International Conference on Advanced Materials and New Technologies in Modern Materials Science 2015

IOP Conference Series: Materials Science and Engineering
Volume 116

Tomsk, Russia
9 - 11 November 2015

ISBN: 978-1-5108-2275-7
ISSN: 1757-8981

Printed from e-media with permission by:

Curran Associates, Inc.
57 Morehouse Lane
Red Hook, NY 12571

Some format issues inherent in the e-media version may also appear in this print version.

Printed by Curran Associates, Inc. (2016)

For permission requests, please contact the Institute of Physics
at the address below.

Institute of Physics
Dirac House, Temple Back
Bristol BS1 6BE UK

Phone: 44 1 17 929 7481
Fax: 44 1 17 920 0979

techtracking@iop.org

Additional copies of this publication are available from:

Curran Associates, Inc.
57 Morehouse Lane
Red Hook, NY 12571 USA
Phone: 845-758-0400
Fax: 845-758-2633
Email: curran@proceedings.com
Web: www.proceedings.com

Table of contents

Volume 116

International Conference on Advanced Materials and New Technologies in Modern Materials Science 2015
9–11 November 2015, Tomsk, Russia

Accepted papers received: 15 February 2016
Published online: 3 March 2016

Preface

Papers

IOP Conf. Series: Materials Science and Engineering **116** (2016) 011001 doi:10.1088/1757-899X/116/1/011001

Preface

The International Conference on Advanced Materials and New Technologies in Modern Materials Science was held at Tomsk Polytechnic University, 9-11 November, 2015. The conference was organized by major Siberian institutions: National Research Tomsk Polytechnic University, Institute of Strength Physics and Materials Science SB RAS, Russian Materials Science Center, National Research Tomsk State University, Novosibirsk State Technical University, Novosibirsk, Russian Federation.

The Conference brought together academics, professionals and young researchers to discuss the current issues of the development and manufacturing of construction materials, improving their properties, as well as the creation of new materials based on new ideas about the nature of the behavior of a solid body under external impact.

The mains topics of the conference were as follows:

- Advanced Materials and Engineering Technologies
- Problems of Strength, Viscosity and Fatigue Durability of Modern Structural Materials
- Surface Hardening and Protective Coatings

More than 110 papers of young and experienced researchers from Russia and abroad were submitted which covered the topics of the conference. 39 selected papers were recommended for publication in IOP Conference Series: Materials Science and Engineering.

The Organizing Committee would like to thank leading scientists in the field of modern materials from Russia, India and Belarus, which made review reports on papers of the young participants of the conference.

We also appreciate the financial support from the Russian Foundation for Basic Research.

The Organizing Committee of AMNT 2015

Dr. Boris Zenin, TPU

Anna Bogdan, TPU

Peer review statement

All papers published in this volume of *IOP Conference Series: Materials Science and Engineering* have been peer reviewed through processes administered by the proceedings Editors. Reviews were conducted by expert referees to the professional and scientific standards expected of a proceedings journal published by IOP Publishing.

Selective deposition of polycrystalline diamond films using photolithography with addition of nanodiamonds as nucleation centers

V V Okhotnikov[1, 2], **S A Linnik**[1], **A V Gaidaichuk**[1], **D V Shashev**[3], **G Yu Nazarova**[3], **V I Yurchenko**[3]
[1] Tomsk Polytechnic University, Tomsk, Russia
[2] Tomsk State University of Control Systems and Radioelectronics, Tomsk, Russia
[3] Research Institute of Semiconductor Devices, Tomsk, Russia

E-mail: vvohotnikov@yandex.ru

Abstract. A new method of selective deposition of polycrystalline diamond has been developed and studied. The diamond coatings with a complex, predetermined geometry and resolution up to 5 μm were obtained. A high density of polycrystallites in the coating area was reached (up to $32 \cdot 10^7$ pcs/cm2). The uniformity of the film reached 100%, and the degree of the surface contamination by parasitic crystals did not exceed 2%. The technology was based on the application of the standard photolithography with an addition of nanodiamond suspension into the photoresist that provided the creation of the centers of further nucleation in the areas which require further overgrowth. The films were deposited onto monocrystalline silicon substrates using the method of "hot filaments" in the CVD reactor. The properties of the coating and the impact of the nanodiamond suspension concentration in the photoresist were also studied. The potential use of the given method includes a high resolution, technological efficiency, and low labor costs compared to the standard methods (laser treatment, chemical etching in aggressive environments).

1. Introduction

Nowadays, the interest to the field of application of various new and better materials in semiconductor electronics is actively growing. One of the promising materials seems to be a diamond. This material is different from the others by a large number of parameters, which are extreme and necessary for semiconductor electronics [1], such as the highest thermal conductivity of all solids (900-2300 W/m·K), chemical and radiation resistance, and a wide bandwidth of the optic radiation from the deep ultraviolet to the far infrared. Pure diamond is a good electrical insulator (specific resistance $\sim 10^{16}$ Ω·cm at room temperature), during doping of diamond its specific resistance can change over a wide range from 10 to 10^6 (Ω·cm), that turns it into a wide-gap semiconductor with a 5.4 eV band gap. The coefficient of thermal expansion (CTE) of the diamond is one of the lowest and at room temperature it can be compared to CTE of invar ($0.8 \cdot 10^{-6}$ K) [1, 2].

Despite these advantages, there are a lot of technological problems in the mass use of diamond in electronics, which complicate its application and impose restrictions on its use in industrial processes. Due to the nature of synthesis and high complexity of processing a monocrystalline diamond, it is difficult and low-tech to create the electronic structures with a complex geometry based on it, especially with the size applicable in micro- and nanoelectronics [1, 3]. Therefore, the application of the diamond in the form of polycrystalline films, which can be deposited onto homo- and getero-substrates, seems very promising [4].

The main problem for their use is the lack of the advanced technology of their selective deposition, which is extremely necessary for the construction of microelectronic devices. Most of the existing methods of creating the necessary topology of semiconductor structures are almost inapplicable for diamond. Wet etching of structures using direct and lift-off stencils is complicated by the fact that using highly aggressive etchants at high P-T parameters is unacceptable in semiconductor electronics. Using the lift-off stencils is limited by the fact that the high density of the diamond does not allow the diamond layers to delete over the etched material, and only the cavities under the film are etched off [5]. The surface laser treatment of the diamond films can be more applied to the ablation treatment (roughness alignment) and cutting, because of high power applied and low controllability of the depth [6]. Ion etching in different atmospheres is hardly applicable because of low efficiency and long duration of the etching process [7].

The use of high purity natural diamond is very expensive and time consuming, and there only tiny crystals in small quantities can be obtained. The use of synthetic diamond enables to simplify the process and to reduce its cost, but its manufacture still requires high-purity diamond substrates, and crystal growth is very limited in size and geometry. In addition, its further processing is complicated, because the use of the conventional methods of etching, grinding, etc., is ineffective due to resistance of diamond to various influences [1]. Using the polycrystalline diamond enables to obtain the films on the substrates with a similar structure of the crystal lattice that enables to avoid the use of expensive and small diamond samples. At the same time, this type of diamond, though slightly ranks below classic diamond, i.e. monocrystalline one, still provides the parameters unapproachable to the other materials, combining it with a higher technological effectiveness of its production and application [1, 3].

This paper proposes a radically different method of growing polycrystalline diamond films, which enables the selective deposition of the films with predetermined parameters of geometry and shape.

2. Experimental

Using the conventional method of producing the growth nucleation centers due to the ultrasonic treatment of silicon wafers in the nanodiamond suspension [8] does not enable to obtaine a required pattern without using a complex and non-technological further processing. The proposed method of obtaining a pattern by creating nucleation centers only in the required areas by applying a photoresist with an addition of nanodiamonds and creation of the required topology using photolithography enables to determine the film geometry before its growth.

In our work, we studied the films obtained by the conventional method and the films obtained with photoresist with different concentration of diamond nanoparticles by weight. The photoresist was prepared by adding 5/10/15 (% by weight) suspension of nanodiamonds (3% by weight). That enabled to produce photoresists with a concentration of nanodiamonds of 0.15/0.30/0.45 (% by weight), respectively.

In the technological process we used: silicon wafers (SW) of 76 mm in diameter, 350-400 μm thick and (111) orientation, photomask (PM) with a suitable pattern, a positive photoresist (SP-051 SU-05) mixed with diamond nanocrystals in the required proportions (0.15, 0.30, and 0.45 % by weight).

The first phase was chemical cleaning of the wafers from contamination in ultrasound. Then, the photoresist was applied using centrifugation, which caused the formation of the photolayer. After that, SW was dried in an oven at a temperature of 70-90°C for solidification of the photoresist. The wafer was aligned with PM followed by layer exposure at the AMK 2104.16 by UV with λ = (250-440) nm. The photolayer was developed in alkaline solution (KOH) and then further developed using plasma treatment in the nitrogen atmosphere. And at the final stage the obtained silicon wafer was finally dried in the oven at 120-140°C.

IOP Conf. Series: Materials Science and Engineering **116** (2016) 012001 doi:10.1088/1757-899X/116/1/012001

Table 1. The main operating parameters

Temperature of filament	Temperature of wafer	Working pressure	Deposition rate	Distance to wafer
2000±50°C	800±25°C	20±1 Torr	0.5 μm/h	10 mm

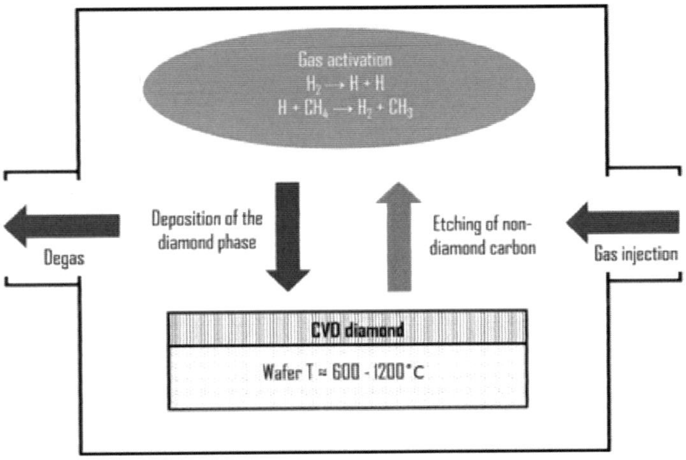

Figure 1. Diagram of the diamond deposition

During the film deposition, we used a specialized CVD reactor. The reactor is based on a well-known method of "hot filaments" [9, 10]. In this method, the gas mixture (H_2/CH_4) is activated thermally using hot tungsten filaments. The temperature of tungsten filaments (0.2 mm in diameter) was 2000±50°C (measured using the infrared pyrometer) in the reactor. The pressure during the film deposition was 20±1 Torr at a methane concentration in hydrogen 2 %. The distance from the wafer to the filaments was constant at 10 mm.

3. Results and discussion

The phase composition and the purity degree of the synthesized diamond films were determined using the most accurate method, i.e. Raman spectroscopy (NanoScan Technology Centaur I HR spectrometer). Figure 2 shows the Raman spectrum obtained in the area of the solid film. It is close to the spectrum characteristic to monocrystalline diamond. The line width, which characterizes sp^3-carbon (1333 cm^{-1}), is less than 50 cm^{-1}, which indicates the absence of amorphous sp^3-carbon. The lines of the parasitic inclusions typical for polycrystalline diamond [11] (lines of trans-PA (1450 cm^{-1}) and G band of graphite (1560 cm^{-1})) are also not observed in the spectrum. Thus, the deposited films consist of high-quality diamond material and the photoresist does not make any significant impurities in them.

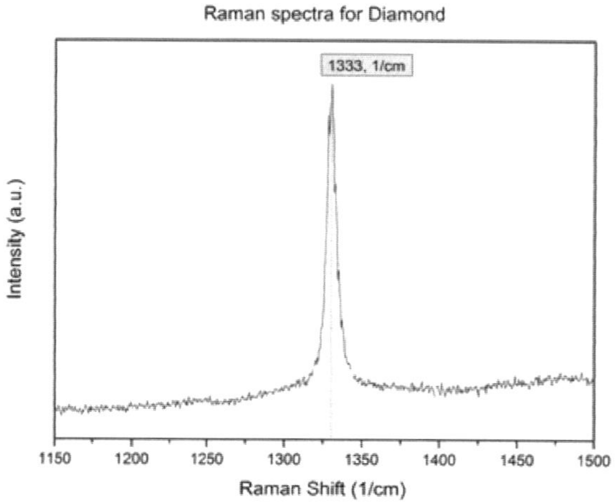

Figure 2. The Raman spectrum of the deposited diamond films

The topology and uniformity of coatings was analyzed using scanning electron microscopy (FEI Quanta 3D SEM) without adding additional conductive layers. The micrographs in Figure 3 show that the method enables film selective deposition, despite the presence of a certain number of "parasitic" crystallites formed in the area, where nanodiamond-containing photoresist was removed before deposition. The creation of these crystallites can be explained by the fact that when photoresist was washed off in the non-exposed areas, some nanodiamonds still remain on the silicon and during deposition they serve as nucleation centers for "parasitic" crystallites. This effect increases with increasing the concentration of nanodiamonds in the photoresist that is demonstrated by the comparison in Figure 3 (a, b, and c) (the content of the nanodiamond is 0.15, 0.30, and 0.45 wt.%, respectively).

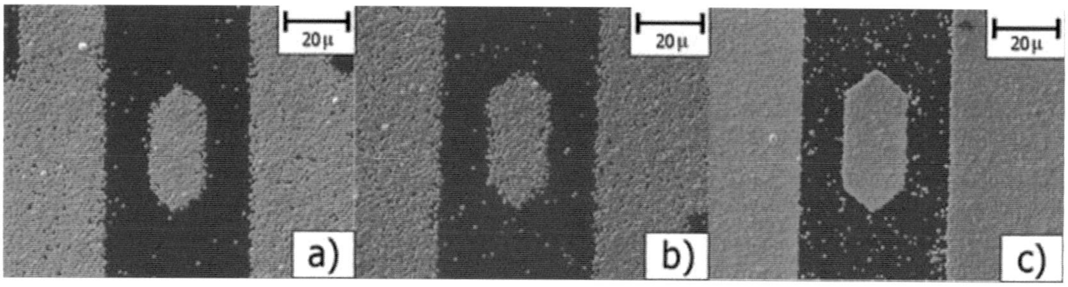

Figure 3. SEM micrographs of selectively deposited diamond films on monocrystalline silicon with different mass content of nanodiamond in the photoresist: a) 0.15%; b) 0.30%; c) 0.45%

ng

uniformity. Thus, the comparison of the film surfaces on the samples prepared with nanodiamond content of 0.15 and 0.45 wt.% (Figures 3a and c) shows that the average concentration of the crystallites is significantly different. The samples, where the selective seeding was carried out at a low content of nanodiamonds, take much more time to form a uniform film, and hence the average crystallite size in such films is significantly higher, that seriously reduces the resolution of this method of selective deposition.

Table 2. Diamond film concentration and uniformity versus the amount of nanodiamond particles in the photoresist

Mass fraction of nanodiamond in photoresist, %	Average concentration of crystals in film	Average concentration of crystals in "clean"	Average uniformity of film coating	Average uniformity of "clean" area

	$(pcs/cm^2) \times 10^7$	area $(pcs/cm^2) \times 10^6$	(%)	surface (%)
0.15	7.98	1.06	91.73	1.22
0.30	9.88	1.94	94.86	1.86
0.45	31.9	5.71	100	2.01

Table 2 shows the comparative statistical data on the dependences of the film uniformity and the diamond crystallite concentration on the amount of the nanodiamond particles in the photoresist. As it is seen, the higher concentration of the nanodiamond produces the films using smaller crystallites and a more accurate pattern, but the number of "parasitic" crystals in the "clean" area increases proportionally.

4. Conclusion

During this work the polycrystalline diamond films of high purity were obtained on the monocrystalline silicon wafers and at the predetermined surface geometry by selective deposition of photoresistive layers with nanodiamond particles using a conventional industrial photolithography. The obtained films, already at this stage, can be applied as a protective layer for devices, which require the coatings with improved properties of strength and chemical resistance. It is possible to simplify the cutting and processing of the required diamond films through selective pattern. Increase in the concentration of nanodiamond in the photoresist increases the uniformity of the films and decreases the average size of the crystallites in them, but also causes an increase in density and "parasitic" diamond nuclei in the areas, which should be free from films. Thus, we can conclude that this technique requires further development to completely eliminate the formation of the unwanted nucleation centers to be widely applied in microelectronics as layers of electronic component of the device (dielectric, semiconductor, etc.).

Acknowledgment

This work was supported financially by the Russian Foundation for Basic Research (research project № 16-32-60018 Mol_a_dk, research project № 16-32-00008 Mol_a), the Foundation for Assistance to Small Innovative Enterprises (research project № 0015685 "UMNIK 15-6") and the Megaproject of Tomsk Polytechnic University "Materials for extreme conditions".

References

[1] Asmussen J, Reinhard D K 2002 *Diamond Films Handbook* (Marcel Dekker, Inc. New York. Basel)

[2] Matsumoto S, Sato Y, Tsutsumi M, Setaka N 1982 Growth of Diamond Particles from Methane–Hydrogen Gas *Journal of Materials Science.* **17** 3106-3112

[3] Davidson J L, Ellis C, Ramesham R 1989 Selective Deposition of Diamond Films *Journal of Electronic Materials.* **18** 6

[4] Linnik S A, Gaydaychuk A V 2015 Synthesis of multilayer polycrystalline diamond films using bias-induced secondary nucleation *Materials Letters.* **139** 389-392

[5] Zhou Z J, Weidong M W, Yongjian T Jun. 2013 Fabrication of Diamond Microstructures by Using Dry and Wet Etching Methods *Plasma Science and Technology.* **15** 550-555

[6] Komlenok M S, Kononenko V V, Ralchenko V G, Pimenov S M, Konov V I 2011 Laser Induced Nanoablation of Diamond Materials *Physics Procedia.* **12** 37–45

[7] Leech P W, Reeves, G K, Holland A July 2001 Reactive ion etching of diamond in CF4, O2, O2 and Ar-based mixtures *Journal of Materials Science.* **36** 3453-3459

[8] Lin C, Chang H, Ben Dao M, Chiang H H, Chen W 2014 Fabrication of Antireflection Nanodiamond Particle Film by the Spin Coating Deposition Technique *International Journal of Photoenergy.* **14** 878

[9] Sharda T, Bhattacharya S 2003 *Encyclopedia of Nanoscience and Nanotechnology* (American Scientific Publishers. California. USA)

[10] Hu J, Foord J S, Holt K B 2007 Hot filament chemical vapor deposition of diamond *Phys. Chem.* **9** 5469–5475

[11] Linnik S A, Gaydaychuk A V, Baryshnikov E Y 2015 Deposition of polycrystalline diamond films with a controlled grain size by periodic secondary nucleation *Materials Today: Proceedings* in press

Microstructure, mechanical properties and machining performance of hot-pressed Al2O3 - ZrO2 - TiC composites

M Grigoriev[1,2*], N Kotelnikov[3], S Buyakova[1,2,3], S Kulkov[1,2,3]

[1] Institute of Strength Physics and Materials Science SB RAS, 12/4, pr. Akademicheskii, Tomsk, 634055, Russia
[2] Tomsk Polytechnic University, 30, Lenin Avenue, Tomsk, 634050, Russia
[3] Tomsk State University, 36, Lenin Avenue, Tomsk, 634050, Russia.

*E-mail: MVGrigoriev@yandex.ru

Abstract. The effect of adding nanocrystalline ZrO_2 and submicron TiC to ultrafine Al_2O_3 on mechanical properties and the microstructure of the composites developed by hot pressing was investigated. It was shown that by means of hot pressing in an argon atmosphere at a sintering temperature of 1500 °C one can obtain the composites Al_2O_3 - ZrO_2 - TiC with a fine structure and minimal porosity. It has been shown that in material a multi-scale hierarchical structure is formed, which possesses high physical and mechanical properties: the hardness and fracture toughness was 21.5 GPa and 5.2 MPa*m$^{1/2}$ respectively, the modulus of elasticity was 500 GPa and bending strength was 390 MPa. Tests on composites' cutting properties were carried out on interrupted cutting of hardened steel. All samples had wear of the cutting edge in the form of chips of the grain rear working surface of the tool, but the cutting tool which was made by Al_2O_3 - 10% ZrO_2 - 10% TiC had a minimum width of wear.

1. Introduction

The development of ceramic composites with high hardness and chemical inertness, able to withstand prolonged exposure to corrosive environments, high pressure, shock, and temperature, with high resistance to brittle fracture and fracture toughness is the main problem of materials science now [1]. This problem is very important, for example, in the cutting of high-strength tempered steel due to the non-uniform loading of the tool, resulting in the tool life being greatly reduced [2]. Conventional cutting tools such as hard-alloys cannot be used in this case due to their lack of toughness and bending strength. One of the most promising materials for the role of such cutting tools is ceramic composites [3,4].

Alumina - TiC composites are widely used as cutting tool inserts due to their ability to machine at higher speeds than cemented carbides and their superior hardness, toughness and strength compared to alumina [5]. They are commonly known as 'black ceramics', having a composition of 70% alumina and 30% TiC, has a high hardness of about 22GPa but insufficient toughness - not more than 4 MPa*m$^{1/2}$[6]. Alumina - zirconia composites are also used as cutting tool inserts where zirconia toughens the alumina matrix by stress induced tetragonal to monoclinic martensitic phase transformation. The hardness and fracture toughness of Al_2O_3- ZrO_2 composites is only 18 GPa and 5 MPa*m$^{1/2}$ respectively [7]. It is well known that addition of either oxide or non-oxide additives improve mechanical properties of alumina. In this context, addition of both oxide and non-oxide additives may be an attractive option as it may impart the beneficial effects of both the additives in the resulting composites. It becomes even more attractive if the oxides and non-oxides are nano sized as the majority of the nano particles may remain at grain boundaries and interact with cracks resulting in interesting features not observed in conventional composites.

The purpose of this research is to obtain an oxycarbide composite based on alumina and additives of zirconia and titanium carbide, in order to study its structure and physico-mechanical properties and test cutting properties.

2. Materials and experimental procedure

This study involves commercial Al_2O_3 powders (USA) with average particle size 4.7 μm and purity 99 %; TiC powders (Russian manufacture) with particle size 5 μm and purity 98.0% and ZrO_2 powders (Grade: TZ-3Y-E, Tosoh Corp., Japan) with nominal particle size 30 nm and purity 99.7%.

Figure 1. SEM micrograph Al_2O_3 powders (A), ZrO_2 (B), TiC (C).

The mixture of powder was prepared as follows: water suspensions created individual components mixed with each other with a magnetic stirrer, followed by suspension sonicated. The deposition of the resulting composition was produced by flocculation of particles from the solution by raising the PH level, followed by vacuum drying. The resulting composite mixture is shown in table 1.

Table 1. - The ratio of components in composites

Samples	Content		
	Al_2O_3, %	ZrO_2, %	TiC, %
AZT-1	85	10	5
AZT-2	80	10	10
AZT-3	70	10	20
AZT-4	60	10	30
AZT-5	75	20	5
AZT-6	70	20	10
AZT-7	60	20	20
AZT-8	50	20	30

Ceramic composites were prepared by hot pressing in an argon atmosphere at a sintering temperature of 1500°C, with a pressing pressure of 50 MPa. The holding time was 10 minutes.

X-ray diffraction data were obtained using an X-ray diffractometer with CuK radiation, grain size and elemental analysis of the composites were carried out on a LEO EVO 50 (Zeiss, Germany) scanning microscope.

Densities of sintered samples were determined by Archimedes' method with distilled water. For hardness measurements sintered samples were polished by diamond paste up to 1 μm grains and indentations were made using a 5 kg load; an average of ten indents was tested. Vickers hardness was determined by using the formula (1).

$$Hv = 1.854 \cdot P/d^2 \qquad (1)$$

where P – load, N; d – diagonal indentation, μm.

Fracture toughness was determined by the formula (2).

$$K_{1c} = 0.035 \cdot \left(H \cdot a^{1/2}\right) \cdot \left(E\varphi / H\right)^{0,4} \cdot \left(l / a\right)^{-0,5} \qquad (2)$$

where H – hardness, GPa; E – Young's modulus, MPa; a – half diagonally indentation, μm; l – the crack length from the corner indentation, μm; φ - constant.

Nanoindentation was performed using a G200 nanoindentor, the load was 100 mN. On the curve load/displacement of the indenter was calculated the modulus of elasticity of the composites.

The flexural strength was measured using the testing machine GP-DLC 30 kN, and was calculated using the formula (3).

$$\sigma_{bend} = 1.5 \cdot P \cdot l/b \cdot h^2 \qquad (3)$$

where: P – load, N; l – distance between supports, mm; b – the width of the sample, mm; h – sample height, mm.

Tests of composites' cutting properties were carried out on interrupted cutting of hardened steel. The material of hardened steel 40X had a diameter of 34mm. Cutting conditions: speed of rotation was 150 m/s; feed 0.1 mm; removal depth of 0.3 mm.

3. Results and discussion

Elemental analysis of the fractured surface of the Al_2O_3 – 10% ZrO_2 – 20% TiC composite (Figure. 2,a) has shown that all phases are distributed randomly (Fig. 2, b-d), but titanium carbide has a higher grain size. Measuring of the average grain size of the individual components has shown that alumina has 1.5 microns; zirconia - 0.8 microns; titanium carbide - 2.5 microns, thus the average grain size of the individual components in the structure of the composites is not significantly higher than the average particle size of the initial powders.

XRD analysis showed that the alumina in composites is in α- modification (corundum), zirconia is cubic and tetragonal modifications and titanium carbide has a NaCl-cubic lattice.

Mechanical properties of sintered composites are shown in table 2. As one can see from the table the best combination of hardness and fracture toughness was achieved in the material containing 10% zirconia, thus the optimum combination of properties was achieved in the composite AZT-3, for other contents of titanium carbide the hardness and fracture toughness are lower.

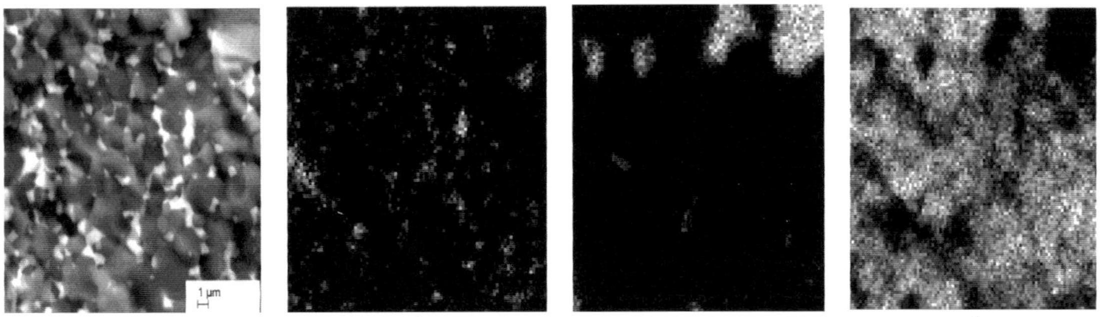

a *b* *c* *d*

Figure 2. The fractured surface of nanocomposite Al_2O_3 - 9% ZrO_2 - 20% TiC (a); element distributions in ZrLα (b) TiKα (c) and AlKα (d) X-rays.

The results of mechanical properties of AZT 1 - 4 samples are shown in table 3. As one can see that the best combination of hardness and fracture toughness is achieved in the compositions comprising 10% zirconia. The hardness and fracture toughness in the bulk and on its surface of materials differ by about 10%: the hardness at the surface is smaller than in the bulk, and conversely there is higher fracture toughness. Thus the optimum combination of properties is achieved in the composite AZT-3. The addition of titanium carbide than 20% lowers the fracture toughness and hardness. A typical crack indentation for sample AZT-3 is shown in Figure 3.

Such differences in the properties at the surface and in the volumes are due to contact of the mixtures with a graphite mold, whereby additional carbonization of oxides at the surface can occur.

Figure 4 shows that the optical image of the edges of cutting tools were made from AZT-2, AZT-3, AZT-4 materials and a control sample manufactured from silicon-aluminum oxide-nitride (SiAlON) with the same sizes. Average widths of wear of the cutting edge, calculated on several optical images for each sample, depending on the composition are shown in Table 3.

As one can see all the samples have wear of the cutting edge in the form of chips of grain on the rear working surface of the tool, but the cutting tool which was made from AZT-2 has a minimum width of wear. It is shown that the width of the cutting edge of the industrial tool SiAlON greater than the same values of AZT-2 and AZT-4 but chipping grains on the cutting surface were not observed.

Table 2. - Mechanical properties of the composites

Samples	Density, g/cm^3	Relative density	Hv, GPa	K_{1c}, MPa*m$^{1/2}$
AZT-1	4.03	0.97	18.6	5.04
AZT-2	4.16	0.99	19.8	5.44
AZT-3	4.26	0.99	21.4	5.68
AZT-4	4.36	0.99	19.3	5.16
AZT-5	4.32	0.99	13.3	5.82
AZT-6	4.35	0.99	17.7	5.44
AZT-7	4.44	0.99	16.0	5.41
AZT-8	4.55	0.99	16.5	5.83

Table 3. Hardness and fracture toughness of the composites on the surface and in the bulk. Bending strength.

Samples	Hv, GPa Vikers		Nanohardness		K_{1c}, MPa*m$^{1/2}$		σ_{bend}, MPa	Width of edge wear, µm
	Surface	Bulk	Hv, GPa	E, GPa	Surface	Bulk		
AZT-1	18.6	20.6	24.9	460	5.04	4.42	438	-
AZT-2	19.8	21.3	28.3	503	5.44	5.12	393	80
AZT-3	21.4	19.0	25.8	502	5.68	4.29	407	140
AZT-4	19.3	21.6	27.9	479	5.16	5.08	365	115
SiAlON	19.6	-	-	345	5.82	-	-	127

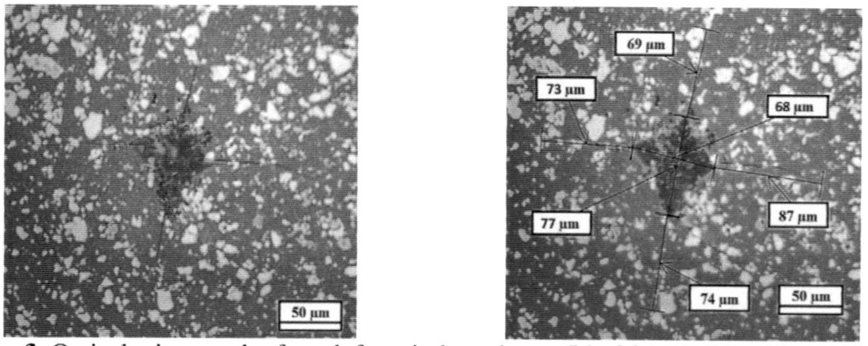

Figure 3. Optical micrograph of crack from indentation at 5 kgf for $70Al_2O_3$-$10ZrO_2$-20TiC

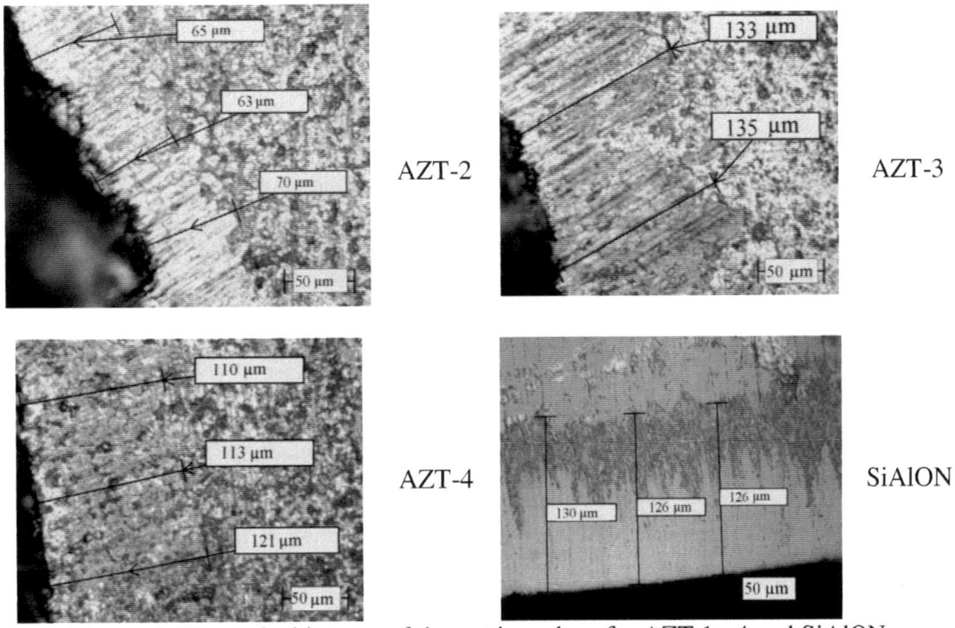

Figure 4. The optical images of the cutting edges for AZT 1 - 4 and SiAlON.

4. Conclusion

1. It was shown that by hot pressing in an argon atmosphere at a sintering temperature of 1500 °C can be obtained the composites Al_2O_3 - ZrO_2 - TiC with a fine structure and minimal porosity.

2. XRD analysis has shown that in the sintered composites phase the content is the same as in initial mixtures. Scanning electron microscopy shows a uniform distribution of the components in a matrix of alumina.

3. The best combination of mechanical properties is shown in the composite Al_2O_3 - 10% ZrO_2 - 20% TiC; its hardness and fracture toughness was 21.3 GPa and 5.12 MPa*m$^{1/2}$ respectively; the elastic modulus was 500 GPa and the bending strength was 390 MPa.

Acknowledgement
This research was conducted with partial financial support by the project #14.607.21.0056-RFMEFI60714X0056. Elemental analysis and images composite structure obtained on a LEO EVO 50 (Zeiss, Germany) in the "NANOTECH" ISPMS SB RAS.

References

[1] Ercin Cura M, Kim S H, Muukkonen T, Varjus S, Vaajoki A, Soderberg O, Suhonen T and Hannula S P 2013 *Ceram Int.* **39** 2093-2105
[2] Kumar A S, Durai A R, Sornakumar T 2003 Machinability of hardened steel using alumina based ceramic cutting tools *Int. J Refract Met Hard Mater* **21**109-117
[3] Shaw M C 2005 Tool materials metal cutting principles. 2nd ed. (New Delhi: Oxford University Press) pp. 307–348
[4] Grigoryev M V, Buyakova S P and Kul'kov S N 2013 *Refractories and Industrial Ceramics* **11/12** 20-25
[5] Grigoryev M V, Molchunova L M, Buyakova S P and Kul'kov S N 2013 *Russian Physics Journal* **7 2** 206-210.
[6] Zhang Y, Wang L, Jiang W, Chen L and Bai G 2006 *J. Eur. Ceram Soc.* **26** 3393-3397.
[7] Savchenko N L, Korolev P V, Melnikov A G, Sablina T Y and Kul'kov S N 2008 *Fundamental problems of modern materials science* pp. 94-99 [in Russian]

Multilayer Coating Formation at the Deposition from Plasma

S A Shanin, A G Knyazeva
National Research Tomsk Polytechnic University, 634050, Tomsk, Lenin Avenue 30, Russia.

E-mail: shanin_s@mail.ru, anna-knyazeva@mail.ru

Abstract. The numerical experiment was carried out for the process of the coating composition formation during deposition from plasma. The chemical reactions between elements are taken into account. The nonuniform composition of the coating is determined by various transfer processes, including diffusion under stress action. To find the stress field the equilibrium problem was solved numerically because all physical and mechanical properties depend on composition. Stress field has been also obtained nonuniform.

1. Introduction

The increase in physical-mechanical properties of protective coatings allows refining the operating characteristics of cutting materials. Structure and properties of the coatings depend on numerous factors [1,2]. Therefore, it is necessary to study the regularities of properties evolution varying the deposition conditions. The coatings from nitride of transition metals, especially TiN [3,4], are widespread, because their properties are unique. For example, TiN - coating elastic module reaches up 466 ± 15 GPa [1]. Material per se (chemical composition) does not ensure the wear resistance growth. Deposition mode, configuration of engineering accessory; preliminary etching or alloying etc determine the coating and interface structure [2].

The experimental results demonstrate that nanohardness is seriously affected by the residual stress [5]. Residual stresses depend on properties of substrates, surface finishing methods and processing parameters [6,7] It was shown that the residual stresses in the coatings deposited with ion assistance depend on the thickness of the coating and the flux and energy of the bombarding ions [8]. It was marked in [9] that residual stresses can arise from three sources: thermal, intrinsic and lattice mismatch. However, there are many other causes for stresses in multicomponent and multi layered coatings. The simplest analytical evaluations of stresses in the coated materials were made in [10].

Since experimental investigations are very expensive and labor-consuming, the mathematical modelling can help in the study of physical phenomena which effect coating structure and properties. Simple model of thickness change of the coating is contained in [11]. Physical and chemical phenomena do not taken into consideration in this model. Monte Carlo method have been used, for example, in [12,13] to understand the fundamental phenomena. There are many papers where finite elements commercial programs are applied for coated material modeling. Limitations of finite elements analysis were discussed for example in [14]. Multi physical model of coating formation is presented in [15]. And multi layered coating growth with chemical conversion and residual stresses evaluation is described in [16]. Numerical experiment had allowed revealing the numerous causes for inhomogeneous stresses field in the coating for the system Ti-Cr-C.

Here we numerically investigate the regularities of the formation of multi layered coating on the iron substrate.

2. Problem formulation

We assume that coating grows at the deposition from plasma with the composition changing due to successive arrival of ions Ti,N and Al,N into the chamber. The interval between arrivals of different ion is given. The stage of ion redistribution in chamber is not considered. Cylindrical specimen made from iron rotates around their axis that provides the same plasma composition along the lateral surface. General mathematical formulation of the problem is presented in [15] and includes three sub-problems: thermal-diffusion, kinetical and mechanical ones. Total mathematical model takes into account the various cross-effects additionally to fundamental transfer mechanisms for the heat and species. This leads to the appearance of additional term in the equations for the heat and mass fluxes. Really, when four elements

$$Ti, N, Al, Fe$$

can present in growing coating (with the mass concentrations C_i, $i = 1,2,3,4$ correspondingly), following summary chemical reactions

$$4Fe + N_2 = 2Fe_2N;$$
$$8Fe + N_2 = 2Fe_4N;$$
$$2Al + N_2 = 2AlN;$$
$$2Ti + N_2 = 2TiN$$

should be taken into account. Therefore, the chemical compounds $Fe_2N; Fe_4N; AlN; TiN$ will be contained in the coating and in the transient layer. C_i are their concentrations, $i = 5,6,7,8$. In this case, the defining relationships for the elastic body (temperature is more less then melting temperature) turn to

$$\sigma_{ij} = 2\mu\varepsilon_{ij} + \delta_{ij}\left[\lambda\varepsilon_{kk} - Kw\right]$$

where

$$w = 3\left[\alpha_T(C_k)(T - T_0) + \sum_{k=1}^{4}\alpha_k(C_k - C_{k0}) + \sum_{k=5}^{8}\alpha_k'(C_k - C_{k0})\right],$$

where λ, μ are Lame coefficients; $K = \lambda + 2\mu/3$ is the bulk module; α_T is the thermal expansion coefficient, α_k and α_k' are the concentration expansion coefficients for diffusing elements and for compounds. Index '0' relates to unstrained state.

In accordance with the irreversible thermodynamics, the diffusion fluxes include the terms corresponding to the usual diffusion and to the diffusion of species due to presence of concentration gradients of other species; to the thermal diffusion and diffusion (mass transfer) due to stress (or/and strain) gradient action:

$$\mathbf{J}_k = -\rho\sum_{i=1}^{3} D_{ki}\nabla C_k + B_k C_k \nabla\sigma_{kk}^{e} - C_k D_{kk} S_{Tk}\rho\nabla T;\ \sum_{i=1}^{4}\mathbf{J}_i = 0,\ k = 1,2,3$$

where $B_k = \dfrac{D_{kk}^0 m_k}{RT}(\alpha_k - \alpha_4)$ are transfer coefficients under stress gradient action; D_{kk}^0 - are self-diffusion coefficients; m_k are molar masses. Partial diffusion coefficients depend on the structure and composition of the substance

$$D_{kj} = D_{kk}^0\left(g_{kj} - g_{kn} + \frac{C_k m_k}{C_n m_n}(g_{nn} - g_{nj})\right),$$

where g_{kj} are thermodynamical coefficients; $n = 4$.

Cross terms appear in the equation for the heat flux:

$$\mathbf{J}_q = -\lambda \nabla T - \sum_{i=1}^{3} A_i \nabla C_i + A_q \nabla \sigma_{kk}^e,$$

where

$$A_i = \sum_{j=1}^{3} D_{ji} Q_j^*,\ i = 1,2,3;\ A_q = T \sum_{k=1}^{3} D_{kk}^0 C_k S_{Tk} (\alpha_k - \alpha_4);$$

Q_j^* are the transfer heats.

Other equations stay the same.

3. Results

Thermal-physical properties were taken in [17] and are contained in the Table 1.

Table 1. The properties of the individual substances

Paremeter	N	Fe	Ti	Al	AlN	TiN	Fe2N	Fe4N	Dimension
λ	0.026	80.1	21.9	203.5	30.14	36.2	67.3	60.1	Vt/(m·K)
ρ	0.8	7.87	4.54	2.7	3.12	4.93	6.68	7.6	·103, kg/m^3
C_p	29.1	25.14	25.1	24.35	30.1	34.23	98.449	29.5	J/(mol·K)
m	14	55.84	47.8	26.98	40.98	61.8	125.6	237.2	·10^{-3}, kg/mol
E	0.14	190	112	70	310	261	217	200	GPa
ν	0.24	0.28	0.26	0.28	0.3	0.29	0.3	0.29	
α_T	2	4	2.8	3.1	3.4	3.2	3.8	4.7	·10^{-6}, K^{-1}

Formal-kinetical parameters used for the calculations are presented in the Table 2. They were found based on chemical thermodynamics similarly to [18].

Table 2. The kinetical parameters for the reactions

Compound	Activation energies, J/(mol K)	Pre-exponential factors	Reaction heat KJ/mol
Fe_2N	80508	2.9·108	-7.54
Fe_4N	79225	3.67·108	-21.78
TiN	200204	2.46·1014	11.64
AlN	319000	3.69·1018	9.63

Since thermal-physical properties of the coating depend on the composition changing in the time, the effective properties

$$\lambda(C) = \sum_{\kappa=1}^{8} \lambda_\kappa \cdot C_\kappa,\ \rho(C) = \sum_{\kappa=1}^{8} \rho_\kappa \cdot C_\kappa,\ E(C) = \sum_{\kappa=1}^{8} E_\kappa \cdot C_\kappa$$

of the growing coating depend on the space coordinate and on the time. Therefore, as opposed to [15, 19], here the sub-problem on mechanical equilibrium was solved numerically.

Below we present the illustrations to some interested result: due to changing composition of plasma, the coating forms with inhomogeneous structure. Due to the diffusion and chemical reactions the transient zones between the layers contain the chemical compounds (Figure 1). This leads to the diffusion retardation, and Fe2N presents only in the layer between the coating and substrate. All elements are distributed not uniformly.

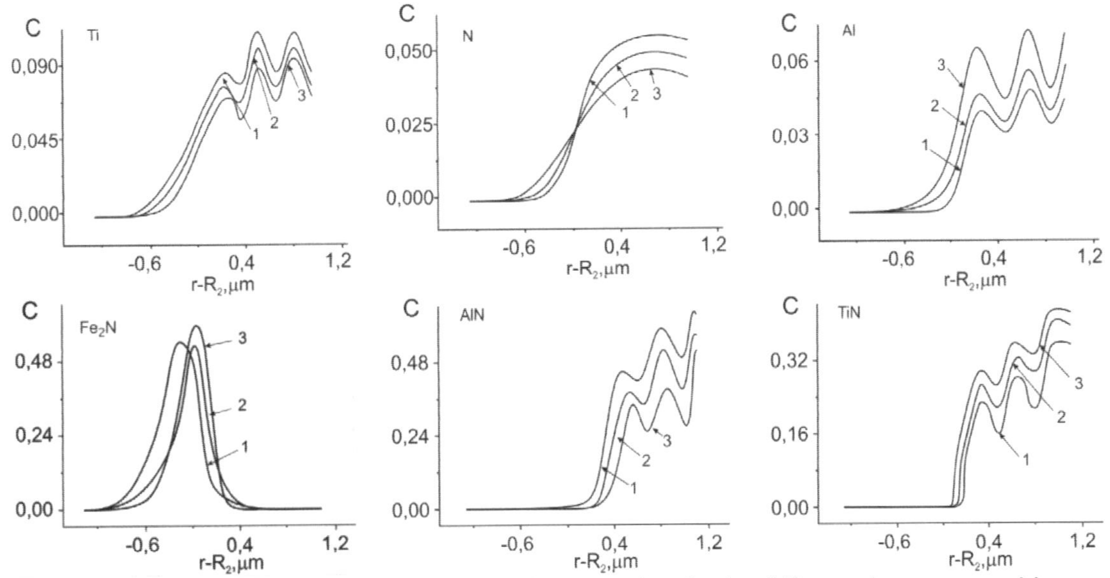

Figure 1. Concentrations of some substances in the coating for the different plasma compositions: 1. – y1=0.5, y2=0.4; 2. – y1=0.55, y2=0.45; 3. – y1=0.6, y2=0.5;

Not uniform stress field in the coating connects immediately with the concentration distribution is shown in the Fig. 2. Obviously, radial stresses equal to zero in the free surface. Axis and tangential stresses are discontinuous in interface.

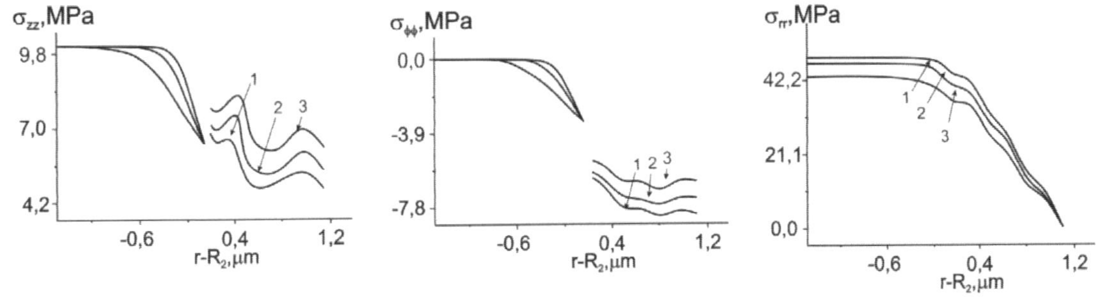

Figure 2. Stress tensor components in the coating depending on the plasma composition. The curves numbers correspond to the curves numbers in the Figure 1

4. Conclusion

The features of the model for multi layered coating formation were described. The coating composition was investigated depending on plasma conditions. It was detected that nonuniform stress field is determined by coating composition.

Acknowledgement

The work was supported by the Ministry of Science and Education of Russia for 2014-2016 No. NIR - 11.815.2014/K

References

[1] Jílek M Jr, Jílek M, Mendez M F, Mayrhofer P H and Veprek S 2014 *J. Thin Solid Films* **556** 361–68

[2] Sarakinos K, Alami J, Konstantinidis S 2010 *J. Surface & Coatings Technology* **204** 1661– 84
[3] Shanyong Zhang, Weiguang Zhu 1993 *J. Materials Processing Technology* **39** 165–177
[4] Bavadi R and Valedbagi S 2012 *J. Materials Physics and Mechanics* **15** 167–172
[5] Hernández L C, Ponce L, Fundora A, López E and Pérez E 2011 *J. Materials* **4** 929–940;
[6] Colombo D A, Echeverría M D, Moncada O J, and Massone J M 2013 *J. ISIJ International* **53** 520–26
[7] Dobrzański L A, Skrzypek S, Pakuła D, Mikuła J and Křiž A 2009 *J. Achievements in Materials and Manufacturing Engineering* **35** 162–68
[8] Solovyev A A, Sochugov N S, and Oskomov K V 2010 *J. The Physics of Metals and Metallography* **109** 111–19
[9] Thomas R and Jagannadham K 1999 *JCPDS-International Centre for Diffraction Data* 443–54
[10] Clyne T W, Gill S C 1996 *Journal of thermal spray technology* **5** 401–18
[11] Panjan M, Čekada M, Panjan P 2010 *J. RMZ – Materials and Geoenvironment* **57** 317–30
[12] Moarrefzadeh A 2012 *J. Wseas transactions on applied and theoretical mechanics* **7** 106–11
[13] Cho J, Terry S G, LeSar R and Levi C G 2005 *J. Materials Science and Engineering* **391** 390–401
[14] Limarga A M, Vaßen R and Clarke D R *J. Applied Mechanics* **78** 1–9
[15] Shanin S A and Knyazeva A G 2014 *Journal of Applied Mechanics and Technical Physics* **3** 192–204
[16] Shanin S A, Knyazeva A G and Kryukova O N 2014 *International Conference on Mechanical Engineering, Autornation and Control Systerns* (MEACS)
[17] http://materials.springer.com;
[18] Kryukova O N and Knyazeva A G 2014 *Russian Physical Journal* **57** 113–17
[19] Shanin S and Knyazeva A 2014 *J. Advanced Materials Research* **1040** 565–70

Surface modification of additive manufactured Ti6Al4V alloy with Ag nanoparticles: wettability and surface morphology study

E Chudinova[1], M Surmeneva[1], A Koptioug[2], A Sharonova[1], K Loza[3] and R Surmenev[1]

[1] Tomsk Polytechnic University, Institute of Physics and Technologies, 30, Lenina ave., Tomsk 634050, Russia
[2] Mid Sweden University, Akademigatan 1, Östersund SE-831 25 , Sweden
[3] Inorganic Chemistry and Center for Nanointegration Duisburg-Essen (CeNIDE), University of Duisburg-Essen, Essen 45117, Germany

E-mail: rsurmenev@mail.ru

Abstract. In this work, the use of electrophoretic deposition to modify the surface of Ti6Al4V alloy fabricated via additive manufacturing technology is reported. Poly(vinylpyrrolidone) (PVP)-stabilized silver nanoparticles (AgNPs) had a spherical shape with a diameter of the metallic core of 100±20 nm and ζ -potential -15 mV. The AgNPs- coated Ti6Al4V alloy was studied in respect with its chemical composition and surface morphology, water contact angle, hysteresis, and surface free energy. The results of SEM microphotography analysis showed that the AgNPs were homogeneously distributed over the surface. Hysteresis and water contact angle measurements revealed the effect of the deposited AgNPs layer, namely an increased water contact angle and decreased contact angle hysteresis. However, the average water contact angle was 125° for PVP-stabilized-AgNPs-coated surface, whereas ethylene glycol gave the average contact angle of 17°. A higher surface energy is observed for AgNPs-coated Ti6Al4V surface (70.17 mN/m) compared with the uncoated surface (49.07 mN/m).

1. Introduction

Titanium alloy foams have significant application potential due to their light weight and exceptional, isotropic mechanical properties as well as high corrosion resistance [1]. These include multiple structural applications in aerospace, aeronautics and automotive systems which take advantage of density-compensated strength and stiffness at elevated temperatures [1]. Murr et al. [2] have also recently demonstrated that Ti6Al4V cellular foams produced by additive manufacturing (AM) using electron beam melting (EBM®) technology have innovative applications in the "next generation biomedical implants".

The AM allows to create customized implants individual for the patients and clinical cases[3]. These applications are utilizing strongest advantages of the AM technologies such as freedom of component shapes, possibilities of computer optimization of the manufactured-to-be component functionality and properties, and good value for money in manufacturing one-off or small series of products [4]. These benefits are already recognized by medical implant manufacturers and practical surgeons[5]. There are certain medical applications, such as acetabular cups mass-produced in

Ti6Al4V by EBM®, which have been successfully implemented for years [6, 7]. There is also a wide variety of biomedical implants manufactured by EBM® technology that are under investigation, as for example, customized hip and knee implants or bone grafting, including craniofacial and maxillofacial replacements [2, 6, 8].

Electrophoretic deposition can be used to fabricate well-distributed particles layers on metallic surfaces [12]. The aim of this study is to generate the silver nanoparticles assembly on the as-manufactured Ti64 surfaces via electrophoretic deposition and investigate the wettability and surface free energy of the developed surface.

2. Materials and methods

2.1. Samples preparation
All the samples used were fabricated using Ti6Al4V ELI powder in ARCAM A2 EBM® machine (Arcam AB, Mölndal, Sweden) [13]. The Arcam machine also makes it possible to produce implants with designed porosity and cellular lattice surface structures that stimulate the osteointegration in the single process procedure[7]. All samples were coin-like discs 10 and 30 mm in diameter with flat surfaces. These discs were manufactured in a single batch with their surfaces normal to the melt layer h. The 3D-scaffolds were made in two different batches with identical parameter settings and same diamond-shaped basic lattice cells. All samples were carefully blasted in the ARCAM powder recovery system using the same precursor powder. Electrochemical etching was carried out at 30° C in the ethanol and 2-propanol solution of the aluminum chloride and zinc chloride with the stabilized dc current density 0.2 A/cm^2[14].

2.2. Synthesis and deposition of silver nanoparticles
The surfaces of as-manufactured Ti6Al4V samples were modified by electrophoretic deposition of silver nanoparticles. Wet chemical synthesis and characterization of Poly(vinylpyrrolidone) (PVP)-stabilized silver nanoparticles were reported elsewhere [15]. Prior to electrophoretic deposition, the silver nanoparticles were dispersed in ethanol. Electrophoretic deposition was carried out at a constant voltage of 50 V for 30 min. The distance between the two electrodes was maintained at 3 mm. The concentration of AgNPs in the dispersion was 60 mg L-1. PVP-stabilized silver nanoparticles had a spherical shape with a diameter of the metallic core of 100±20 nm and ζ -potential of -15 mV.

2.3. Characterization
The morphology and composition of the surfaces were determined using an ESEM Quanta 400 FEG scanning electron microscope (SEM), equipped with energy-dispersive X-ray spectroscopy unit (EDX; Genesis 4000, SUTW-Si(Li) detector) operating in a high vacuum.

Contact angle analysis was performed with an optical contact angle apparatus (OCA 15 Plus Data Physics Instruments GmbH, Germany), using the SCA20 software (Data Physics Instruments GmbH, Germany). The contact angle of water in air was measured by the sessile drop method. A minimum of 10 droplets (2 µL, 5 µL·s-1) of water, 3 droplets of diiodomethane or ethylene glycol were examined for each sample, and the resulting mean values of the contact angles were used for the calculations. The surface free energy was calculated using Owens-Wendt-Rabel-Kaelble (ORWK) method. Three different media (water, diiodomethane and ethylene glycol) were used for these calculations.

3. Results and discussion
Figure 1 illustrates the morphology of the as-manufactured Ti6Al4V alloy samples. Because powder-bed process is used for manufacturing, component surfaces have certain amount of partially-connected (fused) precursor powder grains (60-100 µm in diameter), which resulted in a relatively rough surface morphology. As clearly visible from the images metal surface contains a number of powder grains strongly and rather loosely attached to the bulk metal (Figure 1) providing a roughness with the spectrum ranging from nanometers to about 100 µm, a feature quite common for the powder bed AM

processes. It has been extensively reported that the microscale morphology significantly influences cell behavior such as adhesion, proliferation and differentiation in vitro and in vivo[16].

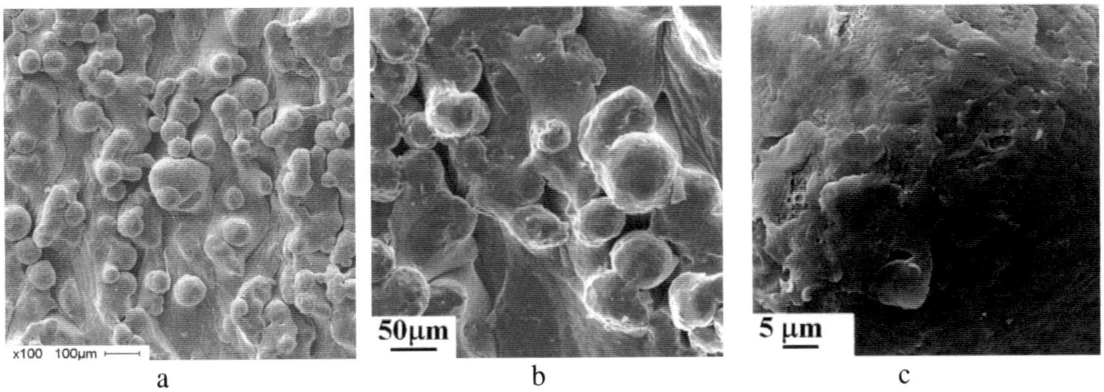

Figure 1. SEM images of the as-manufactured Ti6Al4V sample surface morphology. Samples are additively manufactured using EBM® technology . Images are acquired using SE detector.

Surfaces are also showing distinct roughness at the sub-micron scale (figure 1c).

According to the DLS measurement, the PVP-functionalized AgNPs were negatively charged ($\zeta = -15$ mV) particles with a hydrodynamic diameter of 100 ± 20 nm. The polydispersity index (PDI) was below 0.3, indicating the absence of large agglomerates.

SEM images of the AgNPson deposited on the surface of Ti6Al4V substrate are shown in Figure 2. A uniform distribution of the AgNPs was observed across the surface. The surface morphology of the initial Ti6Al4V substrate was not changed by the deposition of AgNPs. The metallic core diameter of the AgNPs determined from the SEM analysis was 110 ± 30 nm. The ordered AgNPs with well-defined two- (2D) and three-dimensional (3D) spatial configurations are expected to show the homogeneous and sustained release of Ag+ ions from nanoparticles to the surrounding tissue. These ions can reduce or even prevent colonization and subsequent biofilm formation on the surface of an implant. The further deposition of the hydroxiapatite (HA) film on the layer of AgNPs may improve this strategy since the release of silver ions from the AgNPs can be precisely controlled. It has already been shown that the use of HA film can help to control the Ag and Ni ions release [15, 17].

c

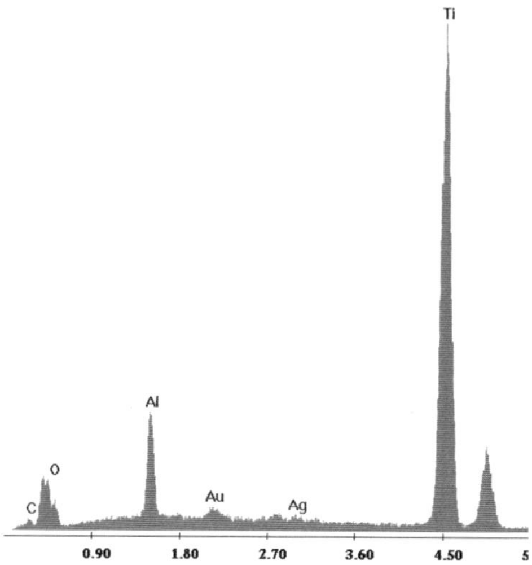

d

Figure 2. SEM image (a-c) and EDX spectrum (d) of the Ti6Al4V alloy coated with AgNPs.

EDX spectrum analysis of the coated samples revealed that the obtained composite contained free silver. EDX analysis revealed silver concentrations of 0.58 at% for the silver nanoparticles deposited on the surface of Ti6Al4V alloy.

A detailed study of the wettability of the scaffold surfaces was done using contact angle sessile drop method and the contact angles for the scaffolds after modification with AgNPs before and after treatment are summarized in Table 1. It should be noted that water contact angle revealed the significant effect from the AgNPs layer. The hydrophobic and super-hydrophobic surfaces can result from the increased surface roughness, for example, this effect occurs naturally on the lotus leaf [18]. The surfaces of these leaves possess a micrometer-level roughness and they are covered with nanosized crystals of wax [18].

The highest contact angles were obtained with water, whereas ethylene glycol gave the lowest contact angles for PVP-stabilized-AgNPs-coated surface. One of the possible reasons for this effect may be due to the lower surface tension of ethylene glycol as compared to water, and as the consequence the AgNPs-coated samples can display a lower wetting resistance against ethyleneglycol [19]. Water (surface tension 72.8 mN m-1), diiodomethane (50.8 mN m-1) and ethylene glycol (48.0 mN m-1) were used as testing fluids. However, this must also be investigated in more details in the future[19]. The surface energy of materials is influenced by some surface characteristics like structure, morphology, roughness, chemical composition and surface charge, however, the correlation among them is not clear.

Figure 3 shows the variation of dispersive, polar and total surface energy of as-manufactured and modified surfaces of scaffolds. The results presented in Figure 3 show that total surface energy of the coated surfaces significantly varied from 29.22 to 70.17 mN/m compared with as-received Ti6Al4V scaffold (49.07 mN/m).

Sample	Contact angle, degrees			Contact angle hysteresis, degrees
	Water	Diiodmethan	Ethylene glycol	
Bare Ti6Al4V	106±1	34±1	42±2	20±3
HA-coated Ti6Al4V	109±2	56±2	80±1	16±1

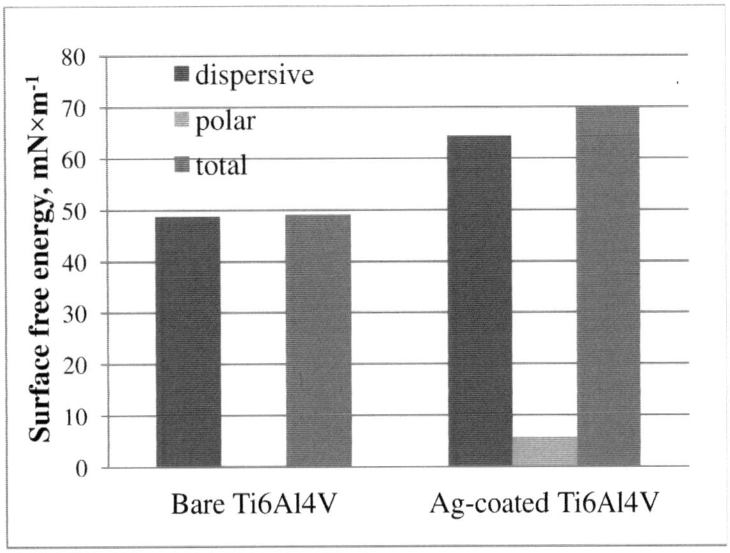

Figure 3. Variation of the dispersive, polar and total surface energy of as-manufactured and AgNPs-coated scaffolds surfaces.

A detailed comparison shows that a higher surface energy is observed for AgNPs-coated Ti6Al4V surface (70.17 mN/m) as compared to uncoated surface (49.07 mN/m). In nanoparticles, surfaces contribute to the chemical potential by creating surface free energy and surface stress [20]. Furthermore, the polar component of the surface energy is significantly higher for AgNPs-coated surface compared to the one of as-manufactured scaffolds. The dispersive component makes a major contribution of 0.1–8% to the total surface free energy for all the samples.

4. Conclusion

The Ti6Al4V alloy prepared via AM using EBM® technology was modified by electrophoretic deposition of AgNPs. The effect of the modification of the metallic scaffolds on the wettability and surface energy has been studied in details. The determined surface free energies of the AgNPs-coated Ti6Al4V scaffolds were significantly different from as-manufactured ones. A detailed comparison shows that a higher surface energy is observed for AgNPs-coated Ti6Al4V surfaces (70.17 mN/m) compared with uncoated ones (49.07 mN/m). The dispersive component made the major contribution to the total surface free energy for all the studied scaffolds. This resulted in the polar component making approximately a 0.1–8% contribution to the total surface free energy.

Acknowledgments
This research was supported by the *Russian Science Foundation* (15-13-00043).

References

[1] Murr L E, Gaytan S M, Medina F, Martinez E, Martinez J L, Hernandez D H, Machado B I, Ramirez D A, Wicker R B 2010 Characterization of Ti-6Al-4V open cellular foams fabricated by additive manufacturing using electron beam melting *Materials Science and Engineering* **527** 1861-1868.

[2] Murr L E, Gaytan S M, Medina F, Lopez H, Martinez E, Machado B I, Hernandez D H, Martinez L, Lopez M I, Wicker R B, Bracke J 2010 Next-generation biomedical implants using additive manufacturing of complex cellular and functional mesh arrays *Philos. Trans. R. Soc.* **368** 1999-2032.

[3] Koptyug A, Rännar L, Bäckström M, Fager Franzén S, Dérand P 2013 Additive manufacturing

technology applications targeting practical surgery *International Journal of Life Science and Medical Research* **3** 15-24.
[4] Cronskär M, Rännar L E, Bäckström M, Koptyug A 2008 Application of electron beam melting to titanium hip stem implants *Proc Intl Conference on Additive Technologies, Vienna, DAAAM International* 1559.
[5] Christensen A, Lippincott A, Kircher R 2007 An Introduction to Electron Beam Melting with Ti6Al4V-ELI for the Orthopaedic Device Industry *BONEZone* **6** 14-17.
[6] Herrera A, Yánez A, Martel O, Afonso H, Monopoli D 2014 Computational study and experimental validation of porous structures fabricated by electron beam melting: A challenge to avoid stress shielding *Materials Science and Engineering* C **45** 89-93.
[7] Thundal S 2008 Rapid manufacturing of orthopaedic implants *Adv. Mater. Process.* **166** 60-62.
[8] Cronskär M, Rännar L, Bäckström M 2013 Production of customized hip stem prostheses - a comparison between conventional machining and electron beam melting (EBM) *Rapid Prototyp. J.* **19** 365-372.
[9] Gristina A 1987 Biomaterial-centered infection: Microbial adhesion versus tissue integration. *Science* **237** 1588-1595.
[10] Lim P N, Chang L, Thian E S 2015 Development of nanosized silver-substituted apatite for biomedical applications: A review Nanomedicine: Nanotechnology, Biology, and Medicine, 2015. **11**: p. 1331–1344.
[11] Loza K, Diendorf J, Sengstock C, Ruiz-Gonzalez L, Gonzalez-Calbet J M, Vallet-Regi M, Köller M, Epple M 2014 The dissolution and biological effects of silver nanoparticles in biological media *J. Mater. Chem.* B **2** 1634-1643.
[12] Boccaccini A R, Keim S, Ma R, Li Y, Zhitomirsky I 2010 Electrophoretic deposition of biomaterials *J Royal Soc Interface* **7** S581-S613.
[13] Standard Specification for Additive Manufacturing Titanium-6 Aluminum-4 Vanadium ELI with Powder Bed Fusion; ASTM F2924–14; American Society for Testing Materials: West Conshohocken, P., USA, 2014.
[14] Tajima K, Hironaka M, Chen K K, Nagamatsu Y, Kakigawa H, Kozono Y 2008 Electropolishing of CP Titanium and Its Alloys in an Alcoholic Solution-based Electrolyte *Dent Mater J.* **27** 258-265.
[15] Ivanova A A, Surmenev R A, Surmeneva M A, Mukhametkaliyev T, Loza K, Prymak O, Epple M 2015 Hybrid biocomposite with a tunable antibacterial activity and bioactivity based on RF magnetron sputter deposited coating and silver nanoparticles *Applied Surface Science* **329** 212-218.
[16] Li L, Xie T 2005 Stem cell niche: structure and function *Annu. Rev. Cell Dev. Biol.* **21** 605-631.
[17] Surmenev R A, Ryabtseva M A, Shesterikov E V, Pichugin V F, Peitsch T, Epple M 2010 The release of nickel from nickel–titanium (NiTi) is strongly reduced by a sub-micrometer thin layer of calcium phosphate deposited by rf-magnetron sputtering *J. Mater. Sci.: Mater. Med.* **21**(4): p. 1233-1239.
[18] Wang C F, Chiou S F, Ko F H, Chou C T, Lin H C, Huang C F, Chang F C 2006 Fabrication of Biomimetic Super-Amphiphobic Surfaces Through Plasma Modification of Benzoxazine Films *Macromol. Rapid Commun.* **27** 333-337.
[19] Tang Z, Hess D W, V Breedveld V 2015 Fabrication of oleophobic paper with tunable hydrophilicity by treatment with non-fluorinated chemicals *J.Mater.Chem.* A **3** 14651 - 14660.
[20] Li W, Ni C, Lin H, Huang C P, S Ismat Shah 2004 Size dependence of thermal stability of TiO2 nanoparticles *Journal of Applied Physics* 96 6663-6668.

The behavior of delaminations in composite materials – experimental results

A S Chermoshentseva[1*], **A M Pokrovskiy**[1] **and L A Bokhoeva**[2]
[1]Bauman Moscow State Technical University, Russia, Moscow, Baumanskaya 2-ya str., 5/1, 105005
[2]East Siberia State University of technology and management, Russia, Ulan-Ude, Kluchevskaya str., 40v, 670013

Email: anechka227@yandex.ru

Abstract. Delamination is one of the most common failure modes of composite materials. It may occur as a consequence of imperfections in the production process or the effects of external factors during the operational life of the composite laminates, such as the impact by foreign objects. This paper presents the results of mechanical tests and the optimum degrees of filling the composite materials (CM) with hydrophobic powder (Tarkosil T-20) depending on the latter mass concentration. The results present test samples of the CM with the underlying interlayer defects. The samples were fabricated of twenty-ply pre-preg (fiberglass or carbon fiber). The industrial grade glass is T-25 (VM) specification 6-11-380-76. The composite materials have nanosized additives in structure. The volume concentration of nanopowders is varying from 0.1% to 0.5%. This kind of research has been done for the first time.

1. Introduction

Laminated composites have high specific strength and stiffness, but they are very sensitive to defects such as delaminations. Their appearance is the result of imperfections in manufacturing technology and of the impact of operational loads. They can significantly affect the strength and rigidity of the structural elements of the laminated composite materials. The initiation, development, and termination of delamination buckling and their growth in a compressed laminate depend strongly on various geometrical parameters, material properties, and loading conditions [1]. One of the features of the laminated process is the local bucking and the bulging of an exfoliated layer with the following defect growth. This can cause the total destruction of the structural element under the load which is significantly lower than the estimated level [1-3].

With the development of nanotechnology it became possible to harden the epoxy resins in layered composite materials due to the introduction of different types of nanomaterials such as nanotubes, nanoparticles, nanofibres into the matrix, which improves their physic-mechanical properties to 20-30% [4]. Currently, the research of the behavior of such materials in products and designs is far from being complete. This paper presents the study of composite materials (CM). Their mechanical properties change depending on the degree of their hydrophobic powder (Tarkosil T-20) filling. As the result we have received the CM test samples with the underlying interlayer defects with the addition of binder nanosized particles (silicone dioxide) with different volumetric concentrations from 0.1% to 0.5%.

2. The methods of experiment

The samples were fabricated of twenty-ply pre-preg (fiberglass or carbon fiber) [5]. The industrial grade glass is T-25 (VM) specification 6-11-380-76. The unidirectional material has the following characteristics: fiberglass - $E_1 = 5.4 \cdot 10^4$ MPa, $E_2 = 1.2 \cdot 10^4$ MPa, $G_{12} = 0.5 \cdot 10^4$ MPa, $\mu = 0.28$. As nano-modifying additives PCM we chose the nanosized silicone dioxide powder Tarkosil - 20 (T-20). A thin strip of teflon film (10 - 100 microns thick) simulates the defect. For research, 30 samples were made with the different mass concentration: C = 0.1%; C = 0.3%; C = 0.5%. For a comparative analysis the control samples (CS) were fabricated without nano-additives, but with the included defect between the layers (10 samples). The data for the samples is given in Table 1.

Table 1. The data for the samples – plate with a through defect

#	Name of nano-particles	Mass concentration of nano-particles, %	L(mm) length of sample	H(mm) thickness of sample	l(mm) length of defect	h(mm) thickness of defect	Number of layers sample	Number of layers in the defective portion	The angle of placing filaments
1	T-20	0.1	230	2	60	0.4	10	2	$(45,-45)_5$
2	T-20	0.3	230	2	60	0.4	10	2	$(45,-45)_5$
3	T-20	0.5	230	2	60	0.4	10	2	$(45,-45)_5$
4	CS	-	230	2	60	0.4	10	2	$(45,-45)_5$

The pre-pregs were made on the basis of fiberglass satin weave with the epoxy binder. The technological process of sample preparation is as follows: the surface of the mold is lubricated with a release coating and is heated. On the surface ten layers of pre-cut pre-preg are laid. Then the mold is placed into hydraulic press and pressurized due to the convergence of molds. The defect models simulating delamination in the samples are manufactured in the following way: they are put between the layers with and without departing yield (depending on the shape of the defect) at the edge of the sample - a thin teflon strip coated with a release layer. The length of the sample between the clamps L = 200mm. We can prevent possible bending jaws on condition that the loading is low. That is why the accepted sample width is 2 mm and the length of jaws from both sides is 3cm. The process of solidification influences the final size of the sample. After the solidification the thickness of each sample was measured. The ratio of the length to the thickness for all samples is approximate (L / H = 60). The boundaries of defects in the samples are outlined with risks. A special installation was designed to carry out the test; special grippers were developed and manufactured. The sample was pinched by upper and lower grips (Figure 1). The pictures of the samples were taken before and after the experiment; the process of loading, deformation and fracture was recorded on video.

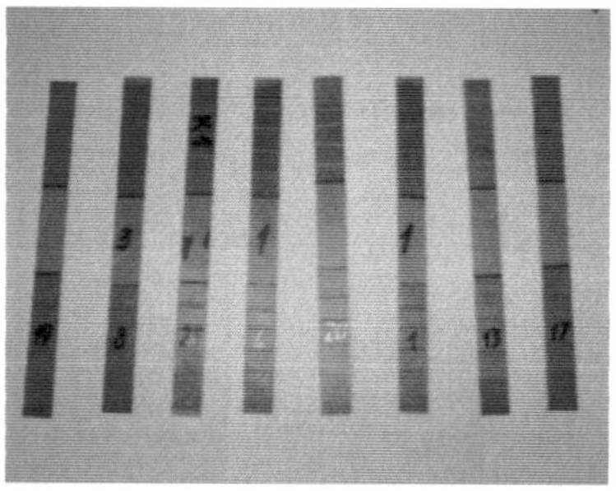

Figure 1. The samples for testing

In order to determine the tensile strength of the samples we used an electromechanical testing machine Instron 5982 (Figure 2). The loading process is performed according to the automatic locking force applied to the sample P on the axial displacement (Figure 3).

Figure 2. The stand for testing

Figure 3. The process of destruction

3. The results of experiment

The sample tests were made with the fibreglass branches. In the process of loading, the automatic recording of dependence of applied forces P on axial movement was made. With the increase of the loading occurs the following: first, a thin layer bucks, then the adhesive bond gets destroyed and the local bulging takes place. The process of destruction is accompanied by the decrease of loads and the bending of the upper, defective, and lower parts and as well as the main part of the plate. The "mixed" form of buckling takes place. At this stage of deformation the process of destruction is accompanied by the sounds of cracking fibers. With the increase of loading, the deformation of all parts of the plate took place. Same experimental results are shown in Table 2: the critical load of the defective area (P_1), the critical load of the defective plate (P_2) and the load during the fracture of the plate (P_3). We received a graphical loading dependence on the sample displacement (Figure 4).

Table 2. The results of experimental research

Type of sample	P_1 (H)	P_2 (H)	P_3 (H)
CS	486.7	458.3	140
CM with the addition T-20 in mass % 0.3	601.7	535.25	268
CM with the addition T-20 in mass % 0.5	576.5	569.25	283.3
CM with the addition T-20 in mass % 0.1	486.9	391	240

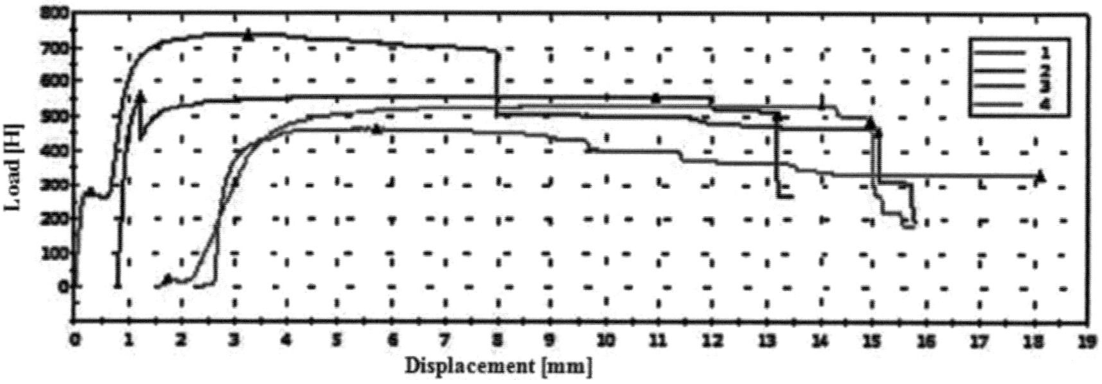

Figure 4. The load versus displacement of composite materials

Figure 3 shows: 1. the composite material with the addition of Tarkosil - 20 in mass % 0.3, 2. the composite material with the addition of Tarkosil - 20 in mass % 0.5, 3. the composite material with the addition of Tarkosil - 20 in mass % 0.1, 4. the control samples.

4. Conclusion

The analysis of the data showed that the introduction of powder Tarkosil in CM with defects leads to the change in the strength of the mixture made on their basis. The best result of P_3 as compared to the control samples was detected in the samples with nanomaterials with T-20 at the concentration of 0.3. Taking into account the results of experimental studies, it should be noted that the maximum load of the samples with defects such as delaminations is less critical than the load of the samples without defects [1]. Although the magnitude of movements is comparable with each other, the experimental data is a bit less than it is theoretical counterpart. It is characteristic of the samples with defects, and without. This fact can be explained as follows:

- The presence of various kinds of initial imperfections, such as the inevitable initial curvature, the heterogeneity of the material.
- During the experiment was very difficult to follow the precise theoretic requirements.

This research will be used to meet the challenges of strengthening the interlayer defects in laminated composites, the appearance of which is the result of imperfections in manufacturing technology and the effects of operational loads, due to the addition of the nanoparticles in the binder [2-4].

Acknowledgement
This research was supported by grant RFFI. Project #15-38-50484 mol_nr "Stability of thin-walled elements of designs from layered composite materials based interlayer defects".

References
[1] Bokhoeva L A 2008 Research of stability the plates with defect in nonlinear statement *J. Mechanical Engineering* **2** 22
[2] Rogov V E, Bokhoeva L A and Filippova X A 2014 Research of interlaminar defects of an elliptic form in element designs from layered materials *J. Advanced Materials Research with the title Material Research and Applications* **875** 624
[3] Bokhoeva L A, Baldanov A B and Rogov V E 2013 Strength of circular samples from layered composite materials based interlayer defects *J. Fundamental problems of modern materials* **10** 87
[4] Brusentseva T, Zobov K, Filipov A, Bazarova D, Lkhasaranov S, Chermoshentseva A and Syzrantsev V 2013 Introduction of nanopowders and mechanical properties of the materials on epoxy resins *J. Nanoindustry* **3** 24
[5] Bohoeva L A, Pnev A G and Chermoshentseva A S 2011 Experimental research of ring samples from layered composite materials with defects *J. Bulletin of the Buryat State University* **9** 230

IOP Conf. Series: Materials Science and Engineering **116** (2016) 012006 doi:10.1088/1757-899X/116/1/012006

Analysis of changes in hardness of a metal surface layer in areas of high stress and methods of determining residual life of parts for mining machines

I E Zvonarev[1*] and S L Ivanov[1]
[1] Department of Mechanical Engineering, Electromechanical faculty, Mining University, 2, 21st line, VO, Saint-Petersburg, Russia

*E-mail: ZVano@mail.ru

Abstract. The methodological bases for determining the energy resource of mechanical transmissions details for mining machines are considered.. Based on the analysis of the accumulation of damage in metal gears, a method of estimating residual life of coarse-toothed wheels by periodically measuring the hardness of the surface layer of the teeth is justified. The regularities in change of hardness of coarse-tooth gear, conditioned by a change in metal strength properties that take into account the micro- and macromechanisms of plastic and elastic deformation, distortion of the metal crystal lattice with formation and movement of vacancies and dislocations. Experimental setup was built and the results of laboratory experiments are given related to the process of destruction of non-standard samples under different loads. Comparison of dimensions and hardness values of the sample allows concluding that a larger deformation corresponds to a greater increase in hardness, their limit value for the material being in the fracture zone. It is established that the detected changes in the local hardness occurs in areas of increased stresses above the limit of proportionality and the work of fracture forces attributed to dislocations density adjacent to the fracture plane expressed in terms of hardness increment is constant.

1. Introduction

Substantiation of timing of the inspection, mining machinery repairs and writing off is carried out by estimating residual life of equipment components and details. Traditional technological and diagnostic procedures of state mechanical transmissions assessment do not enable to estimate the residual life of gears with sufficient accuracy, to assess their technical condition quickly. Today one of the most promising ways to assess residual life of coarse-grained gear mechanical transmission of mining machinery is control of changes in metal surface layer hardness in areas close to the fracture surface.

2. Theory research

Researches on fatigue failure of metals [1, 2], friction and wear in machine parts show that material of the detail is destroyed upon the critical value of the internal energy, which is typical for this detail, is reached, that is each element of a mechanical transmission has its energy source. Production rate of a resource determines duration of the element work. Energy approach to definition of the resource allows to take into account intensity of machine work both in the whole and its individual elements [3, 4].

Value of an energy source is constant for the same type of machines and determined by properties of material of its parts, drive kinematics, its production technology, and it does not depend on the value of load, if it does not exceed the maximum of allowable limit. Residual life of the details and the timing of the limit state proportionally depend on the power loss in transmission elements. Thus, energy source is a property of a machine drive and can be calculated by formula

$$K_K = \int_0^{T_P} \Delta P(t)dt = const\,, \qquad (1)$$

where K_K = the energy source (kW·h); T_P = the resource of machine drive work (h); $\Delta P\ (t)$ = the function of the power losses in the drive for a time t (kW).

However, with this approach the whole drive is seen as «black box», and physical nature of fracture process is not considered, which does not enable to identify effectively causes of possible failure and find potentially dangerous elements. This drawback is eliminated in the course of development of an energy source calculation method [5]. In considering the question of an energy source determination of a gear transmission, taking into account all kinds of destruction, energy source value is determined by formula

$$E = \int_a^{N_{lim}} P'dN = \sum_{i=1}^{n} \frac{N_{lim}^{1-m_i} A_i K_{Bi}}{1-m_i}\,, \qquad (2)$$

where P` – specific power of losses, equal to the work of the dissipative losses in the loading cycle (W); N^i_{lim} – leader-criterion (of the following criterions: $N_{h\ max}$ – number of cycles before the failure by the criterion of the contact strength of the teeth; $N_{f\,max}$ – number of cycles before the failure by the criterion of flexural strength of the teeth; $N_{J\ max}$ – number of cycles before rejection by the wear criterion of the teeth working surface), by which the gear wheel will be broken down; m_i – exponent of the curve energy source equation; A_i – constant coefficient of curve energy source; K_{bi} – coefficient of factors influence for each of the failure criteria: contact chipping, bending strength and wear of active transmission teeth surfaces.

An undoubted advantage of a calculation method for determining an energy source is a possibility of theoretical and experimental evaluation of a resource of such transmission elements as gear wheels, which allows to consider wear phenomena, contact and bending loading of the teeth together, take into account influence of each factor separately for resource of gear wheel in general, depending on the transmission parameters and the properties of each gear wheels.

3. Purpose of research

It is known that the work of fracture is proportional of dislocation density, resulting in a sample under the action of internal stress. Stress, in turn, are reactions to external force. Thus, the specific energy assessment to create a critical density of dislocations in the immediate vicinity of parts fracture plane is possible by periodically measuring the hardness of the surface layer parts.

3.1. Experiments

Experimental investigations of metal surface hardness change in areas of high wear and destruction of mining machines mechanical transmissions parts were carried out at the Department of Mechanical engineering of National university of mineral resources. The measurements were carried out using a universal durometer Zwick ZHU187 (indenter – four-sided diamond pyramid, load – 100 N; soak time – 10 s). Identification of regularities of teeth surface hardness local variation of coarse-grained transmission gears was carried out on the winch gear wheel of career hydraulic excavator lift, worked out about 1.000 machine-hours. Measurements were carried out on 11 teeth wheel (module m 8) at the end faces. Methodology for investigation was providing a partition surfaces of each tooth by 76 sectors, within which the measurements were made in the 3-5 points (see Figure 1).

In comparing the obtained results with the model of the teeth stress-strain state during operation it is revealed similarity of changes in local areas of high hardness with a maximum stress of the tooth deformation at a bend (see Figure 2).

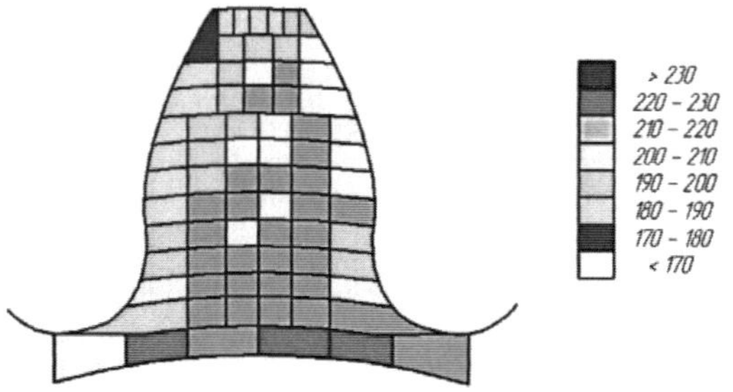

Figure 1. Changes in the surface hardness of coarse-grained gear wheel tooth (module *m* 8 mm, quantity of teeths *z* 19), the nominal hardness is 173HV

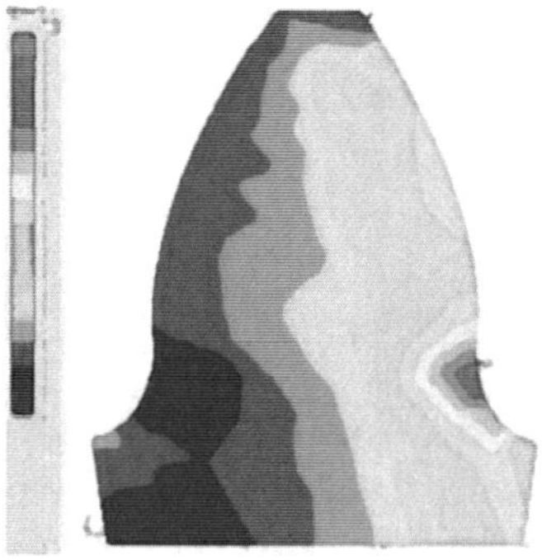

Figure 2. Distribution of principal stresses in the wheel tooth

3.2. Purpose

Identified regularities of hardness change are associated with the process of metals strength properties changing, they allow to take into account micro- and makro-mechanisms of plastic and elastic deformation, causing distortion of the metal crystal lattice with the formation and movement of vacancies and dislocations. Changing of material internal energy density is proportionally to the internal volume of accumulated dislocations. It is not dependent on the loading conditions and it is a physical constant of the material [6].

In working process the gears of mining machines test stress, that causing the different nature of the materials details destruction process. Separation of common energy flow, passing through the gearing in the stress transmission, on the flows, causing the accumulation of the dislocations in the material of the gear wheels, facilitating destruction of the teeth during wear, fatigue failure and fracture of working surfaces of teeth, is difficult [7, 8]. The solution of this problem is possible on the basis of the process researches results of mining machines deformation parts under stress and mechanism of accumulation of dislocations and vacancies in their material.

4. Experimental methods and analysis of results

Investigations of changes in the local hardness in the deformation and destruction of non-standard samples are made by car for static testing Zwick Roell. As a sample for testing was selected a wavy perforated tape – 120.5, made of St10 (see Figure 3). Using the same type of samples with common physical and mechanical properties makes it possible to perform a large number of experiments and high convergence of the results. Periodically changing shape of the samples allows to create different sizes of stress in one force field on the sample. Non-standard wavy perforated tape – 12×0.5 samples is easily reproducible.

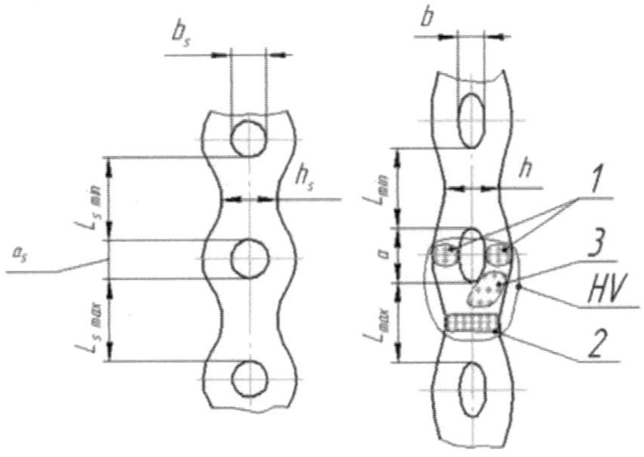

Figure 3. The geometrical dimensions of samples:
a – before stretching; b – after stretching

The first series of five samples was subjected to stretching up to fracture. Average destruction job of five samples line of 210 mm, in static tensile tests on the machine Zwick Roell, is equal to 3.07 J. Battered samples were scanned at the same time, together with the original, undeformed samples from the same batch (five pieces) and a metal ruler (GOST 427-75), chosen as a basis for comparison. Linear measurements were carried out on a computer by the received scanned image in «*Paint.net*» by comparison with the measure in pixels, with subsequent conversion to metric units.

From measurements of geometrical dimensions before and after the fracture it has been found that the dimensions of the ellipse axes *a* and *b* (Figure 3b) changed to 41.7% and -29.7% respectively of the values in the original samples. Hole dimensions of samples from the first from fracture plane vary linearly, approaching in the future to the appropriate sizes of the original sample.

However, the dimensions between axes L_{min} and L_{max}, and the width of the connection element *h* were changed by only 6.9; 3.2 and 0.12% respectively, for the element *h* it is within the accuracy of measurement. It indicates, that the sample in the area of element *h* has not undergone a significant deformation, elongation of the sample out of holes is minimal, and portions of samples in areas of the holes undergone deformation. The dangerous section of the sample is also situated there.

Measurements of destroyed and original samples hardness were carried out together with measurements of geometrical parameters of the samples. The zones of hardness measurements of the surface layer are indicated on Figure 3b:

1. In the line of ellipse axis *b*, and in parallel lines departing from the original till 1.5 mm. The choice of this area is caused by the fact that the destruction of all five samples, subjected to tension, occurred in it. Total 18 measurements for each hole is performed (see 1, Figure 3b).

2. On the line corresponding element *h*, and in parallel lines extending from the original till 1.5 mm. This area is probably dangerous, as it is the smallest in the area *L* between the ellipses. There are only 21 measurements on each bridge (see 2, Figure 3b).

3. On the middle line of the envelope hole. The measurements occurred on three lines, which are equidistant of holes samples internal surfaces and equidistant from each other, on three measurements in the each area, totally nine measurements for each element (see 3, Figure 3b).

The obtained results of measurements were processed using methods of mathematical statistics (see Figure 4).

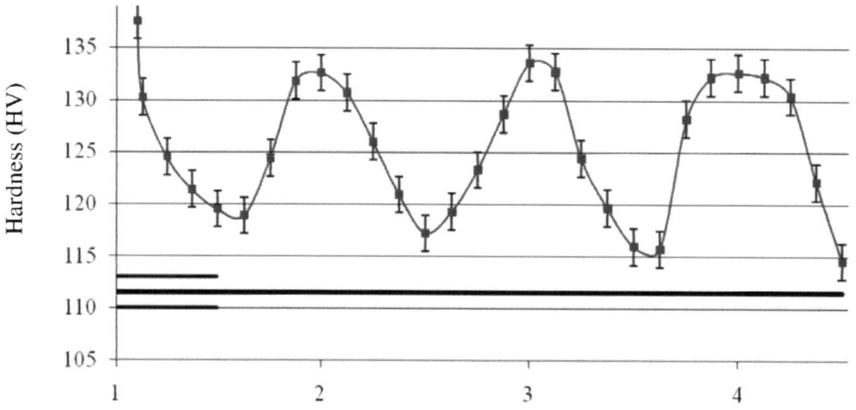

Figure 4. Change in hardness of the samples that were destroyed by car for static testing Zwick Roell compared to the baseline value HV_S = 115.5 HV

Change in the hardness of the sample material from fracture plane to the periphery is wavy contour with maximum values of hardness in the area of perforation (see. 1, Figure 3b) and the minimum in the area of the sample restriction (see. 2, Figure 3b).Comparing changes in the geometric dimensions and hardness values of the sample, it can be concluded that the greatest deformations correspond to a greater increase in hardness, which maximum is in the fracture zone.

At the same time with laboratory experiments on a break it was planned to expose the samples to alternating sign bending. For this purpose the stand (physical pendulum), that represents a simulation model of dissipative energy flows in the tooth during operation of the gear wheel, was projected and made (see Figure 5).

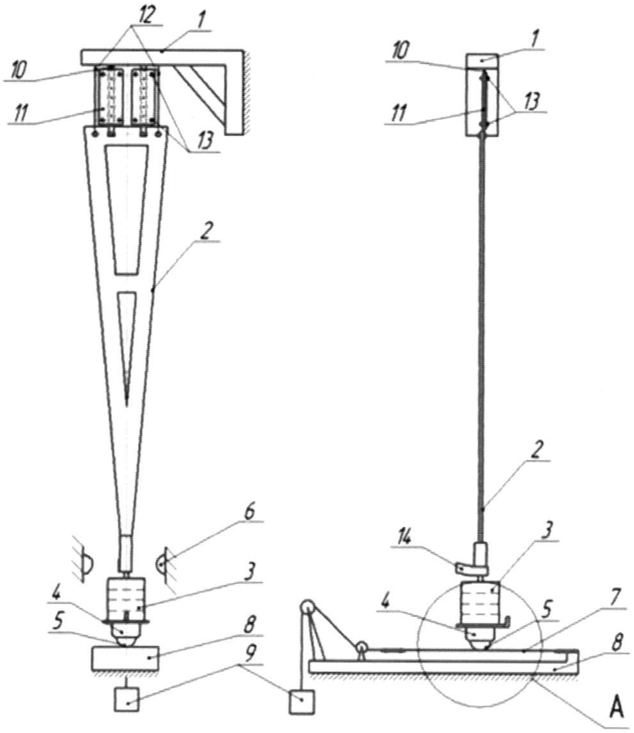

Figure 5. Stand for research of the effects to samples of alternating bending loads

Stand contains pendulum 2, made in the form of an isosceles triangle with an apex angle from 0 to 30°, with a load 3. Grip 4, in which sample, indenter or tool 5 can be installed, is fixed on the pendulum. The sensor of motion controls of the pendulum 6 is connected to the computer by flexible element 7, fixed to the carriage 8, mounted on the lower part of the frame, and provided with a tension device 9 and the gripper 10. The pendulum is rigidly connected to the frame 1 at the point of hanging with possibility of installation between them a sample [9].

Tooth wheel (gear) during operation in engagement experiences bending loads, wear of active tooth surfaces is a consequence of contact loading. Pendulum stand allows to simulate the dissipative losses of separately taken gear wheel of pair, at that the bending stresses of the samples, installed in the grips of the stand, correspond to teeth bending stresses, stretching loads of samples simulate the contact stresses, and the effect of friction in engagement corresponds to the sliding process of the lower roller, mounted on a pendulum, through the surface of the carriage friction element. To estimate the influence on the balance of each elements dissipative loss it is possible to determine their element-wise. At the same time it is possible to combine the dissipative losses in an arbitrary combination.

For fracture of samples by pure bending it is necessary to fix patterns 10 only in the upper part of the frame 1, and the whole structure of the pendulum will be suspended on weightless inextensible threads 13, thereby allowing the sample testing only bending loads during oscillations of the pendulum, which will simulate the bending stresses in the tooth upon engagement in gearing of mining machine.

If the samples 10 to fix both the upper part of the frame 1 and the pendulum 2, the threads 13 are loosened and the entire structure of the pendulum is suspended directly on the samples themselves, thereby additory burdening them by elongation at oscillations of the pendulum and simulating the contact pressures in engagement of wheel transmission.

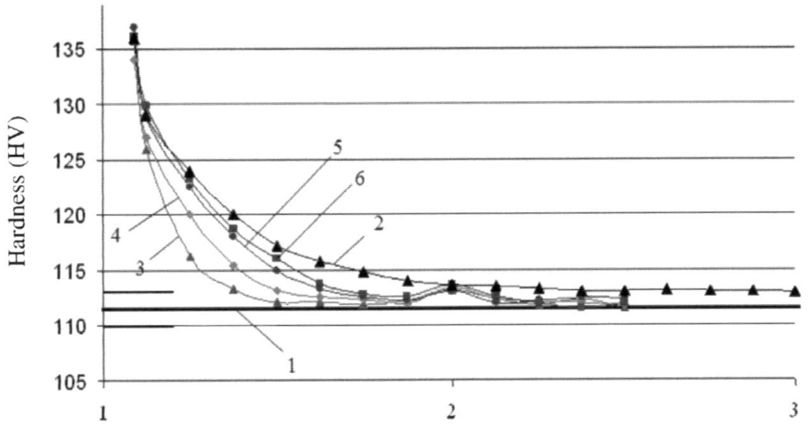

Figure 6. Distribution of hardness along the length of the original sample and destroyed on pendulum stand at an initial deflection angle of 30° 3': 1 – the surface hardness of the initial sample; 2 – the surface hardness of the sample, which was destroyed by the action of bending loads only; 3-6 – the surface hardness of the sample, which was destroyed by the action of bending and stretching axial loads 51.15; 41.15; 31.15 and 21.15 N, respectively

Several series of experiments (seven tests under the same conditions), in which the samples were subjected to loading by pure bending and bending with stretching under the influence of pendulum gravity that is equal to 51.15; 41.15; 31.15; 21.15 N respectively, were produced. The initial deflection of the pendulum was 31° 3'. As a result of the experiments, all the samples were destroyed, at that the number of blocks of variables loading cycles, the number of cycles in a block, the block loading time were fixed and the temperature and humidity to assess the resistance to motion of the pendulum on the air were measured. Thus measured facts was used for further assessment of samples destruction.

Destroyed samples in the pendulum by bend were subjected to measures of the surface layer hardness from the plane of destruction to place of sample fastening in stand by the same method as the samples that were destroyed by stretching (see. Figure 3b). Results of samples hardness measurements were processed by using methods of mathematical statistics (Figure 6).

In the figure 6 the line 2 corresponds to pure bending. Hardness value gradually decreases from the maximum to the nominal value within one link of the sample. Line 3 characterizes the combined effect of bending and stretching loads under the effect of pendulum gravity 51.15 H, change of hardness is more sudden, but there is a small increase in hardness in the area of the first hole from the plane of fracture.

Analysis of the graphs, shown in Figures 4 and 6, allows to draw following conclusions: regardless of the method of loading in the immediate proximity to the fracture plane hardness values are approximately the same, suggesting that failure occurs in the volume of material when it is get the certain accumulation of dislocations in the crystal lattice [10].

Clean work of samples fracture with allowance for dissipation of system for motion in the air was: 1.69 J (51.15 H); 2.288 J (41.15 N); 2.81 J (31.15 N); 3.032 J (21.15 N). There is the effort equivalent to construction gravity of pendulum with load in parentheses. The value of the specific fracture energy, referred to the dislocation density, expressed in hardness value of the surface layer in the fracture zone, can be assumed a constant value within the experimental accuracy, which is 0.023 J, that does not contradict the theory of energy and fracture kinetics [11, 12].

5. Discussions

Despite the fact, that the accumulation of damage and, consequently, an increase in the hardness of the material detail surface layer, is not a necessarily monotonically varying process [13, 14], we can assume, that the process has a certain constant speed during the time interval (or operating time), significantly exceeding cycle of power impact [15]. During the steady flow of damage accumulation process, knowing the actual value of hardness, for example, the end surface of the tooth, identified through diagnostic procedures, and determine maximum hardness value for the fatigue sample of the test material, it is possible to estimate the conditional residual life of the tooth on the criterion of fatigue bending strength:

$$T_{YFP\,\mathrm{res}} = T\left(1 - \frac{\Delta HV_F}{\Delta HV_{MAX}}\right), \tag{3}$$

where T_{YFPres} – the conditional residual life of the gear wheel according to the criterion of fatigue bending strength (h); T – the resource of the transmission, it can be calculated using standard methods or determined experimentally for specific operating conditions (h); ΔHV_F – the increment of the hardness of the surface layer of the tooth end surface in the dangerous section of (HV); ΔHV_{MAX} – the maximum possible (dangerous) increment of the end surface hardness of the tooth in relation to the initial state (HV).

Conditional value of residual life according to the criterion of wear will be determined similarly:

$$T_{YIP\,\mathrm{res}} = T\left(1 - \frac{\Delta S}{[S]}\right), \tag{4}$$

where T_{YIPres} – conditional residual life of the gear wheel according to the criterion of wear (h); ΔS – wear amount at the time of the events on the diagnosis (mm or kg or m^3); $[S]$ – allowable wear (mm or kg or m^3).

Control of surface hardness of the tooth working surface allows to estimate the ratio of the wear and fracture processes according to the criterion of surface contact fatigue of the teeth. If the surface layer hardness of the teeth working surfaces does not change, wear process dominates.

If it is required to estimate the conditional residual life according to the criterion for contact endurance, it should be evaluated similarly to the expression (3), but considering the fact that the process of wear and damage accumulation of contact fatigue occurs within one and the same surface:

$$T_{YHPres} = T\left(\frac{\Delta HV_H}{\Delta HV_{MAX}}\right), \quad (5)$$

where T_{YHPres} – conditional residual life of the gear wheel according to the criterion of the contact endurance (h).

As all three degradation processes take place simultaneously, for coefficient assessing of influence one of the three kinds of gear wheel destruction (expressions (3)–(5)), it is expedient to use the relation

$$K_{Bi} = \frac{T_{YiPres}}{\sum T_{YiPres}}, \quad (6)$$

where K_{Bi} – coefficient characterizing the distribution of the wear types.

Further, in the multiplication of both sides of the expressions (3)–(5) to the corresponding value of coefficient K_{Bi}, at one time with assessment of the three factors influence of the fractures the assessment of residual life is realized directly by fixed diagnostic parameters.

6. Conclusion

From the above mentioned it follows, that the value of the fracture specific work, relegated to the value of the dislocation density, expressed in terms of the surface layer hardness value in the fracture zone, is constant. Thus, the residual life of mining machines details can be determined with high accuracy on the results of change in the surface hardness of the local areas, in particular, the ends of the gear wheels. Perhaps the solution of the inverse problem is the experimental determination of potentially dangerous cross sections by the areas localization of surface layer increased hardness. It should be noted, that the surface hardness of the sample exposed to stresses, varies according with change in the stress state and reaches the maximum in the fracture zone. Accumulation disorders (dislocations) in the sample under uniaxial tension decreases with increasing distance from the plane of the fracture. The value of the sample maximum hardness in the area of the fracture plane does not depend on the method of destruction and determines by limit value of disorders (dislocations) in the sample material.

References

[1] Bushe N A 1996 *Synergetics, structure and properties of materials, self-organizing technologies: theses of the reports* 198-199
[2] Xie Y and Williams J A 1993 *Engin. Depart., Cambridge Uni.* **162-164** 864-872
[3] Black A J, Kopalinsky E M and Oxley P L B 1993 *Proceedings of the Inst. of Mech. Engn. Part C. J. of Mech. Engin. Sci* **207** 335-352
[4] Kolomiitsev M D 1987 *Energy prediction method of mining machines resource.* Notes of Leningrad Mining Institute **117** 69-76
[5] Ivanov S L 1999 *Increase of mining machines transmissions resource based on an assessment of their elements energy loading* (St. Petersburg: State Mining Institute) p 92
[6] Ivanov S L, Zvonarev I E and Shishliannikov D I 2013 *Min. mach. and el. engin.* **11** 28-33
[7] Ibatullin I D 2002 *Bulletin of Samar. state. techn. Univ. Tech. sci.* **15** 123-130
[8] Ibatullin I D 2008 *Kinetics of fatigue damage and fracture of surface layers: a monograph* (Samara: Samar. state. tehn. Univ.) p 387
[9] Zvonarev I E, Ivanov S L, Fokin A S and Semenov M A 2012 *Tech. sci. – Siberian consultant association* 53-62

[10] Ivanov S L 1997 *Mining Journal* **11** 29-30
[11] Gromakovsky D G, Malyarov A N and Samarin Y P 1997 *Abstracts of Papers of the World Tribology Congress* 462
[12] Gekker F R 1999 *Reliability and quality in the industry, energy and transport: collection of intern. conf. works* 19-20
[13] Zvonarev I E, Fokin A S and Ivanov S L 2011 *Processes and means of extraction and processing of mineral resources* 308-311
[14] Chiu Y 1997 Tribology transactions **40** 658-666
[15] Harris T and Yu W 1999 *Transactions ASME Journal of Tribology* **121** 85-89

Composite ion-plasma coatings with nanodisperse reinforced phase: scientific and practical aspects of synthesis

B Brzhozovskii, V Martynov, E Zinina and M Brovkova
School of Electrical and Mechanical Engineering, Yuri Gagarin State Technical University of Saratov, 77 Politekhnicheskaya Ul., Saratov 410054, Russia

E-mail: bmbsar@mail.ru

Abstract. The article describes the main aspects of the synthesis of composite coatings in the surface layer of figurine-shaped product using low-temperature plasma of combined discharge. The example of cutting tools shows the benefits of using the coatings in extreme conditions that occur in machining of materials by cutting.

1. Introduction

During operation, changing the properties of the working surfaces of products for various purposes occurs at the contact areas under the active influence of the external environment. It is the most relevant for the products operating under extreme temperature and power loads, such as cutting tools of complex mechatronic machine tools, precision parts of aircraft drives and others. This results in a continuous change in the shape of the contact surfaces, which leads to a change in temperature and stress distribution in the area of actual contact, and as a consequence, to fatigue and abrasion strength/resistance outage.

A promising way to increase abrasion and fatigue resistance is the creation of technologies and equipment that allow forming structures characterized by dissipative properties in the product surface layer. Special protective coatings may serve as these structures; they are created with the use of single-layer coating deposition techniques (with ion assisting); multilayer coatings with layers of nanometer thickness; multiphase coatings and their combinations, produced by ion-plasma technology [1-4]. It should, however, be noted that the works on producing these coatings in most cases are in the stage of laboratory research and testing. The use of conventional coatings to create dissipative structures is not possible due to the absence of amorphous phase in their structure, and this phase works for permanent local moving of minimum number of particles of minimal size (groups of atoms) in a minimum time without momentum transfer.

In conventional coatings atoms do not have such freedom, so any displacement causes crack formation on the surface, and lattice defects causing delamination and destruction of the coating, reducing its reactivity and mechanical strength, especially during functioning under extreme conditions. In addition, practical implementation of these technologies is associated with the need to create plasma volume which is then transformed and delivered to the treated surface. In this case, most of the energy consumption is spent on the creation of plasma, maintaining its generation conditions and delivery of active plasma particles with a given energy to the object of processing and further scanning them on the surface. The process of forming nanocomposite structure takes only a few percents of the general energy input. In addition, most of the methods that create directional movement of charged particles through the use of various ion source and accelerator systems, cannot modify the geometrically complex parts and products due to their limitations in the machined surface profile (planar and/or cylindrical with no protruding elements). Finally, the methods are expensive and have long processing cycles. The foregoing means that the formation of nano-composite dissipative non-reactive structures with high hardness and resistance in the operation in extreme conditions, and on this basis, providing higher efficiency of the surface layer is an acute scientific and practical problem.

2. The Scientific Basis

It is possible to increase the strength of the product surface layer by its modification with low temperature plasma combined discharge, the formation of which is a sequential flow of the following events [5]:

- application of electrostatic field (displacement potential + 65 ... 200) and microwave electromagnetic field at a reduced level of microwave power (30 ... 90 W) at a pressure of about 300 Pa in a process gas mixture;
- local gas breakdown at the outer boundary of the plasma cloud;
- electron drift to the inner boundary of the plasma cloud;
- formation of compensated flow of charged particles;
- emission of electrons and ions with the inner boundary of the plasma.

To improve the modification efficiency plasma should be formed immediately around the treated surface, which will make it possible to use the maximum amount of energy for proper execution of the synthesis process of the composite coating per se and not on maintaining the conditions of formation and delivery of active plasma particles (electrons and / or ions) to the treated surface as it is implemented in other technologies [[6]-18].

Mathematical description and the analysis of its results, firstly, show that the decisive role in the formation of low-temperature plasma is played by ionization-recombination bulk processes and processes of ambipolar diffusion, and, secondly, allow determining the conditions that ensure plasma formation around the working portion of the article:

- efficient energy deposition in ionization processes under the gas breakdown in ionizing electromagnetic fields and collecting electrostatic field by reducing breakdown voltage using working pressure and easily ionized additives;
- formation of compensated stream of charged particles in the drift space, which results from the Coulomb interaction between charged particles in quasineutral space and prevents separation of charges in their joint motion with the same speed towards the surface of the product.

The validity of the conditions has been confirmed by research into their efficiency, including:

- Recording the emission spectra of the low-temperature plasma under different conditions of its formation, determined by the parameters of the electrostatic and electromagnetic fields;
- The study of the distribution of the spectral line intensities and their correlations, depending on changes in the level of input microwave power;
- Identification of the chemical composition of the gases involved in plasma formation;
- Comparison of the spectra with literature report.

The research results led to the following conclusions:

- Near the boundary of plasma there is a thin layer in which, as a result of ionization, multiply charged ions are formed in a gas mixture of complex composition. The maximum energy of the incident electrons involved in the ionization does not exceed 42 eV;
- Multiply charged ions are formed;
- Changes in external conditions (in the level of input microwave power in particular) entail a change in the concentration of electrons and their temperature, which is accompanied by a change in plasma nonideality (electron and ion temperatures ratio);
- intensity of volumetric recombination processes affects the emission processes at the boundary of the plasma.

Another aspect of composite coating synthesis is connected with the transfer of plasma energy to surface atoms of the workpiece. The results of the mathematical description revealed that the transfer can be performed most efficiently via the surface three-body recombination in which the neutralization of charge carriers by connecting the oppositely charged carriers into neutral molecules at a surface of the product takes place [19, 20[20]]. Unlike ionization, recombination is accompanied by the release of energy, contributing to intense local heating of the surface, with an arbitrarily small kinetic energy of the interacting particles. However, effective recombination process requires directing the energy of the particles in plasma flow to the formation of certain behavior in electrostatic field at its inner boundary: electron deceleration in order to reduce their kinetic energy and ion acceleration to increase their kinetic energy. The control is performed by changing the power input of the electromagnetic field and the placement of the product in the treatment chamber as well as by the value of positive potential, as the first two parameters are responsible for the concentration of charged particles at the outer boundary of plasma, the third is being responsible for the formation of compensated flow in quasineutral space.

Finally, the last aspect of synthesis is the need to form a gradient transition (sublayer) from the coating to the base material (matrix), which is a factor of increasing the adhesion characteristics of parts and products. In the formation of the underlayer, grains of matrix boundary layer must undergo changes similar to the changes occurring in the surface layer, i.e., change of shape. Under these conditions, the melted surface grain layer is able to flow and fill the microvoids, and subsequent cooling fixes a new modified shape working as a binder for unfused grains (Figure 1 a). In operating conditions it will provide additional environmental resistance of parts or products. Otherwise, the inevitable macrodefects, up to chips, occur due to the shift of grains in microcavities (Figure 1 b).

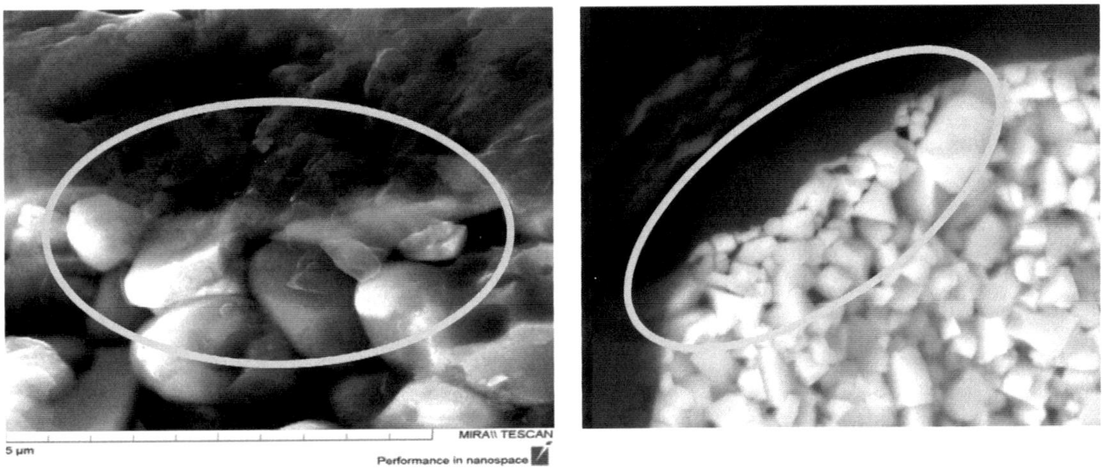

a b

Figure 1. The state of worn matrix of hard alloy products:
a - with a sub-layer; b - without a sublayer.

The results obtained in the study of these aspects became the basis for the creation of equipment and technology for the synthesis of composite coatings on figurine product surface.

The equipment design was carried out to ensure receiving combined discharge and reproducibility of the synthesis results[5]. The high efficiency of microwave energy deposition into the discharge when working at low levels of microwave power allowed simplifying the microwave path without additional protection of the magnetron from reflected power. The equipment power consumption generally does not exceed 2 kW. The positive results of prototype operation (installation 'Chrome') allowed to work out design principles that apply to equipment for both single- and multi-position processing without generality loss.

The main criterion for the technology development was creating identical synthesis conditions on the results of the following actions.

Pre-cleaned and degreased in hot trichlorethylene ClCH=CCl product is installed in a special mandrel fixed in the holder. The surface of the mandrel and a portion of the product surface that are not to be treated are insulated with two layers of thin fluoroplastic tape. Then, the product is placed in the treatment chamber at the coordinates corresponding to its type and size. The processing chamber is sealed and pressurized. The pressure in the camera is decreased to the ultimate level of 10 Pa, and then it is filled with the process gas to 100 Pa, the power supply turns on providing desired potential for the product; the power supply of the microwave energy generator is turned on as well.

By continuously adjusting the anode current of the magnetron microwave power for igniting plasma is fed into the treatment chamber.

Upon processing completion, the time of which ranges from 2 to 20 minutes depending on the size, the product cools in a protective environment, and is removed from the holder after processing chamber depressurization.

3. Practical Aspects

Let us consider practical aspects of the formation of composite coatings through the example of metal-cutting tools for different purposes as a typical figurine-shaped product functioning under extreme temperature and power loads.

The main result of low-temperature plasma impact in the working part of the tool is a significant change in the morphology and physico-mechanical properties of the surface layer (Figure 2): small stripes are virtually absent, scar edges are melted, all microasperities are smoothed. Layer microhardness increases 2-2.5 times, the roughness decreases 1.5-3.0 times, the ohmic resistance is reduced by 6 times. Additionally, the surface is brightened which indicates a change in the chemical composition of the material as a result of its saturation with alloying agents of higher atomic numbers, namely, close to or included into the carbide phase as well as with interstitial atoms of the process gas (nitrogen in this case). The study of the section surface showed the presence of a dislocation structure in the surface layer at a depth of 1.2-1.5 μm to 5 μm, and its absence at a depth of 1.2-1.5 μm. There has been recorded significant reduction in the size and concentration of carbide phase grains in the surface layer (1.8 and 1.74 times, respectively), which is possible only in case of their dissolution, and the change of their shape from a rounded (before) to elliptical (after plasma processing).

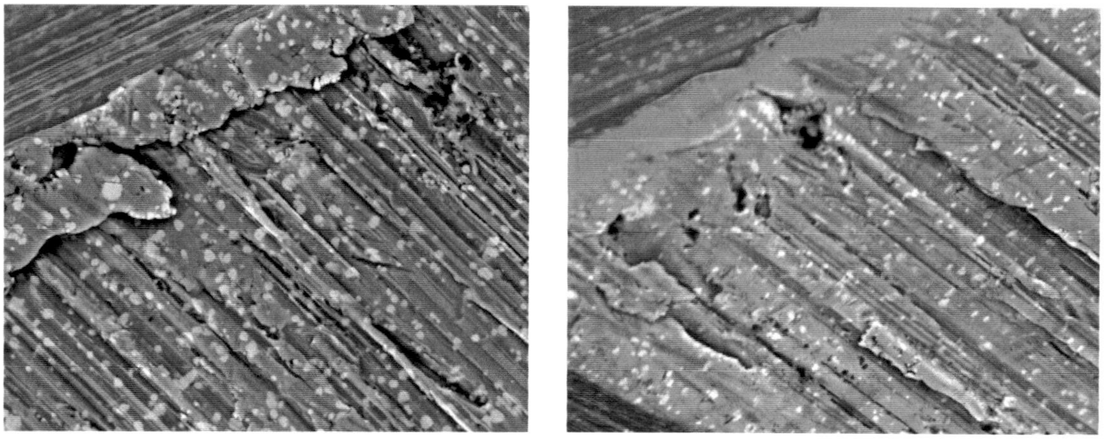

a b

Figure 2. Surface layer of the drill made of steel R6M5 before (a) and after (b) plasma treatment.

Metallographic analysis and electron microscopic examination of the surface showed the presence of structure and phase changes in the surface layer of the tool working part that are not typical for usual thermal processes. The result was the formation of high strength nanoclusters of ~ 50 ... 100 Nm and an amorphous interlayer of ≤10 Nm particles of the same chemical composition as a binder, i.e. composite structure, in the subsurface (to the depth up to 5 μm) (Figure 3). The total depth of the modified layer is 40 μm, while the tool core does not undergo any change, the values of its basic geometric parameters are not changed, distortion is not recorded.

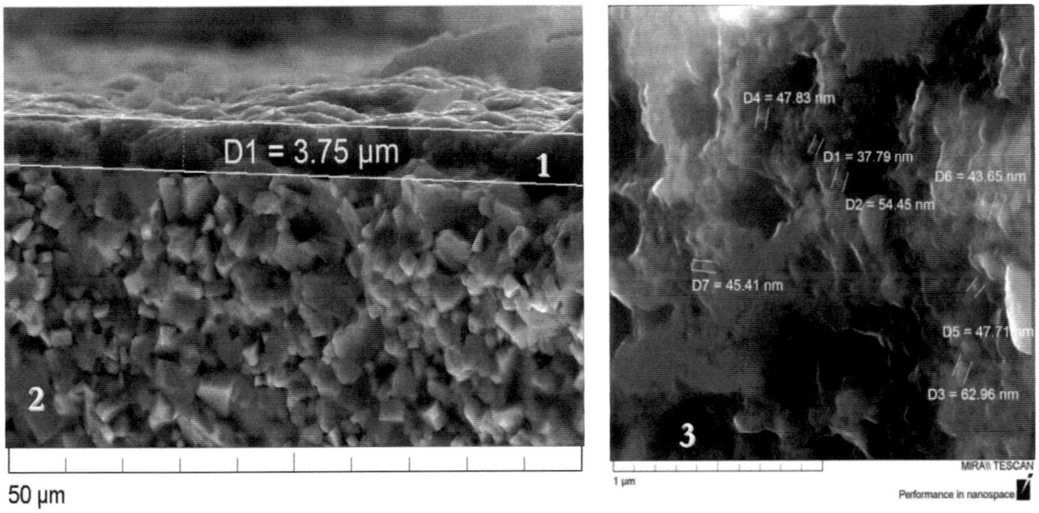

Figure 3. The structure of the modified surface layer:
1 - synthesized composite coating; 2 - matrix;
3 - nanoclusters in amorphous bunch.

The results of the pilot operation of the tool with synthesized composite coating in manufacturing parts from different materials revealed that the average increase in high speed steel tool life time was 3.13 times, the hardmetal tool - 2.35 times, with maximum increase in both types up to 5.0 times. There was also recorded a significant reduction in roughness of the machined surface in AARH parameter: from 2.5-3.5 μm for uncoated tools to 0.80-2.0 μm for coated tools.

The study of the state of the coated tool working surfaces revealed the following. In the cutting process the impact of the cutting waste on the transformed subsurface layers of the tool material causes complex responses associated with both the local surface extension and with its local compression.

The stretching processes are concentrated in the micron surface layers and may have different local areas, determined by the direction of separated material and chips movement. Resistance of the formed coating is determined by the resistance of its nanodispersed reinforced amorphous phase to the tensile stress. The alternation of stretching and compressive effects leads to local changes in the geometry of the cutting edge. The role of the ion-plasma treatment in these conditions is determined by the increasing ability of the coating and gradient sublayer to restructure in the narrow spatial volumes without creating destructive defects. In other words, the synthesized structure acquires the property of self-organization, the mechanism of which is associated with the local transformation of its surface microvolumes as a result of contact processes of interaction with detachable plastic material under the influence and in the direction of the external forces and temperature, particularly at the cutting edge. The result is the formation of a stable streamlined shape of the cutting edge, providing a decrease in friction forces and, consequently, heat generation in the cutting process. Therefore, the structure remains stable and independent of local changes in the surrounding areas even if defects appear on the surface or the cutting edge.

4. Conclusions

The analysis of the materials presented in this paper leads to the following conclusions.

1. The main scientific and practical aspects of the change in the initial properties of the surface layer of figurine-shaped products in low-temperature ion-plasma treatment are associated with the creation of conditions that cause a controlled change in the layer structure and reducing the size of the grain, but in a way that it acquires the properties of increased hardness and ductility that ensure long-term efficiency under temperature exposure or force action during operation.
2. The stability of ion-plasma coating is determined by the ability of its nanodisperse reinforced amorphous phase in the overall situation of growing stress, on the one hand, to resist the tensile stresses on the other - to take compressive stress (ie to perform the dissipative function), since the loss of this ability leads the formation of defects increasing the load on gradient underlayer and further on the matrix.
3. Formation of the composite coating causes a change in the mechanism of wear of working surfaces of figurine-shaped products during their operation: the wear gradually abrases the coating to a state in which defects are formed. Coating self-organization based on the nanodisperse reinforced amorphous phase performance of its dissipative function allows to stabilize the coating purposefully changing the degree and direction of the defect impact and to provide on this basis the most complete performance of their protective functions in relation to the base material of the product without provoking its destruction.

Acknowledgement

The authors express their gratitude to the Russian Science Foundation, through a grant of which (Project No 15-19-00030) the present study was carried out.

References

[1] Grigor'ev S N 2009 Methods for increasing the resistance of cutting tools (Moscow: Mashinostroenie Publishers)

[2] Vereshchaka A S, Grigor'ev S N, Kim V A, Mokritskii B Ya, Altukhova V V and Kirichek A V 2013 High tech tool hardening, *Science Intensive Technologies in Mechanical Engineering,* **6** 19-24

[3] Kurochkin A V 2010 Technological features of multifunctional nanostructured coatings for cutting tools *Vestnik RGATA* **3** 167-172

[4] Migranov M Sh 2005 Investigation of wear-resistant coatings for cutting tools with nanocrystalline structure, *Proceedings of Higher Educational Institutions. Machine Building* **1** 56-62

[5] Brzhozovskii B M, Martynov V V, Zinina E P 2009 Cutting tool hardening through low-temperature plasma combined discharge (Saratov: Saratov State Technical University) [in Russian]

[6] Zinina E P 2001 Low power microwave energy effectiveness for hardening technologies, *Prospects for the development of electronics and vacuum technology for 2001-2006* (Saratov: Scientific conference proceedings) 193-198

[7] Raichuk D Yu 1986 Steel surface hardening by method of plasma hardening with silicon carbide coating, *Improving the quality, reliability and durability of products made of construction, heat-resistant, powder and tool steel and alloys* (Leningrad: Leningrad House of Scientific and Technical propaganda) 73-77

[8] Dashkevich I P, Knyazhevskaya G S, Tomashevich Yu G 1988 Plasma hardening of metal machining tools and die tooling, *New high-frequency technology for machine-building production*, collection of studies (Moscow: VNIIETO) 32-37

[9] Sal'kova S S, Rudman V A 1987 *Experience of using ion nitriding in machine building* (Leningrad: Leningrad House of Scientific and Technical propaganda)

[10] Ryabchenkov E V 1989 *Progressive methods of thermochemical treatment* (Moscow: Mashinostroenie Publishers) 132-142

[11] Butenko O I, Golovchiner Ya M and Skotnikov S A 1982 *Progressive methods of thermochemical treatment* (Moscow: Mashinostroenie Publishers) 122-128

[12] Kaplun V G 1990 Hardening parts of fuel equipment in the glow discharge, *The reliability and durability of machines and structures* **18** 86-89

[13] Nemeshev D F and Shakhnazarov Yu V 1986 *Wear and scaling resistance of steels of different classes after coating ion-plasma bombardment* (Leningrad: Leningrad House of Scientific and Technical propaganda) 63-66

[14] Vasil'eva E V 1989 Influence of the implantation of nitrogen ions and carbon bearing steel for resistance, *Physics and Chemistry of Materials Processing*, **1** 80-82

[15] Guseva N I 1991 Hardening of machine parts by ion implantation, *Problems of mechanical engineering and reliability of machines*, **4** 80-89

[16] Labunov V A and Reise G 1982 Ion-beam sources for the treatment of solid surfaces and produce thin films, *Foreign electronic machinery* **1** 1-45.

[17] Pogrebnyak A D, Remnev G E, Chistyakov S A, Legachev A E 1987 Modifying the properties of metals under the influence of powerful ion beams, *Proceedings of Higher Educational Institutions. Physics*. **1** 52-55

[18] Uglov V V, Kuleshov A K, Fedotova Yu A 1999 Plasma immersion implantation of nitrogen in high-speed steel. The phase composition and mechanical properties, *Physics and Chemistry of Materials Processing*, **5** 18-25

[19] Brzhozovskii B M, Martynov V V, Zinina E P 2014 Effects of low temperature plasma on the surface of complex profile (physical and mathematical aspects), *Current problems electronic engineering (APEP-2014)*, Proceedings of the international scientific and engineering conference, Saratov: SSTU, 244-250

[20] Brzhozovskii B M, Martynov V V, Zakharov O V, Zinina E P, Tsarenko M A 2013 *Innovative ways of increasing the efficiency of cutting tools* (Staryi Oskol: Thin High Technologies Publishers)

Main problems and perspectives of the synthesis of nanocomposite coatings on the surface of complex-shaped components

B Brzhozovskii*, **V Martynov, E Zinina and M Brovkova**
School of Electrical and Mechanical Engineering, Yuri Gagarin State Technical University of Saratov, 77 Politekhnicheskaya Ul., Saratov 410054, Russia

E-mail: bmbsar@mail.ru

Abstract. The main methods for composite coating synthesis in the surface layer of complex-shaped workpiece with dissipative properties have been investigated in this study. The problem of their application can be solved by using low-temperature plasma of combined discharge. The plasma is formed directly around the treated surface, which allows the use of the maximum amount of energy transmitted by its ions for the synthesis process itself. The main mechanism is the synthesis of three-body recombination, in which surface structures are heated to temperatures that change the phase of their constituent atoms. As a result, the original structure of the surface layer of the workpiece is transformed into a composite structure consisting of nanoclusters of ~ 50 ... 100 nm in size and an amorphous layer of particles which acts as binder, and gradient transition from the coating to the base material filling the microvoids.

1. Introduction

The current stage of production development in various industrial sectors is characterized by tighter operating conditions for items made of metals and alloys. This requires continuous refinement of existing methods and development of new manufacturing technologies to improve the product operational reliability in terms of abrasion and fatigue resistance.

Traditionally, increasing fatigue strength and abrasion resistance is ensured by protective coatings [1-7]; however, a promising approach is associated with the creation of technologies and equipment that allow formation of coating structures with dissipative properties due to the presence of the amorphous phase that performs the work of irreversible local movement of minimal number of minimum size particles (groups of atoms) in a minimum of time without momentum transfer. In conventional coatings, atoms do not have such freedom, so, any displacement leads to surface cracks and crystal lattice defects causing delamination and destruction of coating as well as reduced chemical activity and mechanical strength.

Special coatings obtained by ion-plasma deposition (IPD), as well as vapor deposition of nano-dispersed powder materials can act as dissipative structures [8-11]. Studies in this direction are carried out in many countries: Russia, Japan, the USA, Germany, France, Canada, South Korea, Switzerland, Israel, Liechtenstein, Ukraine and Belarus. Coatings produced by IPD methods are homogeneous, poreless and tight fit on the base layers of the workpiece of any configuration. In the course of IPD, plasma condenses on the metal surface with a possible simultaneous plasma chemical reaction to form a compound of the desired chemical composition.

There are several modifications of the IPD method:

1) activated evaporation with chemical reaction (ARE);
2) reactive electron-beam plasma deposition (REP);
3) condensation of matter with ion bombardment (CIB);
4) low pressure plasma deposition (LPPD);
5) hollow cathode deposition (HCD);
6) cathodic arc low temperature separated ion deposition (CALTSID) which lets you create:
- multilayer nano-gradient coatings from separative plasma of arc discharge;
- atomic layer coating by applying microlayers of various components of 1-100 nm thick;
- self-lubricating and anti-friction diamond-like coatings (DLC coatings), i.e. "solid lubricant", which operate in a wide range of temperatures in vacuum, at high and low temperatures on the cutting tool, in ground power plants and aerospace engines;
- wear and corrosion-resistant multilayer ion-plasma coatings for various applications, used in hard alloys working in heavy conditions of interrupted cuts on metal cutting machines.

2. Problems

It should be noted, however, that obtaining these coatings is associated with the solution of the whole set of problems, complex from both technological and technical points of view, which include:

- elimination of the droplet phase accompanying the ion-plasma deposition, which requires substantial modernization of technological equipment;
- realization of a long and multi-stage process cycle which includes, for example, pumping - ionic surface cleaning - ion etching - ion-plasma nitriding - application of microlayer of titanium and aluminum nanolayers 1-100 nm thick - microlayer of titanium nitride and aluminum nitride 1-100 nm thick, with ionic surface cleaning after each microlayer;
- creation of ultra-high vacuum (~ 10-7 mm Hg) in the working chamber, which leads to lower coating speeds and hence a longer cycle time. The latter, in turn, determines the need for special preventive measures for maintaining the equipment in working condition.

In addition to the above, the absence of transition layer in the coatings is responsible for their low adhesion to the base substrate and, as a result, it cannot be used at high temperatures and power loads. The formation of 'coating – substrate' transition layers also requires ion implantation that needs special equipment with high accelerating voltage (~ 50 kV). Finally, these technologies dictate the transformation of the volume created by flaming workpiece and its delivery to the working surface. In this case, most of the energy consumed is spent on the plasma creation, maintaining its generation conditions and the delivery of active plasma particles with a given energy to the subject of treatment with their subsequent scanning on the surface; while the same process of nanocoating formation requires only a few percent of the total energy input.

The use of nano-dispersed powder materials also has a number of negative technological features caused by the resulting powder as a fraction of 60-100 nm when grinding and the special conditions of storage of such powders as well as difficulties of formation of structurally uniform coatings when applying nanodisperse powders by gas-phase synthesis with vapour condensation method.

Thus, the objective difficulties of forming multi-layer nano-coatings make them exclusive, expensive and only available for laboratory research, while their industrial use seem problematic. It leads to the conclusion that the formation of geometrically complex products of chemically inactive nanocomposite dissipative structures with high hardness and stability, especially when operating in extreme conditions in the surface layers presents a critical scientific and practical problem.

3. Solution procedure

It is possible to generate dissipative structures in the workpiece surface layer by exposing it for 1.5 ... 15 min. to a low-temperature plasma formed by superimposed electrostatic microwave electromagnetic field at a reduced level of microwave power (30 ... 90 W) at a pressure of about 300 Pa in a mixture of process gases at a total power consumption of 2.0 kW [12]. The plasma is formed directly around the treated surface, which allows the use of the maximum amount of energy transmitted by its ions for the modification process itself, rather than for maintaining the conditions for the generation and delivery of the active plasma species to the workpiece.

The study of the impact of low-temperature plasma on the surface of the product revealed that its main mechanism is the three-body surface recombination (Figure 1) at which neutralization of charge carriers by combining oppositely charged carriers into the neutral molecules on the surface of the product takes place. Unlike ionization, recombination occurs with the release of potential energy at arbitrarily low kinetic energy of the interacting particles. Mathematically, this process is described by the following relationships:

$$J_{\text{emis}} = \frac{I_e + I_{p_1^+} + I_{p_2^+} + I_{p_2^-}}{S_{\text{bd}}}, \qquad (1)$$

$$Q = n_e n p_{p_1^+} \frac{N_B}{S} \gamma_{p_1^+} \langle U_i + U_a \rangle + n_e n_{p_2^+} \frac{N_B}{S} \gamma_{p_2^+} \langle U_j + U_a \rangle, \qquad (2)$$

where J_{emis} is the compensated flow emitted from the unit surface area of the lower boundary of plasma S_{bd}, which consists of low-energy electrons (I_e), singly charged ions ($I_{p_1^+}$ $I_{p_2^+}$ $I_{p_i^-}$) z: type p_1, p_2 particles and negative ions p_i^-, falling on unit surface area of the product S; Q is the amount of heat transferred to the surface atoms per unit time in the process of recombination of positive and negative monovalent ion; $\langle U_i + U_a \rangle$ and $\langle U_j + U_a \rangle$ are averaged energies. The value $\langle U_i + U_a \rangle$ is greater than or equal to the potential of p_i atoms ionization, and $\langle U_j + U_a \rangle$ is greater than or equal to the energy of formation of negative and positive ions.

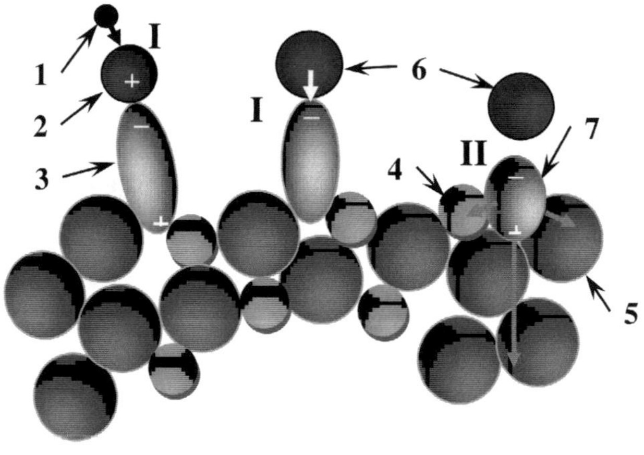

Figure 1. Steps (I-III) of the three-body recombination: 1 - incident electron, 2 - adsorbed ion, 3 - surface atom (dipole), 4, 5 - subsurface atoms, 6 - neutral atom resulting from recombination (during and after transmission of the stored energy), 7 - surface atom displaced as a result of recombination.

The main result of the recombination process is the heating of the surface structures to temperatures changing the phase of their constituent atoms and leading to their displacement in the nature of thermal vibrations. In other words, a layer of the liquid phase (melt) is formed that contains unstable groups of individual disordered atoms and dense metastable atom groupings with higher local resistance and ordering symmetry different from that of the crystal one. Subsequent quenching of the surface under reduced pressure gives fixation (solidifying) of the formed melt. As a result, the original structure of the surface layer of the workpiece is transformed into a nanocomposite structure consisting of (Figure 2):

- nanoclusters of ~ 50 ... 100 nm in size and high strength, and an amorphous layer of particles of the same chemical composition with the size of ≤10 nm, which actss as binder (coating);
- gradient transition from the coating to the base material (sublayer) filling the microvoids.

4. Results and discussion

The results of the structure microhardness measurement and their analysis showed [13] that, firstly, the microhardness increases (Figure 3), and secondly, it is achieved by consolidation of the surface layer (Figure 4), which affects the structure behavior under working conditions. Above all, this applies to operation under conditions of intense temperature and power loads, including alternating stress, for example, of the tool during mechanical or ultrasonic machining. Therefore, to assess the prospects for the practical use of products with nanocomposite coatings it is necessary to study their operation process.

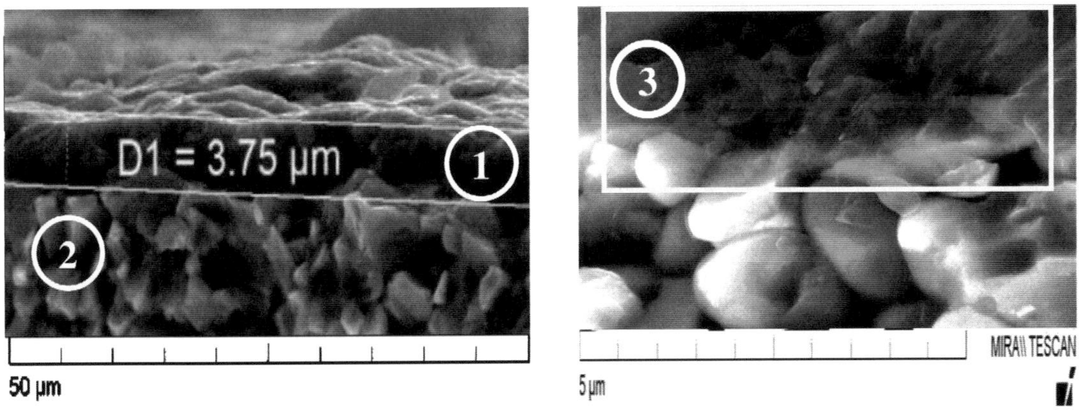

Figure 2. The structure of the working part of the product: 1 - synthetic composite coating; 2 - matrix; 3 – sublayer.

Figure 3. Imprints of the diamond pyramid on the surface of the carbide article at the same load before (a) and after (b) plasma treatment.

Research would be appropriate from the standpoint of the following provisions:

- in an elementary volume of the coating containing some free space in the form of voids between the atoms of the amorphous phase coating and some free space in the form of grain voids of adjacent underlayer, local deformation will be accompanied by an increase in the free volume both in the coating and in the underlayer;
- the coating is exposed to typical for amorphous materials processes of local tension and compression which lead to local surface micro volumes displacement determined by the direction and nature of the applied load;
- the presence of nanoclusters enhances the coating capacity for structural adjustment in the narrow spatial scale;
- underlayer of a high density and, therefore, hardness and having a structure with a minimum of microvoids decreases mobility of the grains of the base material and increases its resistance under external temperature and force pressure.

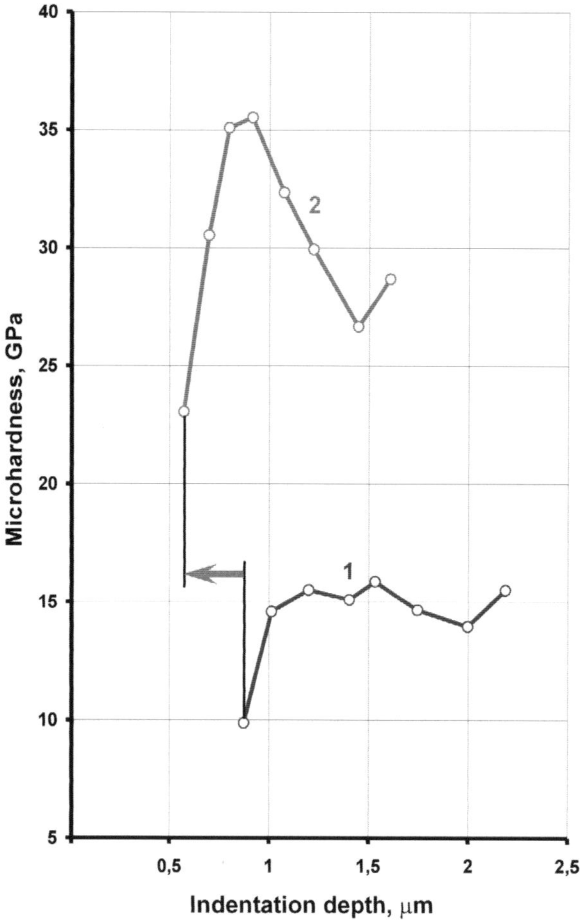

Figure 4. A typical example of the distribution of microhardness in depth of the surface layer of the carbide product before (1) and after (2) composite coating formation: arrow shows the consolidation of surface layer structure.

Taken together, these provisions determine the ability of the nanocomposite coating to perform the dissipative function in the operation of the product.

5. Conclusions

The preliminary results show (Figure 5) that the dissipative functions are carried out by a composite coating structure according to multi-stage scheme, since with the loss of stability of the structures of ~ 200 nm, the structures of ~ 40-70 nm maintain the contact with the amorphous binder, flexibility, the ability to migrate and can be transformed, i.e. form a stable streamlined contact surface. This gives reason to hope that, based on the study, we will be able:

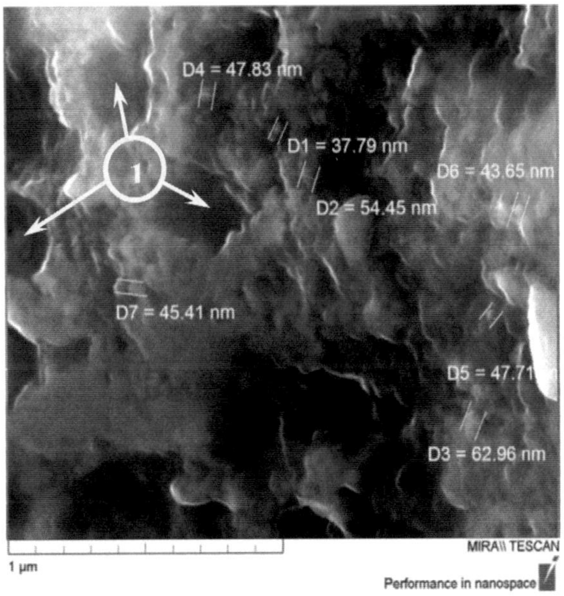

Figure 5. The result of the composite coating structure dissipation functions during operation of the solid alloy workpiece: 1 - voids of submicron structures (~200 nm).

- to establish the interrelations between the physical-mechanical, chemical characteristics of the 'cover – sublayer' structure and the conditions of its synthesis;
- to solve the problems of optimizing the process of synthesis of the nanocomposite coating according to the criteria of intensity and uniformity of the impact of low-temperature plasma on the surface of the workpiece, and the process of use of the product by the criterion of composite coating guaranteed performance of its dissipative functions.

Acknowledgement
The authors would like to acknowledge the assistance of the Russian Science Foundation. The present study was supported by the Russian Science Foundation grant (Project No 15-19-00030).

References

[1] Stolyankov Yu V, Antyufeeva N V, Rasutin E A and Karimova S A 2014 The possibility study of synthesis of layered metal-polymeric composite materials using light gage amorphous alloys, *Composites and nanostructures* **6(1)** 25-31 (in Russian).
[2] Dmitrienko Yu I, Sborshikov S V, Prozorovsky A A et al. 2014 The development of multi-layered polymer composite with a discrete orthotropic filler, *Composites and nanostructures* **6(1)** 32-48 (in Russian).
[3] Popov V S, Sevastyanov V G and Kuznetsov N T 2012 Synthesis of nanostructured SnO_2 coatings using APCVD method with induction heating and volatile precursors, *Composites and nanostructures* **1** 33-43 (in Russian).

[4] Dvornik M I, Ershova T B and Zaitsev A V 2011 Synthesis of nanostructured tungsten-cobalt high-resistance hard alloys, *Composites and nanostructures* **2** 40-49 (in Russian).
[5] Bakhvalov Yu O, Mishenznikov G E, Anisimova A S et al. 2011 Synthesis and study of production of antifrictional high-resistance coating for mechanical engineering using magnetron deposition of chromium and ion implantation, *Composites and nanostructures* **4** 5-13 (in Russian).
[6] Korneeva E A, Skomorokhov A N, Kolobov Yu R, et al. 2011 The structure and properties of "titanium-coating" composites obtained using micro-arc oxidation, *Composites and nanostructures* **4** 32-42 (in Russian).
[7] Pashentsev V 2010 Vacuum ion-plasma deposition of nanostructured coatings, *Nanoindustry*, **5** 26-28 (in Russian).
[8] Grigor'ev S N 2009 Methods for increasing the resistance of cutting tools, Moscow: Mashinostroenie (in Russian).
[9] Vereshchaka A S, Grigor'ev S N, Kim V A, Mokritskii B Ya, Altukhova V V, Kirichek A V 2013 High tech tool hardening, *Naukoemkie tekhnologii v mashinostroenii*, **6** 19-24 (in Russian).
[10] Kurochkin A V 2010 Technological features of multifunctional nanostructured coatings for cutting tools, *Vestnik RGATA*, **3**, 167-172 (in Russian).
[11] Migranov M Sh 2005 Investigation of wear-resistant coatings for cutting tools with nanocrystalline structure, *Izvestiya vuzov. Mashinostroenie*, **1** 56-62 (in Russian).
[12] Brzhozovskii B M, Martynov V V, Zinina E P 2009 Cutting tool hardening through low-temperature plasma of combined discharge, Saratov: Saratov state technical university (in Russian).
[13] Brzhozovskii B M, Martynov V V, Zinina E P, Starinnova V O 2013 Justification and development of a software algorithm for evaluation of the results of plasma modification of the cutting tool working surfaces, *Science Intensive Technologies in Mechanical Engineering*. **5** 41-48 (in Russian).

IOP Conf. Series: Materials Science and Engineering **116** (2016) 012009 doi:10.1088/1757-899X/116/1/012009

Synthesis of positively and negatively charged silver nanoparticles and their deposition on the surface of titanium

A Sharonova[1], K Loza[2], M Surmeneva[1], R Surmenev[1], O Prymak[2] and M Epple[2]

[1] National Research Tomsk Polytechnic University, Tomsk, Russia
[2] University of Duisburg-Essen, Essen, Germany

E-mail: rsurmenev@mail.ru

Abstract. Bacterial infections related to dental implants are currently a significant complication. A good way to overcome this challenge is functionalization of implant surface with Ag nanoparticles (NPs) as antibacterial agent. This article aims at review the synthesis routes, size and electrical properties of AgNPs. Polyvinyl pyrrolidone (PVP) and polyethyleneimine (PEI) were used as stabilizers. Dynamic Light Scattering, Nanoparticle Tracking Analysis, X-ray diffraction (XRD), scanning electron microscopy (SEM), and energy dispersive spectroscopy (EDX) have been used to characterize the prepared AgNPs. Two types of NPs were synthesized in aqueous solutions: PVP-stabilized NPs with a diameter of the metallic core of 70 ± 20 nm, and negative charge of -20 mV, PEI-stabilized NPs with the size of the metallic core of 50 ± 20 nm and positive charge of +55 mV. According to SEM results, all the NPs have a spherical shape. Functionalization of the titanium substrate surface with PVP and PEI-stabilized AgNPs was carried out by dropping method. XRD patterns revealed that the AgNPs are crystalline with the crystallite size of 14 nm.

1. Introduction

Metals have a long history in the treatments of dentistry and orthopedics. Pure titanium is commonly used as artificial joints and implants in both dental and orthopedic clinics because of its biocompatibility and mechanical properties. A major factor that determines the success of dental implantation is osseointegration, which is the stable anchorage of an implant in living bone achieved by direct bone-to-implant contacts [1, 2]. It is commonly known that the implantation of a foreign object into the human body may be rejected. Moreover, a wide range of local tissue reactions, in particular inflammation, giant cell formation and fibrosis can be induced [3]. A promising solution to this problem is formation of an antibacterial bioactive coating on the implant surface, which allows to avoid rejection and speed up the treatment and recovery process [4, 5].

The consequences of implant-associated infection are significant and usually require revision surgery, with removal of the implant and prolonged antibiotic treatment. Various approaches have been investigated to reduce the rate of infection. Two recent strategies are the coating of implants with antibiotics and followed by covalently attaching of antimicrobial nanoparticles to the implant surface. The remarkable mechanism of the antimicrobial activity of Ag nanoparticles (AgNPs) is related to the formation of free radicals and consequent free-radical-induced oxidative damage of the cell membranes of bacteria [6, 7]. The purpose of these bioactive surfaces is to disrupt the metabolic machinery of the microbes or to prevent bacterial adhesion to the implant and, consequently, the development of biofilm [8].

Currently, many methods have been reported for the synthesis of AgNPs by using chemical [9-11], physical [12, 13], photochemical [14, 15] and biological [16-18] routes. Each method has advantages and disadvantages with the common challenges such as production costs, scalability, particle sizes and size distribution. Among the existing methods, the chemical methods have been mostly used for production of AgNPs. Chemical methods provide an easy way to synthesize AgNPs in solution.

A vast number of nanoparticle systems have been reported to be used to prepare highly stable Ag NPs, based on various groups that are capable of binding to metal surfaces: phosphonic acid [19], aminocellulose as a combined reducing and capping reagent [20]. Useful coatings for biomedical application and delivery of AgNPs, based on glutathione, thiols, disulfides, thioethers, thioesters, thiocarbonates, and thiocarbamates are used in the preparation of metal nanoparticles [21]. Poly(vinylpyrrolidone) (PVP) and poly(ethylenimine) (PEI)-stabilized AgNPs are stable in suspension and exhibit little change in size or electrophoretic mobility [22, 23].

The purpose of this study is to find appropriate parameters of AgNPs such as size, ζ - potential, charge for the deposition on metallic substrates. These parameters of AgNPs are important for the development of the structure of scaffolds.

2. Materials and methods

2.1. Samples preparation

To prepare the samples as substrates commercially pure titanium (99.58 wt% titanium; 0.1 wt% oxygen; 0.15 wt% iron; 0.05 wt% carbon; 0.04 wt% nitrogen; 0.08 wt% silicon) plates (10×10 mm) were used. The plates were chemically etched in acid solution containing HF (48%) and HNO_3 (66%) dissolved in the distilled water with the ratio 1:2:2.5 in volume. After acid etching, the samples were ultrasonically washed sequentially in ethanol and demonized water for 10 minutes at room temperature.

2.2. The synthesis of positively and negatively charged AgNPs

The negatively chargedAgNPs were synthesized by wet chemical reduction method of silver nitrate (Fluker, p.a.) using glucose (D-(+)-glucose, Baker) as a reductant and polyvinylpyrrolidone (PVP K30 Povidon 30; Fluka, molecular weight 40000 g mol^{-1}) as a stabilizer. 2 g glucose and 1 g PVP were dissolved in 40 g water and heated to 90 °C. Then 0.5 g $AgNO_3$ dissolved in 1 mL water was quickly added. The synthesis of positively charged, PEI-stabilized, AgNPs was proceeded without use of glucose. The 1g polyethylenimine (PEI; branched, Aldrich, molecular weight 25 000 g mol^{-1}) was dissolved in 40 g water and heated to 90°C. Then 0.5 g $AgNO_3$ dissolved in 1 mL water was quickly added. The dispersion was kept at 90 °C for 1 h, PVP-stabilized AgNPs and then left to cool to room temperature. The particles were collected by ultracentrifugation (3 times, 30 000 rpm, 30 min), redispersed in pure water and collected again by ultracentrifugation. Thereby NO^{3-}, excess glucose and its oxidation products, excess PVP/PEI, and excess Ag^+ were removed. The silver nanoparticles were further redispersed in water.

2.3. Characterization of AgNPs

The hydrodynamic diameter (HDD) of the AgNPs were measured by Dynamic Light Scattering (DLS) using a Malvern Zetasizer Nano ZS and Nanoparticle Tracking Analysis (NTA) using a NanoSight LN 10. Zeta (ζ) potential of the AgNPs were measured by DLS. Both of these methods are based on the rate of Brownian motion to particle size and applicable to analyse the particles of a diameter 10-2000 nm in liquids. DLS technique is used to measure the diffusion of particles and converts this to the size and a size distribution using the Stokes-Einstein relationship. Non-invasive back scatter technology is used to give the highest sensitivity simultaneously with the highest size and concentration range. NTA is a unique method for visualisation. The light scattered by the particles is captured using a scientific digital camera and the motion of each particle is tracked from the frame to frame by the specially developed software. This rate of particles movement is related to a sphere equivalent hydrodynamic radius as calculated through the Stokes-Einstein equation [24-26]. By scanning electron microscopy (SEM) the morphology and size of the AgNPs were estimated using an ESEM Quanta 400 FEG instrument operating in a high vacuum with gold/palladium-sputtered samples. The Grazing-incidence small-angle X-ray scattering (GISAXS) was used to determine the internal structure of studied AgNPs. The phase parameters of AgNPs were investigated by the method of X-ray diffraction in the Bragg-Brentano geometry (Siemens Diffractometer D500), in a direction perpendicular to the diffraction plane, calculated according to the Scherrer Equation [27]:

$$R_{hkl} = k \cdot \lambda(\beta \cdot \cos\theta)^{-1}, \quad (1)$$

where, R_{hkl} - crystallite size in the direction perpendicular to the diffraction plane hkl; $k = 0.9$, λ - wavelength of the analyzing X-ray beam; θ - angle subtended by the diffraction, $\beta = B_{1/2}-b_{1/2}$ - the true width of the line; $B_{1/}$ and $b_{1/2}$ - experimental and instrumental broadening of the width of the diffraction peak at half intensity.

2.4. *Dropping method of AgNPs deposition*

The deposition of PVP and PEI stabilized AgNPs on titanium substrate was done by dropping method. The process of dropping method based on formation of a drop 120 μL of the working solution with the concentration 60 μg/mL and following drying at 55.5°C.

3. Results and discussions

Table 1 contains information about the diameter of the metallic core, HDD as determined through DLS, the PDI and ζ–potential. The HDD, ζ-potential and polydispersity index (PDI) of the nanoparticles were measured by DLS using a Malvern Zetasizer Nano ZS. The results of DLS analysis revealed that PVP-stabilized AgNPs had the ζ- potential of -20 mV, average HDD of 110 nm and PDI of –0.195, indicating the absence of large agglomerates and presence of a monodisperse system. Figure1 illustrated the data describing PVP-stabilized AgNPs with the dispersion time 1 hour. The SEM analysis confirmed the ability to attain a uniform distribution of AgNPs. Thus, SEM images showed that the PVP-stabilized AgNPs had a spherical shape with a diameter of the metallic core of 70±20 nm. It is reported [28] that silver nanoparticles (PVP-stabilized, 70 nm) affected the viability of Staphylococcus aureus colonies. The antimicrobial activity of silver nanoparticles was tested using standard methods which determine the minimum inhibitory concentration (MIC) and the minimum bactericidal concentration (MBC). S. aureus (10^6 cells mL^{-1} in medium) was treated without (S. aureus in pure medium: positive control; pure medium: negative control) or with 50 mg/mL or 30 mg/mL of silver nanoparticles for 24 h under cell culture conditions. Subsequently, the bacteria were plated and incubated for further 24 h at 37 °C on blood agar plates. From the reported experiments and the available literature on the dissolution of silver nanoparticles, they conclude that the particles will dissolve after coming into contact with air, therefore they will become increasingly bactericidal (and cytotoxic) with time.

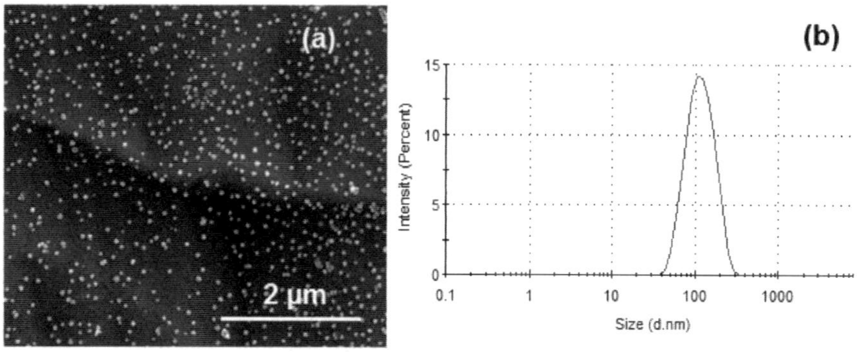

Figure 1. a) SEM image of PVP-stabilized AgNPs, b) DLS estimated distribution size of AgNPs.

The results of DLS analysis of PEI-stabilized AgNPs showed the ζ-potential of +55 mV, average HDD of 110 nm (figure 2). The PDI for PVP-stabilized AgNPs was lower than 0.3, which could indicate that system was monodisperse, however, PDI in case of PEI-stabilized AgNPs was slightly more than 0.3. Moreover, according to DLS of AgNPs obtained using PEI had a multidistribution. Typical SEM image of AgNPs with spherical shape was shown in figure 2. Zhiguo Liu et al. [29] showed that the PEI-functionalized Ag nanoparticles were positively charged. Moreover, the Ag colloids exhibited stronger antibacterial activity in the bactericidal test. Its bactericidal efficiency exceeds the commonly used antibacterial agents such as Erythromycin, chloramphenicol and penicillin as well as $AgNO_3$ solution. These results prove that our synthesis method is efficient to produce a stable PEI-functionalized Ag colloid with excellent antibacterial activity.

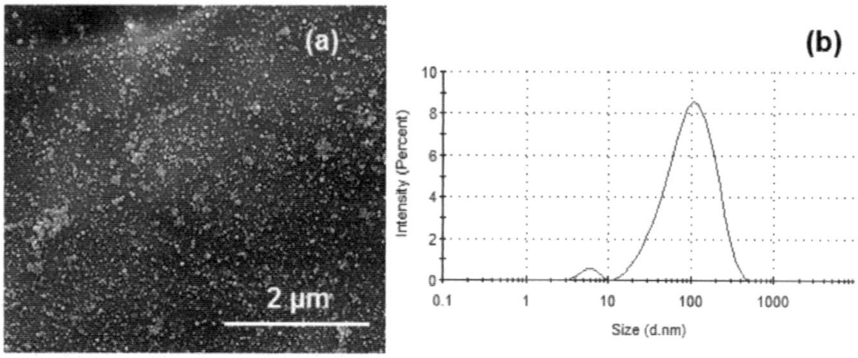

Figure 2. a) SEM image of PEI-stabilized AgNPs, b) DLS estimated distribution size of AgNPs.

Negatively charged AgNPs are monodisperse and have a narrow size distribution. These properties of AgNPs are prospective to develop antibacterial layers on implant surfaces.

Table 1. The properties of silver nanoparticles stabilized with PEI and PVP polymer.

Polymer	Diameter, nm			PDI	ζ–potential, mV
	NTA	DLS	SEM		
PVP	115	110	70	0.195	-20
PEI	30	110	50	0.404	+55

To form an antibacterial interface of metal implant the dropping method was used (120 μL with concentration 60 μg/mL).

The patterns of AgNPs distribution on titanium surface substrate are shown in figure 3.

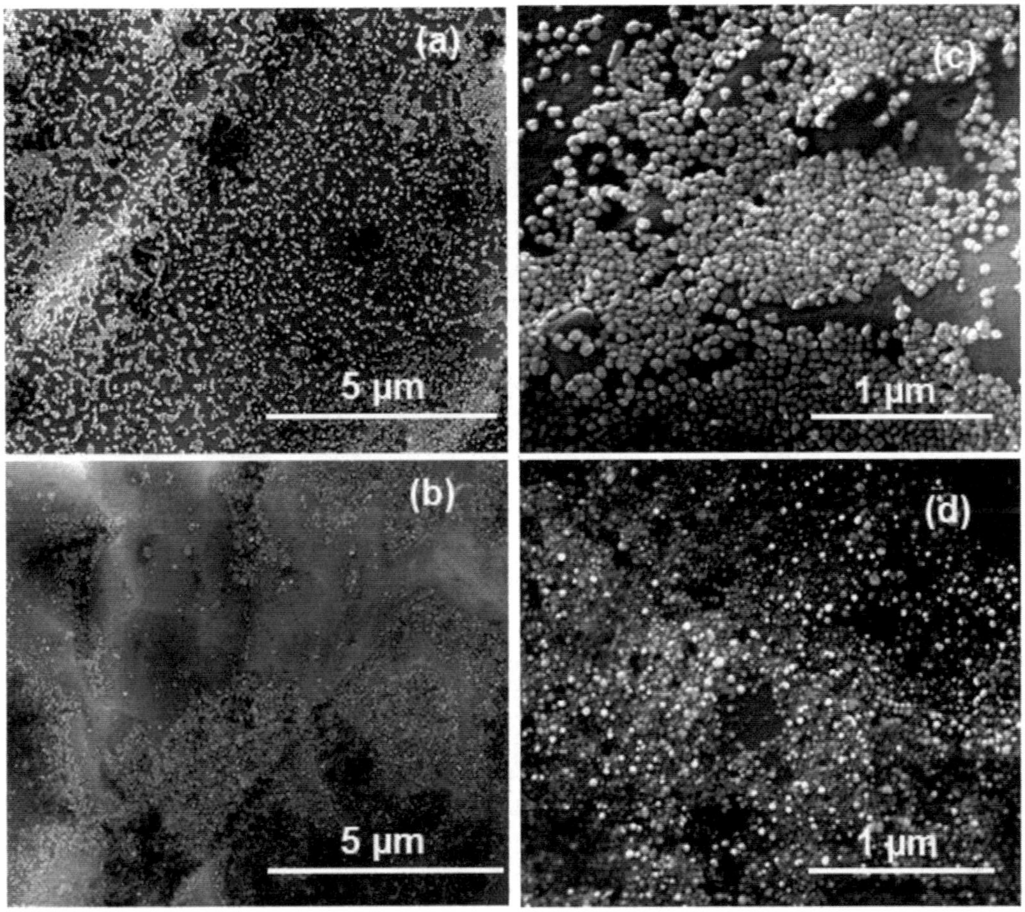

Figure 3. SEM images distribution of AgNPs on the surface of pure titanium by drop method (120 μL with concentration 60 μg/mL) stabilized with a), c) PVP and b), d) PEI.

SEM images revealed that the both types of AgNPs have a spherical shape (figure 3). PVP-stabilized AgNPs were homogeneously distributed over the entire surface. Dropping method has the disadvantage associated with the surface tension and as a consequence the samples are not fully covered by the particles. The particle size of PEI-stabilized AgNPs was smaller than PVP-stabilized AgNPs as it could be seen from the obtained data (Table. 1). In case of positively charged AgNPs the dropping method does not allow a complete and homogeneous coverage of the sample surface. The negatively charged AgNPs were used to create a homogeneous layer of AgNPs on the surface of titanium without any agglomeration. Negatively charged AgNPs were studied by XRD to reveal their structure parameters. In figure 4 and Table 2 the typical XRD patterns are presented.

XRD pattern of AgNPs on the titanium substrate showed the presence of reflexes at 2 Theta angles of 44.3° and 77.3°, which can be indexed to (200) and (311) planes of pure silver (PDF # 04-0783). Since the main reflection of titanium and silver were very close it was decided to use a silicon substrate as a control the presence of silver.

The typical XRD pattern, which illustrates the formation of intense lines of reflection of the Ag, is shown on silicon substrate (figure 4). The presence of peaks at 2 θ values 38.1°, 44.3°, 64.4° and 77.3° corresponded to (111), (200), (220), and (311) planes of silver, respectively. Thus, the XRD pattern confirmed the cubic crystalline structure of silver (JCPDS, 4-0783). The lattice constants calculated from XRD pattern of PVP and PEI-stabilized NPs were found to be a= 4.093 Å and 4.085 Å, correspondently. The crystallite size of AgNPs was found to be 14 and 13 nm in the case of titanium and silicon substrate, respectively (Table 2). The aim of the developed biocomposite is a long-term antibacterial protection against implant associated infections.

Table 2. The lattice parameters and crystallite size of AgNPs on titanium and silicon substrate.

Sample	Lattice parameters		Crystallite size, nm
	a=b, Å	*V, Å*^{3}	
Titanium	4.093	68.52	14
Silicon	4.085	68.22	13

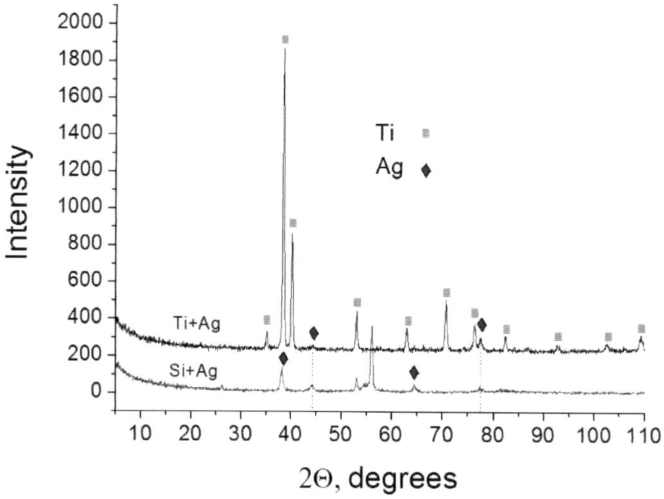

Figure 4. The XRD patterns of AgNPs deposited by dropping method (120 μL with concentration 60 μg/mL) on titanium and silicon substrates.

4. Conclusions

In this study two types of AgNPs were synthesized: PVP-stabilized negatively charged (-20 mV) nanoparticles with the size of 70 ± 20 nm and PEI-stabilized positively charged (+55 mV) nanoparticles with the size of 50 ± 20 nm. According to the SEM data, all the particles had a spherical shape. The dropping method was used for AgNPs deposition on titanium substrate. It was revealed that the dropping method did not allow to get a homogeneous layer of nanoparticles in case of the PVP-stabilized AgNPs. In contrast with the PVP-stabilized AgNPs, the uniform layer of PEI-stabilized AgNPs was formed. No agglomeration of AgNPs after deposition on the titanium substrate was observed. The XRD data obtained for the PVP-stabilized AgNPs on Ti substrate showed the typical peaks of Ag at 2 Theta angles of 44.3° and 77.3° with the coherent scattering region of 14 nm.

References

[1] Hanawa T 2010 *Japanese Dental Science Review. Series* **46** pp 93–101
[2] Guo C Y, Matinlinna J P and Tang A T H 2012. *Int. J. Biomater.* **2012** 381535
[3] IARC Working Group on the Evaluation of Carcinogenic Risks to Humans 1990 *Surgical implants and other foreign bodies* (Lyon: IARC Press)

[4] Lilja M. 2013 *Bioactive Surgical Implant Coatings with Optional Antibacterial Function Digital Comprehensive Summaries of Uppsala Dissertations from the Faculty of Science and Technology 1091* (Room: Acta Universitatis Upsaliensis)

[5] Farooq I, Irman Z, Farooq U, Leghari A, Ali U. 2012Bioactive Glass: A Materials for the Future *World Journal of dentistry* **3** 199–201

[6] Gupta K, Singh R. P., Pandey A, Pandey A 2013 Photocatalytic antibacterial performance of TiO_2 and Ag-doped TiO_2 against S. aureus. P. aeruginosa and E. coli *Beilstein J. Nanotechnol.***4** 345–351

[7] Hwang E T, Lee J H, Chae Y J, Kim Y S, Kim B C, Sang B I, Gu M B 2008 Analysis of the toxic mode of action of silver nanoparticles using stress-specific bioluminescent bacteria *Small* **4** 746–50

[8] Ketonis C, Parvizi J, Jones L C 2012 Evolving Strategies To Prevent Implant-associated Infections *JAAOS* **20** 478–80

[9] Chen S, Zhang H, 2012 Aggregation kinetics of nanosilver in different water conditions *Adv. Nat. Sci.: Nanosci. Nanotechnol.* **3** 035006

[10] Dang T M D, Le T T T, Fribourg-Blanc E, Dang M C 2012 Influence of surfactant on the preparation of silver nanoparticles by polyol method *Adv. Nat. Sci.: Nanosci. Nanotechnol.* **3** 035004

[11] Patil R S, Kokate M R, Jambhale C L, Pawar S M, Han S H, Kolekar S S 2012 One-pot synthesis of PVA-capped silver nanoparticles their characterization and biomedical application *Adv. Nat. Sci.: Nanosci. Nanotechnol.* **3** 015013

[12] Siegel J, Kvítek O, Ulbrich P, Kolská Z, Slepička P, Švorčík V 2012 Progressive approach for metal nanoparticle synthesis *Mater. Lett.* **89** 47-50

[13] Tseng K, Lee H, Liao C, Chen K, Lin H 2013 Rapid and Efficient Synthesis of Silver Nanofluid Using Electrical Discharge Machining J. of Nanomater. **2013** 174939

[14] Christy A J, Umadevi M 2012 Synthesis and characterization of monodispersed silver nanoparticles *Adv. Nat. Sci.: Nanosci. Nanotechnol.* **3** 035013

[15] Sakamoto M, Fujistuka M, Majima T 2009 Light as a construction tool of metal nanoparticles: synthesis and mechanism *J. Photochem. Photobiol. C* **10** 33–56

[16] Sintubin L, Verstraete W, Boon N 2012 Biologically produced nanosilver: Current state and future perspectives *Biotechnol. Bioeng.* **109** 2422–36

[17] Amaladhas T P, Sivagami S, Devi T A, Ananthi N, Velammal S P 2012 Biogenic synthesis of silver nanoparticles by leaf extract of Cassia angustifolia *Adv. Nat. Sci.: Nanosci. Nanotechnol.* **3** 045006

[18] Umadevi M, Shalini S, Bindhu M R 2012 Synthesis of silver nanoparticle using D. carota extract *Adv. Nat. Sci.: Nanosci. Nanotechnol.* **3** 025008

[19] Chaikin Y, Bendikov T A, Cohen H, Vaskevich A, Rubinstein I 2013 Phosphonate-stabilized silver nanoparticles: One-step synthesis and monolayer assembly *J. Mater. Chem. C* **1** 3573–83

[20] Cheng F, Betts J W, Kelly S M, Schaller J, Heinze T 2013 Synthesis and antibacterial effects of aqueous colloidal solutions of silver nanoparticles using aminocellulose as a combined reducing and capping reagent *Green Chem.* **15**, 989–98

[21] Taglietti A, Diaz Fernandez Y A, Amato E, Cucca L, Dacarro G, Grisoli P 2012 Antibacterial activity of glutathione-coated silver nanoparticles against Gram positive and Gram negative bacteria *Langmuir* **28** 8140–48

[22] Römer I, White T A, Baalousha M, Chipman K, Viant M R, Lead J R 2011 Aggregation and dispersion of silver nanoparticles in exposure media for aquatic toxicity tests *J. Chromatogr. A* **1218** 4226–33

[23] Kima K, Lee H B, Lee J W, Shin K S 2010 Poly(ethylenimine)-stabilized silver nanoparticles assembled into 2-dimensional arrays at water–toluene interface *J. Colloid Interface Sci.* **345** 103–108

[24] Chuaha L H, Robertsb C J, Billac N, Abdullahd S, Roslid R, Manickame S, 2014 Using Nanoparticle Tracking Analysis (NTA) to Decipher Mucoadhesion Propensity of Curcumin-Containing Chitosan Nanoparticles and Curcumin Release *J. Dispersion Sci. Technol.* **35** 1201–07

[25] Arzensek D 2010 *Dynamic light scattering and application to proteins in solutions* (Ljubljana: University of Ljubljana Faculty of Mathematics and Physics)

[26] Frisken, Barbara J 2001 Revisiting the Method of Cumulants for the Analysis of Dynamic Light-Scattering Data *Appl. Opt.* **40** 4087–91

[27] Surmenev R A, Surmeneva M A, Pichugin V F, Epple M 2009 RF-magnetron calcium phosphate coatings on materials medical implants *Bulletin of the Tomsk Polytechnic University* **315** 138–141.

[28] Loza K, Diendorf J, Sengstock C, Ruiz-Gonzalez L, Gonzalez-Calbet J M, Vallet-Regi M, Koller M, Epple M 2014 The dissolution and biological effects of silver nanoparticles in biological media *Cite this: J. Mater. Chem. B* **2** 1634–43

[29] Liu Z, Wang Y, Zu Y, Fu Y, Li N, Guo N, Liu R, Zhang Y 2014 Synthesis of polyethylenimine (PEI) functionalized silver nanoparticles by a hydrothermal method and their antibacterial activity study *Mater Sci Eng C Mater Biol Appl.* **42** 31–37

[30] Ivanova A A, Surmenev R A, Surmeneva M A, Mukhametkaliyev T, Sharonova AA, Grubova I Yu, Loza K, Chernousova S, Prymak O, Epple M, 2014 Antibacterial AgNPs/CaP biocomposites *CUET* **2014** 472–474

Shock Wave Treatment of Polytetrafluoroethylene: Structure and Properties

A E Volkova[1], N A Adamenko[1], G V Agafonova[1] and A V Kazurov[1]
[1]Volgograd State Technical University, 28, Lenina Av., Volgograd 400050, Russian Federation

E-mail: mvpol@vstu.ru

Abstract. The paper considers changes of the structure and thermomechanical properties of polytetrafluoroethylene pressing after the explosive treatment depending on the thickness of the pressing. Structural changes and defects in samples, preserved even after sintering, were detected.

1. Introduction

Purposive modification of fluoropolymers to change their structural organization and controlled variation of properties is one of the most urgent tasks in modern polymer materials science which has great practical importance in the development of materials with increased thermal and heat resistance. Currently different processing methods based on the use of external energetic influences: pyrolysis, radiation, laser and ultrasonic processing et al. are developed [1, 2, 3]. Explosive treatment (ET) is promising for polymers and composites on their basis which by the actuation of high speed shock wave in the solid provides the realization of both high pressure (up to 10 GPa) and temperatures (up to 1000 °C). This causes different activating physical (compaction, crushing, deformation and welding of the particles) and chemical (destructively-recombinational and mechanochemical reactions) transformations of the polymer [4-7]. It is determined that the modification of polymer materials, especially by high-energy impacts (including explosive treatment) leads to considerable changes in their structure, physico-chemical and mechanical properties, thereby converting the polymer to an activated state. ET is a multifactorial process, and the intensity of transformations is adjusted by changing of process parameters of the loading and the properties of the treated polymer powder.
This paper presents the results of research of the ET influence on the structural changes effecting on the thermomechanical properties of polytetrafluoroethylene (PTFE).

2. Study subject

The work researched the influence of various ET parameters on the structure formation and thermomechanical properties during the production of PTFE compacts (GOST 10007-80) of various thickness (0.8 and 3.2 mm). ET was realized by the flat loading with the sliding shock front (SF) with a detonation velocity of explosive 1680 ms^{-1}. Pulse of pressure (P) in the pressing powder depends not only on the power of the explosion, but also on the thickness of the pressed material. Increasing the height of the pressed powder with the same explosives reduces the pressure of ET and simultaneously extending its effect (τ). Thus, while ET of pressings 0.8 mm calculated values were P = 0.7 GPa and τ =13 ms (sample #1), and for 3.2 mm - P = 0.3 GPa and τ = 43 ms (sample #2).

3. Methods

The density (ρ) was determined by hydrostatic weighing using the analytical weighing system Shinko HTR-220SE GOST 15139-69. Structural changes were considered by a scanning electron microscope of FEI Versa 3D DualBeam and diffractometer "DRON-3.0" in the radiation of a copper anode CuK_{α}, analyzing the profiles of reflections with 100 and 003 indexes. Correlation of structures and thermal properties was determined by the thermomechanical analysis (TMA) 402 F3 Netzsch Hyperion by standard methods, by detecting deformations caused by penetration of the indenter into the sample under load 1N and heating velocity of 5 °C/min.

4. Results and discussion

The research shown, that the maximum compaction is reached after ET (pressings density is 2.2 – 2.28 Mg/m^3). Subsequent sintering doesn't change the density of the most thin samples (ρ = 2.23 Mg / m^3), and with the increase of thickness of pressing density reduces by 10% in the sample with thickness of 3.2 mm, due to the specialties of the PTFE crystalline phase melting during sintering and shrink ripple.

While studying the morphology (figure 1) of cleavage surface of the sample #2 (P = 0.3 GPa, τ = 43 ms), dense areas were found out (figure 1 a), and at a higher magnification (figure 1 c) - packless short fibrillar areas. In such a case, a level of interparticle interactions and properties reduce.

Samples with smaller thickness #1 (P = 0.7 GPa, τ = 13 ms) have consolidated particles and packless areas are not detected (figure 1 b). PTFE particles are considerably flattened in the direction of shock compression in both cases (figure 1 c, d), and dense formations are similar to the layered structures in fluoroplastic.

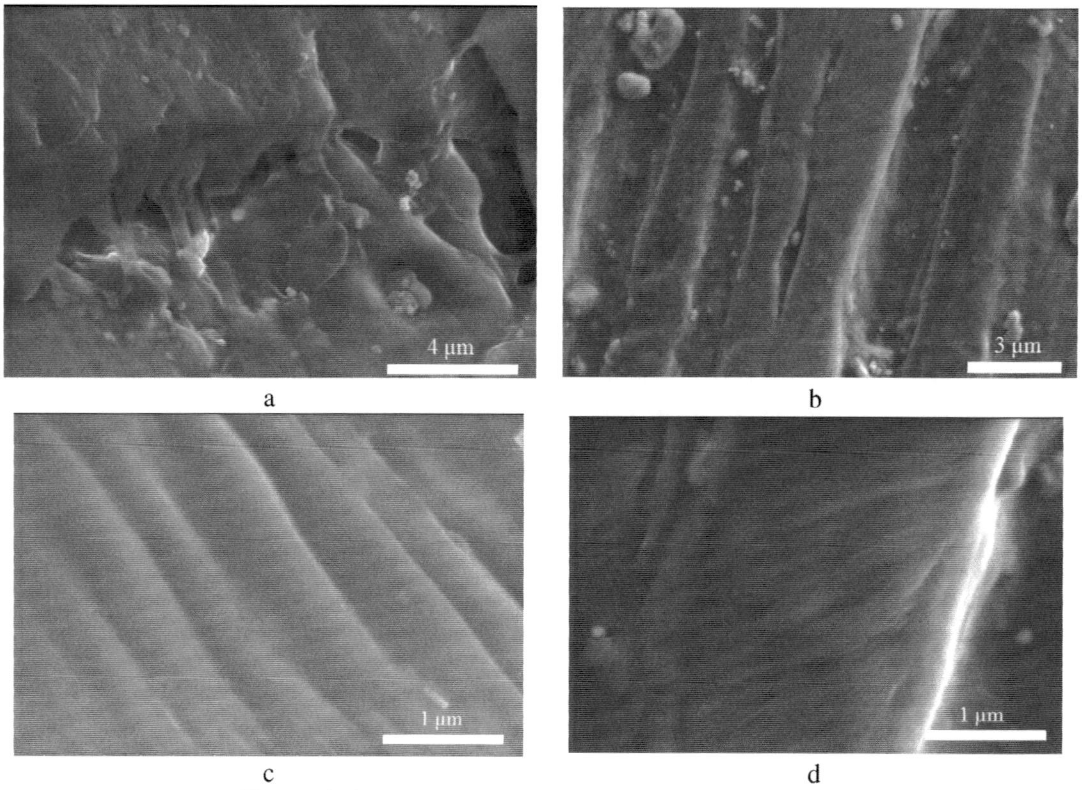

Figure 1. SEM images of PTFE after sintering and ET:
a, b - δ = 3.2 mm, c, d - δ = 0.8 mm

At high magnifications (figure 1 d) sample #1 shows dense oriented lamellar structure that is not found in other samples, which confirms more intense deformation of particles of thin layers in the ET. The energy dispersive analysis indicates the absence of destructive processes in the ET: the elemental composition of the PTFE is almost the same for all samples, and identical to that of the original polymer.

Research of the supramolecular structure of the samples by X-ray analysis showed, that amorphous crystal structure of PTFE remains with a high degree of crystallinity (65-71%), the crystalline peak at $2\Theta = 18°$ and diffuse halo in the region of $2\Theta = 40\text{-}50°$. However, the sample №2 revealed more intense changes in crystal structure in comparison with the thin sample №1. The broadening of the main reflex of crystalline phase (100) is mostly caused by crushing of crystallite size, which decreases more than 2 times from 38 to 15 nm, and the relative strain of the crystal lattice increases by 20% (table 1). Such changes are caused by the fact, that the mechanical resistance of the the polymer has ductile nature, and the effective in the explosive loading is not only the value of the impact pressure, but also pretty much its influence duration, the value of which in the sample №2 is 3 times (43 ms) bigger. The degree of crystallinity of PTFE with increasing of density of pressing under the conventional methods of processing should increase, but for the more dense sample №1 it is below 5% as well as after the radiation exposure of the polymer [8], when an increase of the radiation dose leads to the increase of the density of the material and simultaneously reduces crystallinity χ.

Table 1. The parameters of the crystal structure of PTFE

Thickness , δ (mm)	χ (%)	Structural defects		
		D (nm)	β (mrad)	$\Delta d/d$ (10^{-3} rad)
0.8	65	38	2.92	4.43
3.2	71	15	3.36	5.27

Changes occurring in the morphology and supramolecular structure of PTFE with the ET influence its thermomechanical properties (figure 2, table 2). The results of thermomechanical analysis (TMA) correlate with the results mentioned above. The residual porosity in the sample #2 leads to penetration of the indentor while heating, and its absence leads to expand of hard-grained sample #1. Amplification of intermolecular and interparticle interactions, as well as minimal crystal structure imperfection in sample #1 led to the PTFE melting point elevation from 327 to 340 ^{0}C. Alternatively, the melting point of sample #2 almost corresponds to the melting point of the primary polymer (327 ^{0}C).

Table 2. The thermomechanical properties of PTFE after ET and heat treatment in dependence of the thickness of the sample

δ (mm)	Softening		Relative deformation ε (%) at temperatures (°C)				
	t (^{0}C)	ε (%)	200	250	300	350	390
3.2	325	6	2	3.5	5	12	14
0.8	340	3	-2	-0.5	0.5	22.5	24

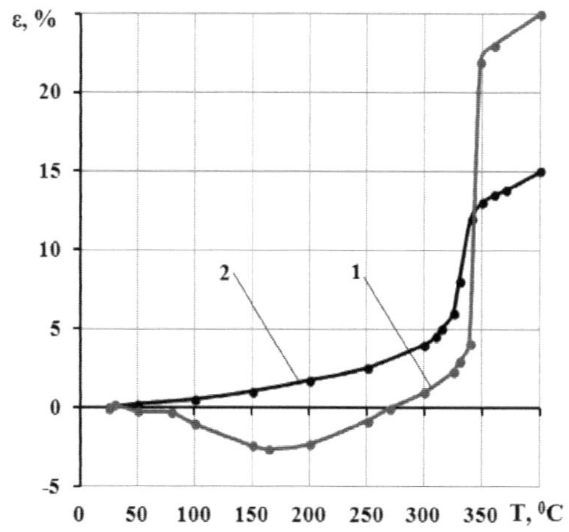

Figure 2. Thermomechanical curves of PTFE compacts after ET and sintering with thickness of 0.8 mm (1) and 3.2 mm (2)

5. Conclusion

Therefore, it is determined that changes in the ET parameters (P from 0.3 to 0.7 GPa) and the pulse duration (from 13 to 43 ms) are accompanied by changes of the structural organizations and allow to adjust the degree of compaction, structure, interparticle interaction in PTFE. Quantitative criteria of parameters of PTFE explosive treatment (P=0.7 GPa, τ = 13 ms) are determined, with the achievement of maximum bodying and minimal crystal structure imperfection, which allowed to produce the material with high heat resistance and melting point of 340 ^{0}C.

Acknowledgements
The investigation was performed by a grant from the Russian Science Foundation (project № 14-29-00158).

References

[1] Hatipov S A and Artamonov N A 2008 *J Rus. chemical. Zh.* **52** 89-97

[2] Kuryavyi V G, Bouznik V M, Ignatieva L N, Zverev G A, Kaydalova T A and Sukhoverkhov S V 2010 *J Mater., technol. and tools* **15** 84-8

[3] Grakovich P N, Ivanov L F, Kalinin L A, Riabchenko I L, Tolstopyatov E M and Krasovsky A M 2008 *J Ros. chemical. Zh.* **52** 97-105

[4] Adamenko N A, Fetisov A V and Kazurov A V 2007 *J Volg. GTU* 240

[5] Ignatieva G N, Zverev G A, Adamenko N A Kazurov A V Sergeev I V, Ustinov A Y, Tkachenko I A and Bouznik V M 2015 *J. Fluorine Chem* **172** 68–73

[6] Adamenko N A, Kazurov A V, Agafonova G V, Ryzhova S M and Gerasimuk A E 2015 *J Proc. Eng.* **113** 418-22

[7] Ignatieva G N, Adamenko N A, Agafonova G V, Gerasimuk A E, Kuryavyi V G, Zverev G A, Merkulov E B, Ustinov A Y and Bouznik V M *J* 2013 *J Vest. DVO RAN* **5** 44-52

[8] Hatipov S A, Kabanov S P, Konov E M and Ivanov S A 2012 *J Vysokomol. Conn. Rev A* **54** 1267-74

Structure Formation Mechanisms during Solid Ti with Molten Al Interaction

L Gurevich, D Pronichev and M Trunov
Materials Science Chair, *Volgograd State Technical University*,
400005, Lenin Avenue 28, Volgograd, Russia

E-mail: mv@vstu.ru

Abstract. The study discuses advantages and disadvantages of previously proposed mechanisms of the formation of structure between solid Ti and molten Al and presents a new mechanism based on the reviewed and experimental data. The previously proposed mechanisms were classified into three groups: mechanisms of precipitation, mechanisms of destruction and mechanisms of chemical interaction between intermetallics and melt. The reviewed mechanisms did not explain the formation of heterogeneous interlayer with globular aluminide particles and thin layers of pure Al, while the present study reveals variation in the solid Ti/molten Al reaction kinetics during various phases of laminated metal-intermetallic composite formation. The proposed mechanism considers formed during composite fabrication thin oxide interlayers between Ti and Al evolution and its impact on the intermetallic compound formation and explains the initial slow rate of intermetallic interlayer formation and its subsequent acceleration when the oxide foils are ruptured.

1. Introduction

One of the main problems of Ti/Al laminated metal composites with intermetallic interlayers (or just metal intermetallic laminates - MILs) production is the demand to obtain layers with proper thickness ratio.

The upper boundary of temperature range, where solid state diffusion in Ti/Al composites results in the formation of aluminides is only 100°C. The growth rate of aluminides is relatively small. The required thickness of intermetallic compound in Al/Ti MILs for possible applications is usually 50-100 μm. The time necessary to obtain intermetallic interlayer of the required width in Ti-Al composite exceeds tens of hours. Thus the efficiency of MIL production is significantly reduced by mentioned above factors.

Investigations on the possibility of intensification of diffusion between Ti and Al melts revealed the kinetics and the steps of diffusion processes [1-3]. However the mechanism of heterogeneous interlayer with globular aluminide particles and thin layers of pure Al formation remains unclear.

The aim of the present study was to analyze the disadvantages of existing hypothesis for the mechanisms of interaction between Al melt and solid Ti and to propose a new explanation which corresponds with experimental data.

2. Materials and Experiments

In order to investigate the evolution of intermetallic formation at the interface of laminated Ti/Al composites, Ti/Al sheet was fabricated via explosion welding and subsequently heat treated.

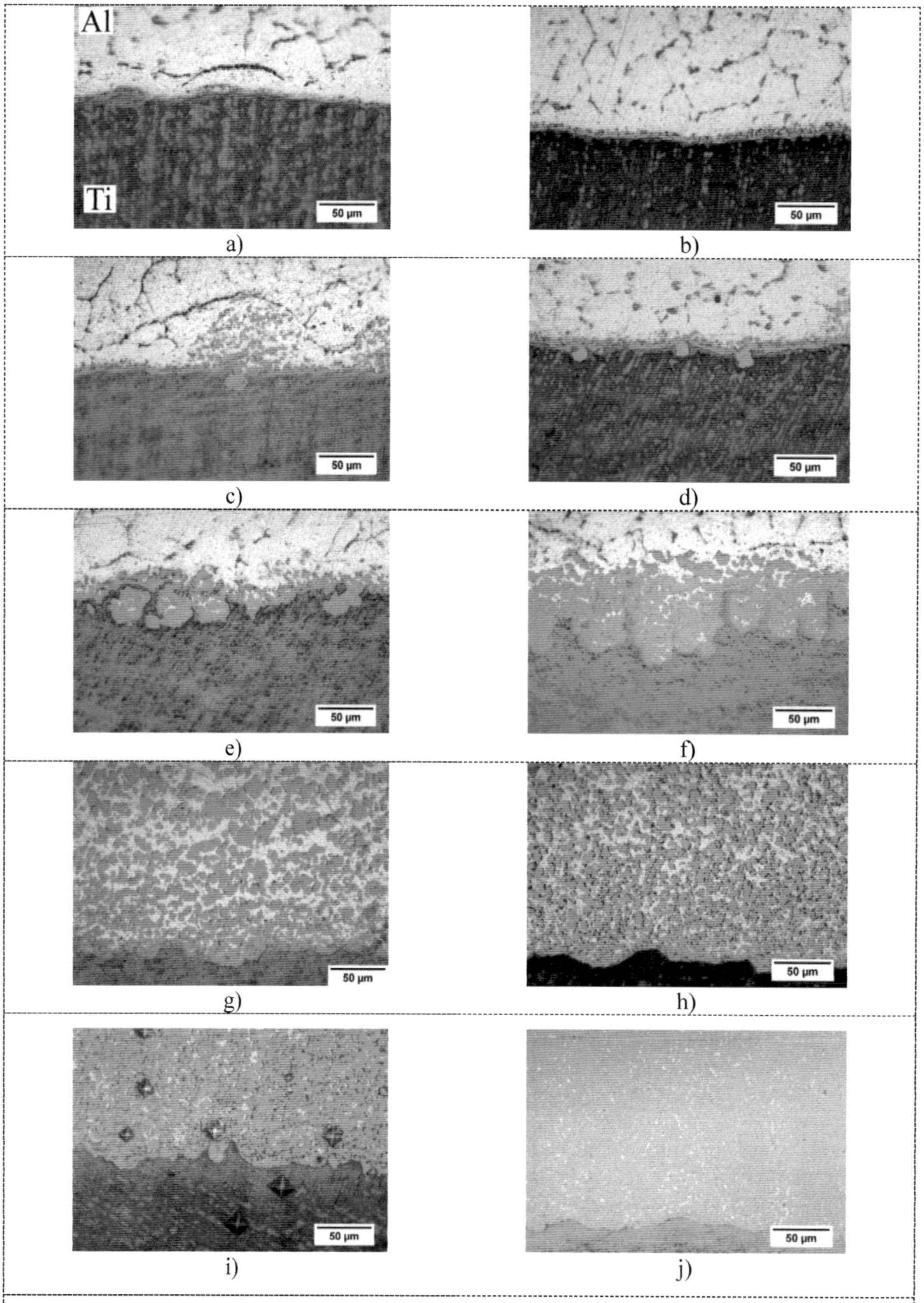

Figure 1. Optical image of the explosion welded Ti/Al composite after heat treatment at 700°C for a) 10 min; b) 20 min; c) 30 min; d) 60 min; e) 90 min; f) 120 min; g) 150 min; h) 180 min; i) 210 min; j) 240 min.

The explosion welding parameters were chosen to avoid molten zone formation. Heat treatment was performed at 675°C, 700°C and 750°C for the range of times between 10 minutes and 4 hours.

The obtained specimens were studied by means of optical microscopy (optical microscope Olympus BX-61) and X-Ray diffraction patterns (diffractometer Dron-3).

The microstructures of the Ti/Al interface after explosion welding and subsequent annealing at 700°C is shown in figure 1. Microstructural changes, presented in figure 1 are similar to those obtained during heat treatments at 675°C and 750°C. The heat treated Ti/Al specimens diffraction patterns analysis identified the presence only of $TiAl_3$ intermetallic compound in the whole range of investigated temperatures and times.

As shown in figure 2 the initial formation rate of the uniform intermetallic layer is rather slow for all investigated heat treatment temperatures. The slow intermetallic formation process is followed by the acceleration of the intermetallic growth rate. The acceleration is achieved by the growth of the interlayer composed of spherical particles (figure 1 c-f).

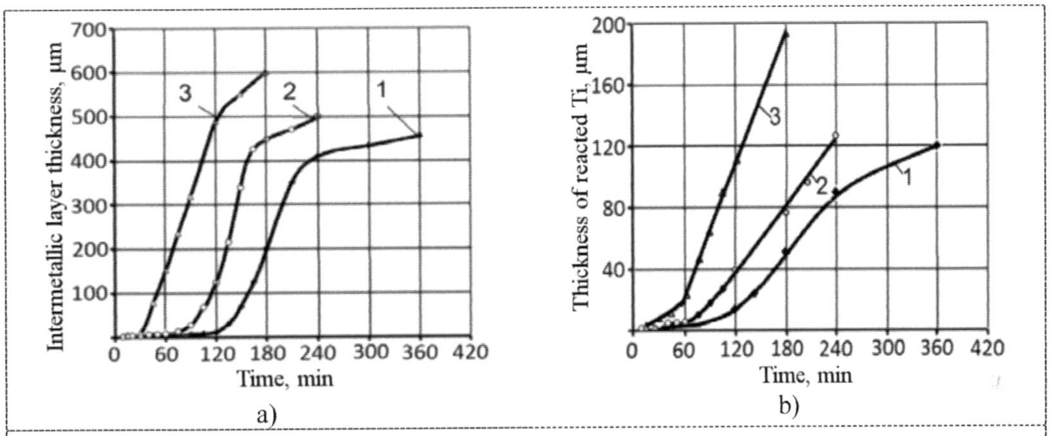

Figure 2. The evolution of the total thickness of intermetallic layers (a) and total thickness of the Ti (b), involved into the reaction during heat treatment at 1 – 675°C, 2 – 700°C, 3 – 750°C

In order to explain the observed feature of the initial slow intermetallic growth rate with its subsequent acceleration, the consideration of solid Ti molten Al interaction mechanism is required.

3. Solid Ti/Molten Al interaction mechanism

The formation of two-phase interlayer with high content of aluminides between Al melt and solid Ti was previously observed during Ti brazing with Al solder. The proposed mechanisms of aluminide formation can be divided into three groups: mechanisms of precipitation, mechanisms of destruction and mechanisms of chemical interaction between intermetallics and melt.

Mackowiak J. and Shreir L. L. [4] proposed the three stage mechanism of precipitation, which included:

- the formation of saturated solution as a result of dissolution of Ti in Al;
- precipitation of the porous $TiAl_3$ layer, which is permeable for molten aluminum;
- the destruction of the continuous interaction layer as a result of tensile stress.

Subsequent investigations on the interaction between solid Ti and, surrounding it Al melt, have brought authors to reconstruct their theory. The stage of formation of saturated solution as a result of dissolution of Ti in Al was excluded, while new stages were proposed [5].

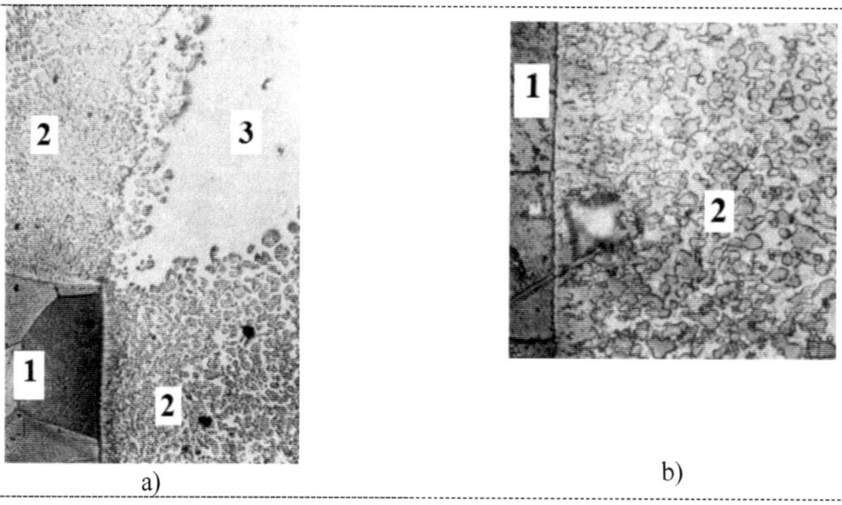

Figure 3. Optical image of the interface and the interaction layer of the specimen obtained through immersion of Ti in molten Al at 850°C, a) Ti - lower left hand corner, b) cylindrical Ti-Al surface, 1 – Ti, 2 – interaction layer, composed of $TiAl_3$ in Al matrix, 3 – interaction layer gap [5].

The four-stage mechanism of precipitation was proposed instead of the previous theory. The four stages were:

1. chemical reaction between solid Ti and Al melt, which resulted in the formation of thin molten aluminum previous layer;
2. the repulsion of precipitated aluminides from Ti boundary, caused by tensile stress in the reaction layer. The tensile stress occurred due to volume differences of $TiAl_3$ layer and saturated Ti;
3. the continuous interaction layer rupture. The rupture occurred in the corners of subjected in melt cylindrical Ti specimen and was caused by tensile stress exceeding $TiAl_3$ strength (figure 3);
4. concurrent precipitation and breakdown leading to the production of a cruciform pattern

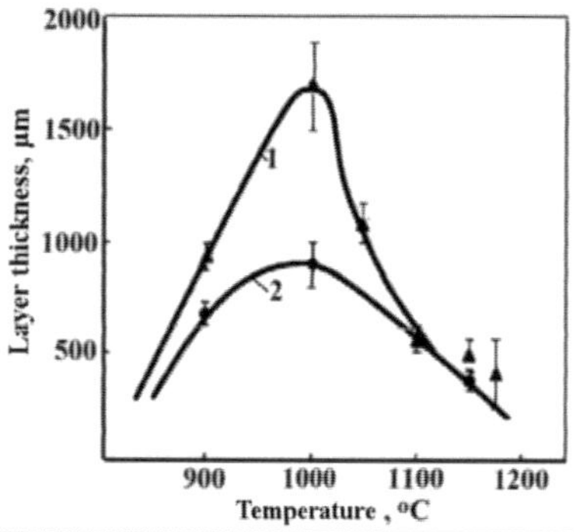

Figure 4. Thickness of biphase zones formed in 0.5 h on cold-worked (1) or recrystallized (2) niobium as a function of temperature [6]

Slama G. and Vignes A. [6], based on the studies of microstructures of reaction layers, formed during the interaction between solid Ti and Nb with Al melt, proposed that the formation of two-phase layer, which contained crystallized Al and the reaction products of Al, can be explained by the rupture of continuous Al layer on the solid Ti boundary. The rupture is caused by tensile stress, which occurs due to volume differences in reacted Ti (Nb) and intermetallic.

The fact that solid Nb - molten Al system has a critical temperature below which the two-phase interaction layer is formed, while above which only the continuous intermetallic layer on solid Nb is formed, acknowledges the hypothesis of Slama G. and Vignes A. The authors proposed the emergence of stress exceeding the yield point of Nb on the interface between materials above the critical temperature. The stress leads to the deformation and hardening of Nb.

However due to recrystallization of Nb the hardening does not result in the Nb to achieve the strength higher than of aluminide and thus does not lead to the rupture of the aluminide. The complex relation of continuous aluminide thickness on the temperature of melt on the surface of Nb (figure 4) was explained by coexistence of two features:

1. the increase in the speed of reaction between solid Nb and Al melt. The stress value increased during the reaction which contributed to aluminide rupture.
2. the decrease in the yield point of Nb and acceleration of recrystallization processes, which contributed to stress relaxation.

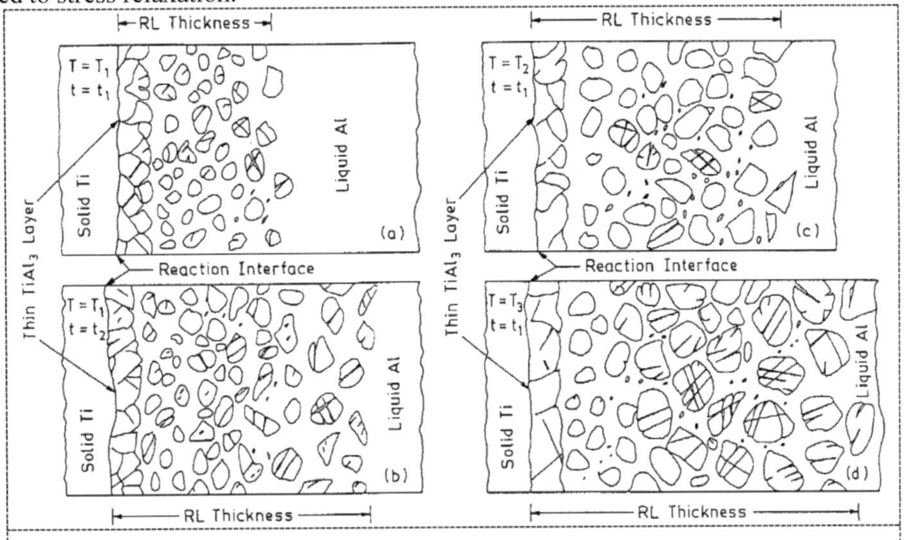

Figure 5. A schematic diagram of the microstructural evolution in the reaction layer at different reaction temperatures T_1, T_2 and T_3 and time t_1 and t_2 ($T_1<T_2<T_3$ and $t_1<t_2$) [7]

The authors did not observe critical temperature in solid Ti - Al melt system even at 1200-1300°C above which the constant growth of continuous intermetallic layer could occur. The microhardness tests and SEM identified the emergence of solid solutions of Al in Ti in front of the $TiAl_3$-Ti interface under the studied temperatures. The microhardness of the solutions was 2.7-5.2 GPa. High hardness value of the emerged Ti(Al) blocked the evolution of plastic deformation and thus decreased the stress value in the formed aluminide layer.

Subsequent studies attempted to complement the proposed by Slama G. and Vignes A. mechanism. Thus Sujata. M., Bhargava S. and Sanga S. [7] agree with the statement that $TiAl_3$ layer, formed on solid Ti, experiences elaborate stress state due to high difference between density value of Ti and $TiAl_3$ and the stress value increases with the growth of aluminide thickness or with the acceleration of aluminide growth rate.

Due to the limited plasticity of $TiAl_3$, the layer can undergo crack and fracture if the critical thickness is achieved due to the critical stress. Fragments separated from the solid Ti layer uncover the surface on which the chemical reaction between the solid Ti and molten Al resumes.

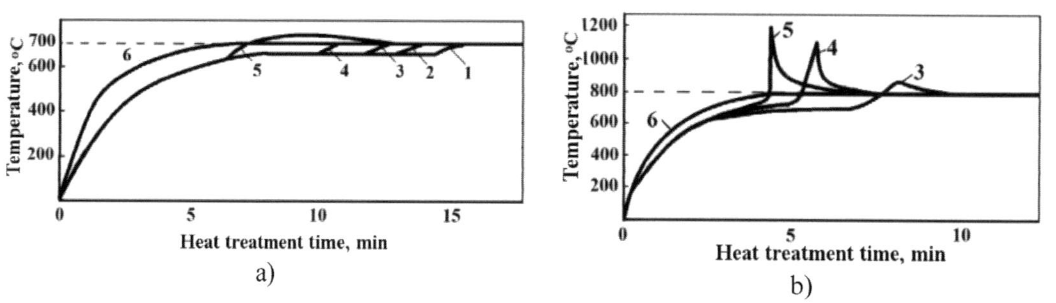

Figure 6. The evolution of temperature of Ti-Al compacts of various chemical compositions during powder agglomeration a) – at 700°C b) – at 800°C; 1 – 5 at. % Ti; 2 – 10 at. % Ti; 3 – 12.5 at. % Ti; 4 – 15 at. % Ti; 5 – 20 at. % Ti; 6 – temperature of the dilatometer tube [8]

The figure 5 demonstrates the interaction layer microstructure evolution at various temperatures and reaction times. In order to explain the evolution of the interaction layer, Sujata, M. et al. [7] assumed the produced in the interaction layer heat leads to the convective heat flux in the melt occurring. The heat is produced as a result of exothermal reaction of $TiAl_3$ formation and leads to the temperature growth of reaction products and adjacent Al melt. The flow of molten Al, caused by thermal gradient, carry separated $TiAl_3$ fragments away from the reaction layer. Cracks in the separated particles contribute to their subsequent fragmentation during thermal flow transportation.

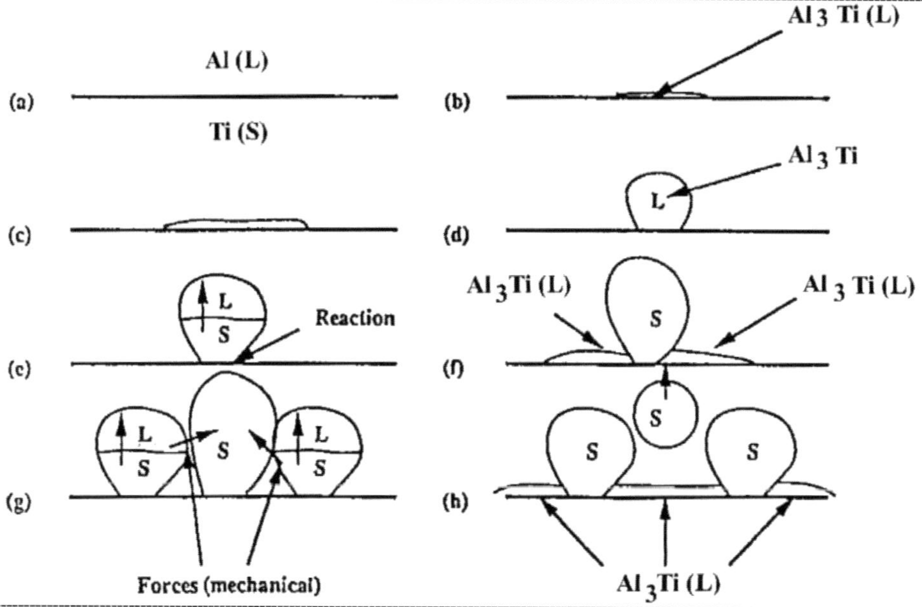

Figure 7. Sequence of events in the synthesis of Al3Ti spherules at the solid Ti–liquid Al interface: (a) through (c) nucleation and growth of thin layer, (d) spheroidization as a result of interfacial energy, (e) solidification and reduced reaction kinetics, (f) adjacent Al3Ti regions form in newly exposed interface, (g) spheroidization and solidification of adjacent Al3Ti regions, and (h) expulsion of spherule [9]

The results of Ti-Al liquid phase heat treatment dilatometric investigations [8] contribute to the importance of thermal effects role. Short-term temperature growth of Ti-Al compacts subjected to temperatures higher than 770°C was observed due to exothermal process of aluminide formation. The temperature of compacts was 400°C higher than the furnace temperature (figure 6).

The proposed mechanisms cannot explain observed in this study modification in the kinetics of 2-phase layer formation. As seen in figure 2, during this modification the uniform intermetallic layer growth in the bond between Ti and Al stops and the continuous interlayer with spherical intermetallic inclusions starts growing.

Harach D. J. and Vecchio K. S. [9] proposed the third group of mechanisms of reaction layer with spherical $TiAl_3$ inclusions formation. The mechanisms are based on the idea of molten intermetallic compounds formation. As shown in figure 7 the reaction starts at the interface of Al/Ti.

The surface tension dominates the gravitational forces and the reactions products are in the form of spheres. The reduction in contact area between Ti and Al leads to the decline of the reaction rate which results in the spherical particles solidification. Additional intermetallic nucleuses occur and grow along the Ti/Al interface during the solidification process.

Due to the spherical particles are constrained in space, the process of solidification leads to the emergence of loads, which extrude the first spherical particle into the liquid and uncover hidden interphase area.

Such process is apparently possible only at temperatures above 800°C when, due to significant exothermal effects, the temperature in specific areas can exceed the $TiAl_3$ melting point (1395°C). However two phase layer was observed in the range of heat treatment temperatures between 675°C and 750°C (figure 1). In this temperature range the two phase layer formation reaction rate is rather low and temperature increases very slightly.

4. The new solid Ti/molten Al interaction mechanism idea

Taking previous studies and experimentally obtained results into consideration the new mechanism of the aluminide formation processes can be proposed. Ti and Al have a high affinity for oxygen. As a result of hot rolling or explosion welding the Ti-Al interface has oxygen layers with highly defective structure.

Such oxygen layer discontinuities can occur during the deformation caused by the difference in plasticity values of oxygen layers and metals. Subsequent heat treatments of the bimetal can also lead to oxygen layers rupture due to the difference in thermal expansion coefficient values.
When the bimetal is treated above Al melting point, the Ti/Al interface represents a multilayer system: molten Al; solid Al oxide Al_2O_3; solid Ti dioxide TiO_2 or monoxide TiO; solid Ti. Thus molten Al does not experience uniform contact with Ti, the solid Ti- molten Al contact is located only in some areas with fractured oxide layers [2].

The transfer of titanium and aluminum atoms through the oxide interlayers, caused by the difference in chemical potentials of Ti and Al, results in the formation of local areas with solid solutions of aluminum in titanium as well as the dissolution of diffused titanium atoms in the melt. The rate of diffusion processes of Ti and Al atoms is decreased by oxygen layers. Thus the atomic transition is carried out mainly by diffusion through isolated channels in the areas with ruptured oxygen layers and leads to the formation of local regions of supersaturated solid solution Ti(Al). The intermetallic nucleuses of various compositions stochastically form in the local regions. The most thermodynamically possible composition $TiAl_3$ has the smallest size of nucleus.

When the size of the formed nucleuses exceeds the critical value (figure 8), the possibility of their growth is increasing. The ratio of intermetallic growth along the TiO_2 solid -Ti solid interface is the highest among all directions. This contributes to the diffusion in the direction with injections of Al atoms.

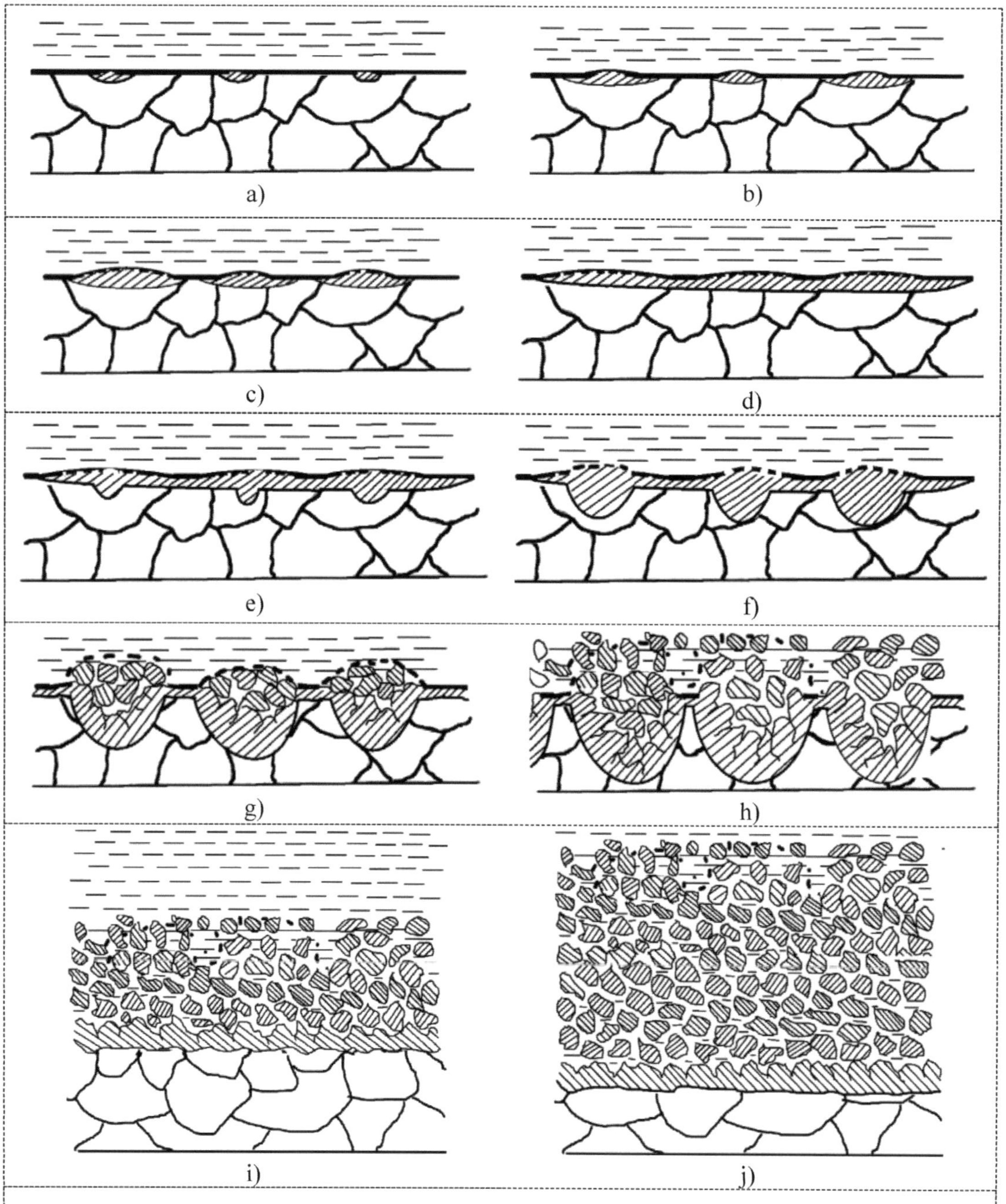

Figure 8. Mechanisms of Ti-Al structure transformation during heat treatment at temperatures higher than Al melting point: a) – Al nucleus formation, b, c, d) – transversal growth of thin interlayer parts, e, f) – formation and accelerated growth of intermetallic phase nucleus in the channels between oxide layer parts, g, h) – intermetallic fragmentation, i, j) – uniform intermetallic layer formation, the layer is covered with molten Al.

The growth of aluminide fragments ends up with the formation of the uniform layer at the interface. Ti atoms, which penetrated through oxide layer ruptures dissolve in the Al melt until the solubility limit of Ti in Al (0.2-1.3% at 670-750°C) is reached. The subsequent Ti atoms supply results in the formation of arbitrarily located crystals of $TiAl_3$ phase, which is driven away to grain boundaries by expanding nucleus during the solidification and crystallization processes.

Uniform distribution of the intermetallics across the volume of the melt can be explained by higher speed of transportation processes in the melt compared to the diffusion of Ti through the oxide layers. The growth of aluminide layer volume causes the oxide layer fragmentation, which leads to the increase in the contact area between Ti intermetallic and Al melt. The formation and growth of intermetallic phase nucleus takes place in the regions of Al melt - Ti intermetallic contact. The kinetics of the process was previously described by Kolmogorov [10]:

$$X(t) = 1 - \exp\left[-\int_0^t I(t')V(t',t)dt' \right], \tag{1}$$

where *X(t)* – the forming phase volume fraction, *I(t)* – the rate of new phase centers formation, *V(t',t)* – volume at time t of the nucleus, formed at *t'*.

$$V(t',t) = gR^D(t',t),\ R(t',t) = \int_{t'}^{t} u(t)dt, \tag{2}$$

where *u(t)* – nucleus growth ratio, *R(t',t)* – nucleus radius, *D* – the dimension of space, *g* – geometrical multiplicator: g=2,π,4 π /3 for D=1,2,3 respectfully.

The formation of intermetallic compound in regions constrained by reacted Ti causes the subsequent oxide layer between Ti and Al melt rupture. This leads to the increase in internal stress and contributes to the fragmentation of intermetallic layer. The emerged intermetallic fragments are then pressed out of reaction volume.

The Rehbinder effect [11] contributes to the fragmentation of the formed intermetallic layer. The effect consists in the adsorption reduction in strength, strain relief and decrease in strength in solids.

The intermetallic $TiAl_3$ has ionic bond; the liquid Al melt wetting of the intermetallic is lower than pure Al and Ti wetting. However $TiAl_3$ contact angle is significantly lower than 90° due to its metallic properties. The more the crystal area is covered by Al melt, the less surface energy the crystal has. Such feature facilitates the diffusion of Al melt across the surface of the forming intermetallic crystal. The micro cracks expansion is contributed by the melt which diffuses across the juvenile surfaces with active centers. The surfaces occur at the beginning of small cracks. The oxide layer rupture, as well as fragmentation of intermetallic compound, accompanied by emerged capillaries, significantly facilitates the Al melt transport to the reaction surface, which contributes to the acceleration of local intermetallic growth and consequently uniform interlayer covered with melt formation.

The rupture of the diffusion barrier (oxide foil) and fragmentation of the intermetallic layer between Ti and Al melt contributes to the increase in thickness value of the two-phase reaction products (aluminum spherical particles, covered with Al melt) layer. The reaction products are constantly separating from the intermetallic layer of critical thickness value. The separated spherical particles are driven away from the interaction layer by newly formed particles. The increase in temperature results in the decrease of viscosity of the melt and contributes to the acceleration of intermetallic formation reaction and thus leads to thermal gradient growth. Heat flows transport Al spherical particles which results in their number reduction in the two-phase interaction products layer.

The proposed mechanism explains the initial slow formation of the uniform intermetallic layer on the Ti/Al interface with subsequent acceleration of the intermetallic growth and formation interaction during the growth of interlayer composed of spherical particles.

5. Conclusion

1. Based on the comparison between mechanisms of precipitation, destruction and chemical interaction between intermetallics and melt, and analysis of the evolution of the microstructure of Ti/Al bimetal treated above the Al melting point, the new explanation of the bond area processes was proposed. The explanation considers the oxide layers at the Ti/Al interface.
2. The explanation for the conversion of the intermetallic growth rate from slow to high has been proposed. The impact of the Rehbinder effect and the oxide layer rupture was considered.

Acknowledgements
The present study was carried out according to the assignment of Ministry of Education and Science of the Russian Federation No 2015/16 project No 1183 (2015).

References

[1] Gurevich L M, Trykov Yu P, Zhorov A N and Arisova V N 2005 *Izvestija VolgGTU* **3** 9-12.
[2] Gurevich, L M, Trykov Yu P, Zhorov A N, Gurulev D N and Loktushin V A 2008 *Zhurnal funktsional'nykh materialov* **4** 153-157.
[3] Gurevich L M, Trykov Yu P, Arisova V N, Kiselev O S, Kondrat'ev A Yu and Metelkin V V 2009 *Izvestija VolgGTU* **11** 5-11.
[4] Mackowiak J and Shreir L L 1956 *Acta Metallurgica* **4** 556-557.
[5] Mackowiak J and Shreir L L 1959 *Journal of the Less Common Metals* **1** 456–466.
[6] Slama G, Vignes A 1971 *Journal of the Less Common Metals* **24(1)** 1-21.
[7] Sujata M, Bhargava S and Sangal S 1996 *ISIJ International* **36(3)** 255-262
[8] Romanov G N, Tarasov P P, D'yachkovskiy P K, Savitskiy A P and Martsunova L S 2006 *Izvestiya Tomskogo politekhnicheskogo universiteta* **309** 114-119.
[9] Harach, D.J. and Vecchio, K.S. 2001 *Metallurgical and Materials Transactions A* **32(6)** 1493-1505.
[10] Kolmogorov A N, 1937 *Izv. AN SSSR. Ser. matem* **1** 355-359.
[11] Andrade E N and Randall R F Y 1949 *Nature* **164** 1127.

IOP Conf. Series: Materials Science and Engineering **116** (2016) 012012 doi:10.1088/1757-899X/116/1/012012

Contact hardening of Al interlayer in laminated Mg/Al composites during compressive and tensile loading

L Gurevich, D Pronichev and M Trunov
Materials Science Chair, *Volgograd State Technical University*,
400005, Lenin avenue 28, Volgograd, Russia

E-mail: mv@vstu.ru

Abstract. The study presents result of the FEM simulation of Mg alloy/Al/Ti alloy composite under tensile and compression loads. The simulation revealed the strength of Al interlayer at its various thickness values.

1. Introduction

Fusion welding of Al alloys with Mg alloys results in the formation of brittle intermetallic compound at the interface between metals [1]. Therefore explosion welded Al/Mg bimetals are usually applied to join Al with Mg. Mg alloy is fusion welded to Mg layer of the bimetal, while Al alloys are joined with Al layer. Laminated Al alloy–Al–Mg alloy joints can be utilized in the −196°C–+100°C temperature range, if the Al–Mg alloy interface is not overheated during the fusion welding. Strength of the joint is determined by Al interlayer which prevents diffusion and intermetallic formation at high temperatures and performs the plasticity buffer function during explosion welding.

The contact hardening effect is observed when the relative thickness ratio of the interlayer decreases. The relative thickness ratio is given by $\chi = h/d$, where h denotes the interlayer thickness in mm, d – specimen diameter. Stress state FEM simulation under tensile loads of such composites results converged with experimentally obtained data [2]. Currently there is little knowledge on the behaviour of the composites during compression testing; however utilization of the materials under compression conditions is also possible.

The aim of the present study was to investigate the behaviour of laminated Al alloy-Al-Mg alloy joints under compression and tensile loads.

2. Materials and methods

Abaqus FEM simulation software was applied to carry out investigations. 6 mm diameter Al alloy–Al–Mg alloy specimens were loaded transversally to the bond. The hardening of the materials was considered using Johnson-Cook model [3], which gives the hardening function of the material:

$$\sigma_Y = \left(A + B \cdot \varepsilon_p^n\right)\left(1 + \ln\frac{\dot{\varepsilon}_p^n}{\dot{\varepsilon}_0}\right)\left[1 - \left(\frac{T - T_r}{T_m - T_r}\right)^m\right], \tag{1}$$

where ε_p – accumulated plastic strain, $\acute{\varepsilon}_p$ – plastic strain rate, $\acute{\varepsilon}_0$ - reference strain rate, T – current temperature, T_r – room temperature, T_m – melting point, A, B, C, n and m – model constants.
Material rupture was also considered using Jonson-Cook model [4], which establishes the rupture when $D=1$, where:

$$D = \frac{1}{\varepsilon_f} \sum_i \varepsilon_p^i, \tag{2}$$

where ε_f – the equivalent strain to fracture, given by:

$$\varepsilon_f = \left[D_1 + D_2 \exp\left(D_3 \cdot \frac{p}{\sigma_{ef}} \right) \right] \times \left(1 + D_4 \ln \frac{\varepsilon_p}{\varepsilon_0} \right) \times \left[1 + D_5 \left(\frac{T - T_r}{T_m - T_r} \right) \right], \tag{3}$$

where $D_{1,2,3,4,5}$ – material constants, ε_{ef} – effective stress, p – pressure in the considered element.
The Jonson-Cook model parameters for researched materials are given in the tables 1 and 2. The deformation rate impact was not considered in this study.

Table 1. Johnson Cook constitutive model parameters for materials used in this study [5].

Material	A, MPa	B, MPa	m	n	ε_0, sec^{-1}	T_m, K	T_r, K
Mg alloy	218.3	704.6	0.93	0.62	1	773	293
Al	60.0	6.4	0.859	0.62	1	933	293
Al alloy	100.0	380	1.04	0.28	1	773	293

Table 2. Fracture constants for materials used in this study [5].

Material	D_1	D_2	D_3	D_4	D_5
Mg alloy	0.178	0.389	–2.246	0	0
Al	0.071	1.428	–1.142	0.0097	0
Al alloy	–0.35	0.6025	–0.4537	0.206	7.2

The simulation was carried out in the relative Al interlayer thickness ratio range from χ_{Al}=0.67 (4 mm) to 0.041 (0.25 mm). The thickness of Mg alloy and Al alloy layers was 10 mm. Mesh size in the loading direction was 1/20 of the thickness of the each considered layer. Mesh size in the parallel to the bond direction was 1/125 of the cylinder's perimeter.

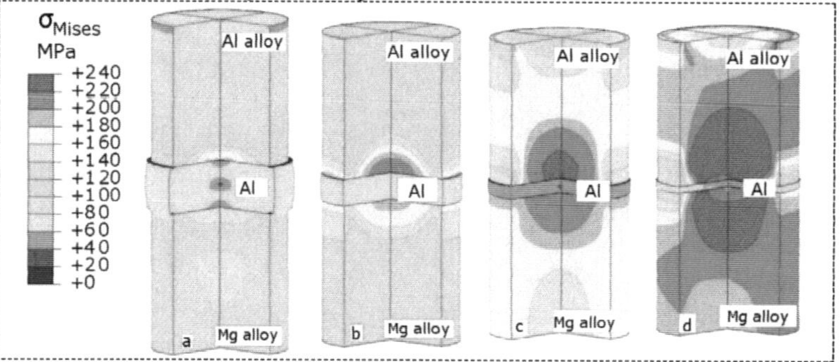

Figure 1. von Mises stress distribution in Mg alloy–Al–Al alloy composite during compressive loading at the moment of the first rupture: a – χ_{Al}=0.67; b – χ_{Al}=0.33; c – χ_{Al}=0.17; d – χ_{Al}=0.08

The extension/compression rate was 2 mm/s.

3. Results and Discussion

The simulation revealed significant difference in the deformation and rupture behaviour of the composite base layers both for various interlayer thickness values and for loading conditions.

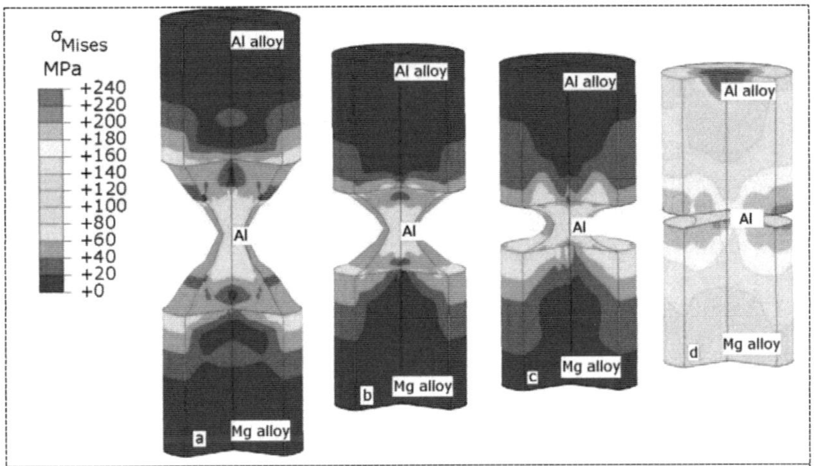

Figure 2. von Mises stress distribution in Mg alloy–Al–Al alloy composite during tensile loading at the moment of the first rupture: a – χ_{Al}=0.67; b – χ_{Al}=0.33; c – χ_{Al}=0.17; d – χ_{Al}=0.08

Al interlayer ruptured in the whole range of investigated χ_{Al} values and both under tensile and compressive loads. The initial Al rupture was observed in the adjacent to base metals area. The reduction in Al interlayer thickness contributed to von Mises stress growth in Al and Mg alloy layers. The maximum Mises stress was observed in adjacent to the bond interface central parts of base materials during compression loading (figure 1) and in the peripheral parts of base materials during tensile loading (figure 2).

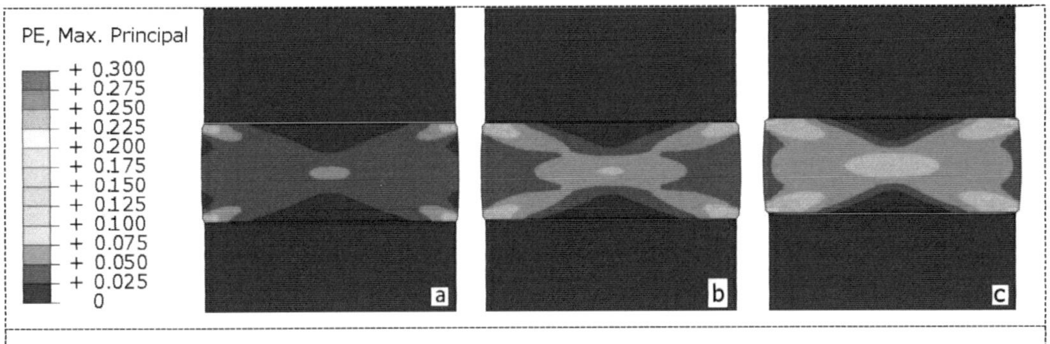

Figure 3. Plastic deformation distribution in 4 mm thick Al interlayer of Mg alloy-Al-Al alloy composite during compressive loading when the total deformation is equal to: a – 0.21 mm; b – 0.28 mm; c – 0.35 mm

The simulation revealed the non-uniform strain distribution in the Al interlayer. When the interlayer thickness was 4 mm, peak deformation was observed in the peripheral zones of the cylinder in the adjacent to base materials areas, as well as in the central part of the interlayer along its axis (figure 3). The increase in the specimen strain rate results in the strain value and size growth of the particular regions.

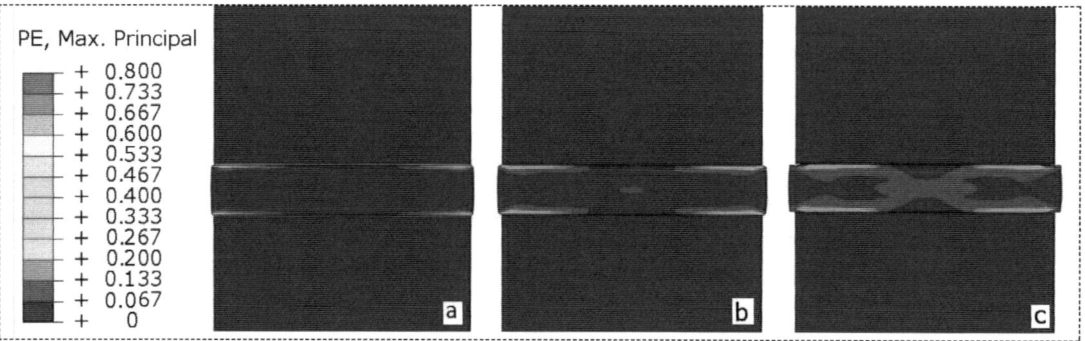

Figure 4. Plastic deformation distribution in 2 mm thick Al interlayer of Mg alloy-Al-Al alloy composite during compressive loading when the total deformation is equal to: a – 0.135 mm; b – 0.18 mm; c – 0.225 mm

The decrease of the interlayer thickness down to 2 mm changes the deformation localization: peak interlayer deformation is observed along the bond in adjacent to Mg and Al alloy base layers (figure 4). The deformation of the central part of the interlayer perpendicular to the bond starts only when the total deformation of the specimen exceeds 9% of the Al interlayer thickness.

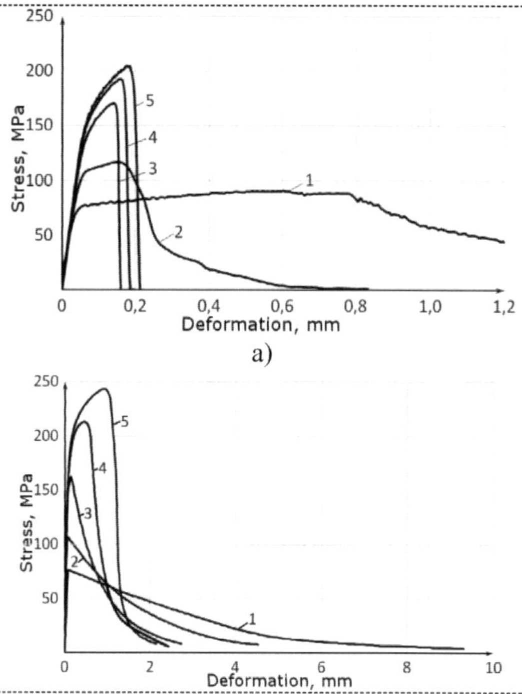

Figure 4. Stress-strain curves obtained during FEM simulation of Mg alloy–Al–Al alloy composite compressive (a) and tensile (b) loading at various values of χ_A: 1 – χ_{Al}=0.67; 2 – χ_{Al}=0.33; 3 – χ_{Al}=0.17; 4 – χ_{Al}=0.083; 5 – χ_{Al}=0.041

The obtained during the simulation stress-strain curves for both loading methods are presented in figure 5. The peak stress values both for compression and tensile loading remain on the same level,

however the strain values are significantly higher for tensile loading (figure 5). Compression leads to the 20-25% Al interlayer deformation.
Thus the compression of composite materials with soft interlayers can lead to the rupture at relatively small deformation values.

4. Conclusion

1. FEM simulation of the tensile and compressive loading revealed significant differences in the deformation and rupture behavior of the composite both due to various interlayer thickness values and loading conditions.
2. For the whole range of investigated interlayer thickness values and for all loading conditions rupture was only observed in the Al interlayer.
3. The variation in loading direction does not significantly influence the critical stress values, but the acceptable strain rate is approximately 10 times lower for compression loading.

Acknowledgements
The study was supported by Russian Science Foundation grant No 14-29-00158.

References

[1] Li Y, Liu P, Wang J and Ma H 2007 *Vacuum* **82(1)** 15-19
[2] Gurevich L, Trykov Y, Pronichev D and Trunov M 2014 *WSEAS Transactions on Applied and Theoretical Mechanics* **9** 275-281
[3] Johnson G R and Cook W H 1983 *Proceedings of the 7th International Symposium on Ballistics* **21** 541-547
[4] Johnson G R and Cook W H 1985 *Engineering fracture mechanics* **21(1)** 31-48
[5] Giraud E, Rossi F, Germain G and Outeiro J C 2013 *Procedia CIRP* **8** 522-527

IOP Conf. Series: Materials Science and Engineering **116** (2016) 012013 doi:10.1088/1757-899X/116/1/012013

Eddy Current Defectoscope for Monitoring the Duralumin and Aluminum-Magnesium Alloys

S Dmitriev[1], L Dmitrieva[1], V Malikov[1], A Sagalakov[1]
[1]Altay State University, Faculty Physics and Technology, Barnaul, Lenin Street, 61, Russia

E-mail: osys@me.com

Abstract. The system developed is based on an eddy-current transducer of the transformer type, and is capable of inspecting plates made of duralumin and aluminum-magnesium alloys for defects. The measurement system supports absolute and differential control modes. The system was tested on a number of duralumin and aluminum-magnesium plates with internal flaws located as deep as 5 mm under the surface. The article provides data that demonstrates a link between the response time and the presence of defects in similar structures at a signal frequency of 1000 Hz.

1. Introduction

Existence of defects in the product structure or the object critical parts may substantially affect their service life. In the course of many years, great attention has been paid to the development and application of non-destructive methods, which would be able to predict the service life or the risks connected with further exploitation of the object in question [1].

Eddy current testing, which forms part of the electromagnetic testing methods, is a non destructive testing (NDT) method that relies on electromagnetic induction to detect discontinuities in conductive material. Discontinuities such as geometrical changes, variations in material properties relating to conductivity and permeability and the presence of defects, both surface¬breaking and subsurface, can be detected.

When a conductive material is placed close to an alternating magnetic field, there will be an oscillating electrical current induced in the conductive material due to Faraday's law of induction. These induced eddy currents will produce a secondary alternating magnetic field which is in the opposite direction to the original alternating magnetic field. The presence of any discontinuity or defect in the material will disturb the eddy current flow; hence, the secondary magnetic field will be disturbed.

The application of methods and means of nondestructive eddy current testing is possible to search for defects in products made of any electroconductive materials. This measurement method allows to investigate each instance in the series of products in the factory, if it is necessary.

The most informative parameter in this method is the stress introduced into the measuring coil of the primary sensor, the eddy-current transducer.

This quantity depends on the composition and structure of the material, the type and parameters of flaws, their structure, the frequency of an alternating magnetic field which induces eddy currents, and on technology factors such as the construction and parameters of the transducer, clearance between the transducer and material surface, and other [2].

An important place in modern civil and military industries occupy duralumin and aluminum-magnesium alloys. Such alloys are the main structural material in aircraft and space exploration - thanks to a good combination of strength and lightness. These alloys are widely used in the high-speed trains production (e.g., trains, Shinkansen (Japan) and in many other industries. Duraluminium is also used in many other branches of modern industry: in the electrical industry, in the chemical and food industry, as this alloy is resistant to most chemicals and products. Also, duralumin is used in the ventilation systems production, in radio engineering and building. For example, the alloy of D16AM grade is used in extreme low temperature conditions. Duralumin D16T is flexible and therefore it is used in shipbuilding and furniture manufacture.
Defects in these alloys can occur both at the stage of casting and other parts production, for example, as a result of defective welding. It is well known, in recent years the usage of aluminum-magnesium alloys in welded structures has increased).

2. Specifications, material choices and design

Despite our great experience in using eddy-current transducers (ECT) in the sphere of nondestructive control, several important aspects still need to be studied nowadays. In particular, the most vital problems are: carrying out local contactless measurements combined with controlled object scanning; price reduction in sensor engineering; making the control locality more accurate and the quick allocation of the frequency range, the most proper to scan the control object.
Along with the considerable price reduction of the device, the virtualization of the generator and signal receiver makes it possible to mark out the frequency range, at which the process of scanning of the control object gets very effective. The effectiveness is reached due to the quick varying of the current frequency, given by the virtual generator to the ECT. Thus, researchers can define the frequency, at which changing the amplitude of the signal on the measuring winding (when it is passing through the imperfect area) is at a maximum. Besides, frequency varying enables us to find the depth of the defect hiding, because the depth of the electromagnetic field distribution of the ECT directly depends on the frequency of the field sent to the actuating winding.
The worked out conception of the virtualized measuring device (transformer) enables a researcher to make a great number of measurements in different spheres with the help of one sensor only. To these spheres belong fault detection of conducting materials and stratified compos, thickness measuring, profilometry, tension measuring of permanent and variable magnetic fields and conductivity measuring of unferromagnetic materials and some others.
Producing such devices demands making a special technological line, due to that the final price of the device will increase significantly. To make the price lower we came to the conclusion that we should replace the expensive hardware blocks with the software for personal computers (PC). As a result, out device consists of a vortex-current transformer (ECT) only, which is connected to the sound card of the PC, managed with the special software [3].
The software manages the voltage giving to the generative transformer winding, it also reads out values of the voltage from the measuring winding in conventional units. Later these units are transferred into the values of conductivity taking into account the preliminary calibration.
In general the scheme of the sensor is represented in Figure 1.

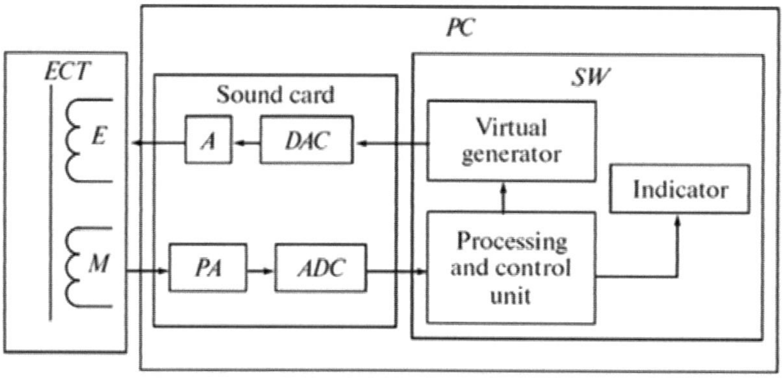

Figure 1. Block diagram of the probe.

The digital signal from the virtual generator is input to an analog-to-digital converter (DAC), a sound card, after which the analog signal through the power amplifier (A) is fed to the exciting winding transducers. Passing through the excitation winding (E), sine wave creates an electromagnetic field which induces a voltage in the measuring winding (M) in eddy-current converter, which depend on the parameters of the object of control.This signal is fed to the microphone input of the sound card and then through the preamplifier (PA) to the input of an analog-to-digital converter (ADC) sound card. The analog signal is converted to digital one and transferred to the processing and control unit of the software (SW). The processing and control unit fixes the level of the digital signal in conventional units corresponding to the difference between the voltages on the measuring windings.

The principle of operation of the ECT is based on the alternating magnetic field, localized in the controlled object with the help of the oxide pyramid shaped core. The shape of the core is conditioned with the necessity to localize the magnetic flux from the actuating (generator) coil.

Exciting subminiature transmitter winding consists of 10 turns, and its diameter is 0.13-0.12 mm. Measuring winding consists of 130 turns and its diameter is 0.05-0.08 mm.

In order to minimize the influence of the exciting windings on the received signal, compensating winding is included into the circuit. The compensating winding is connected to the measuring winding so that the voltage of the exciting windings can be subtracted. The compensating winding consists of 20 turns. Copper wire 5 μm thick is used to the wind the turns. The windings are wound on pyramidal shape core. The core is made of HM3 2000 ferrite with initial magnetic permeability of the value of of μ_{max}=500.

A copper wire, that is 0.005 mm thick, is used to reel the turns. The windings are wound round the pyramid shaped core. The scheme of the subminiature ECT is represented on the Figure 2.

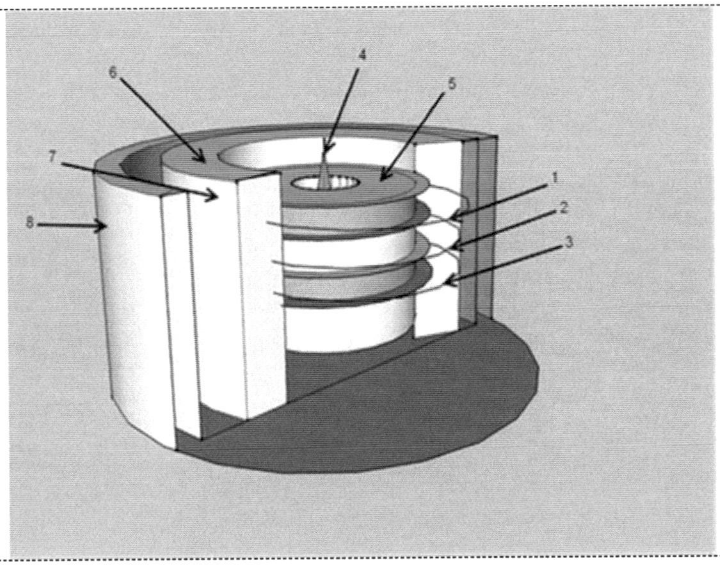

Figure 2. The scheme of the ECT.

The ECT has 1) an actuating winding 2) a compensation winding 3) and a magnetic conductor 4) which is placed inside the cylindrical platform 5) with the tracks for the windings, cut on the outer side 6) which is impregnated with a compound at 200 degrees centigrade in order to prevent the windings from destruction while the ferrite screen is being put on 7) which is meant for the magnetic field localization on the controlled object. From the outside the sensor is covered with a corundum washer 8 that protects the core 4 from contacting with the controlled object.

3. Experimental results

The worked out ECT enables the researchers to investigate the defects rather effectively. These defects may be found in the duralumin and aluminum-magnesium alloys. To demonstrate the operational capability of the suggested method we used plate of the duralumin. Plate thickness was 5.3 mm.

As a model defects used slits having a thickness of 0.25 mm. The defect was located at a distance of 1-5.3 mm from the probe in the plate of the duralumin.

Figures 3, 4 shows the picture observed when the probe moves above medium inside which a defect is located. The signal level from the measuring winding characterizes conductance values on the studied area.

Two major inspection modes are normally used viz. absolute and differential, with the older type reflection or bridge probes used for more specialised applications.

The absolute inspection mode utilise a single inspection coil with an internal instrument balance. Absolute mode is sensitive to variations in temperature and lift-off and is generally used to find large volume indications in tube inspection applications or generalised wall loss areas on thin conductive samples. Surface crack detection is also performed with the signal amplitude used to quantify the indication depth. The results of the experiment when scanning in absolute inspection mode presented in Figure 3

The differential inspection mode utilise two probes in close proximity to one another and comparing one surface area with another. It is very sensitive for finding small indications and the effect of probe wobble and lift-off is also reduced since both coils are subjected to the same movement or distance influence. The results of the experiment when scanning in differential inspection mode presented in Figure 4.

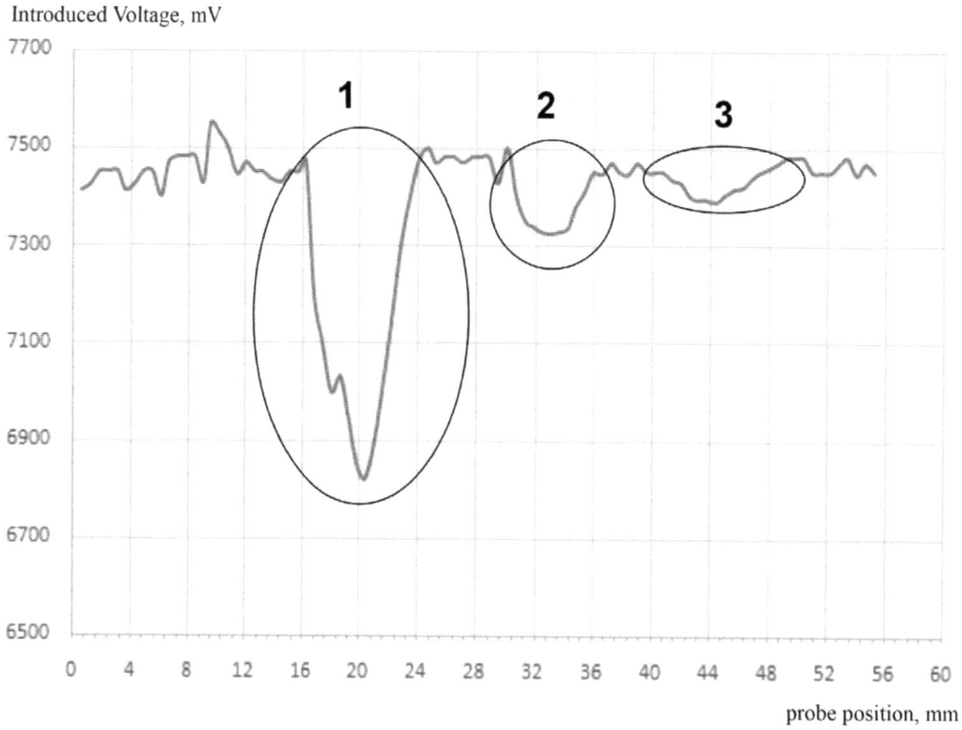

Figure 3. Results of scanning plate. Absolute inspection mode.

For the 1000 Hz basic operating frequency, the voltage introduced into the measuring winding was 7450 ± 30 mV. Domain 1 in the graph, in which the voltage level drops to 6850 mV, correspond to the defects. This change in the signal amplitude is 8% from the signal level corresponding to the defect free area of the sample. Domain 2 and 3 in the graph, correspond to the deep-lying defects.

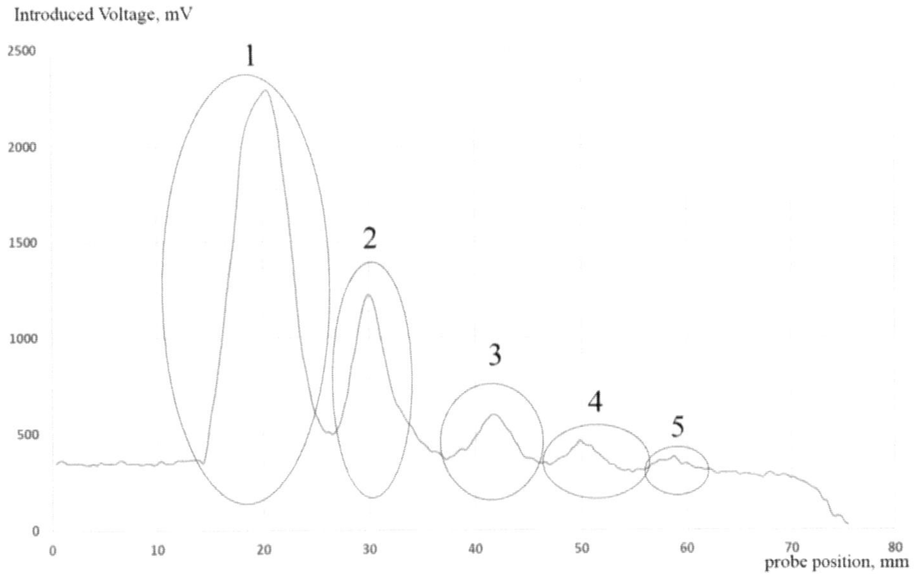

Figure 4. Results of scanning plate. Differential inspection mode.

For the 1000 Hz basic operating frequency, the voltage introduced into the measuring winding was 6000 ± 30 mV. Domain 1 in the graph, in which the voltage level drops to 3750 mV, correspond to the defects. This change in the signal amplitude is 37.5% from the signal level corresponding to the defect free area of the sample. . Domain 2 in the graph, in which the voltage level drops to 4950 mV. This change in the signal amplitude is 17.5% from the signal level corresponding to the defect free area of the sample. Domain 3, 4 and 5 in the graph, correspond to the deep-lying defects.

4. Conclusion

Summing it up, the experiment results demonstrate great capabilities of the eddy current method when the defects, hidden in the metal depth, need to be studied. Earlier the eddy current control method could be used to investigate only surface defects (such as cracks, cuts and other examples of metal surface discontinuity), now, due to using subminiature ECTs and special software, it is getting possible to localize the magnetic field in a small zone of the controlled object and to achieve a high degree of the (magnetic) field penetration depth into the investigated object. The only condition is to choose a proper field frequency, created by the actuating winding.

5. References

[1] Korenska M, Manychova M, Pazdera L, Topolar L. 2014 Monitoring of the Damage in Stone Blocks by Means of Non-Destructive Methods. *Key Engineering Materials*. **592-593**. pp 545-548

[2] Malikov V, Dmitriev S, Sagalakov A, and Ishkov A. 2014 Subminiature Eddy Current Transducers for Studying Metal-Dielectric Junctions. *Instruments and experimental techniques*. **57**. **No.6**. Pp. 751-754

[3] S F Dmitriev, A V Ishkov, V N Malikov, A M Sagalakov and A O Katasonov 2014 Non-destructive testing of the metal-insulator-metal using miniature eddy current transducers. *IOP Conference Series: Materials Science and Engineering*.**71**

The empirical evaluation of thermal conduction coefficient of some liquid composite heat insulating materials

M V Anisimov[1], V S Rekunov[1], M N Babuta[2], Nguyen Thi Hong Bach Lien[2]
[1] Tomsk State University of Architecture and Building, Tomsk, Solyanaya sq. 2, Tomsk 634003, Russia
[2] National Research Tomsk Polytechnic University, Lenina St. 30, Tomsk 634050, Russia

E-mail: teploproekt@mail2000.ru

Abstract. We experimentally determined the coefficients of thermal conductivity of some ultra thin liquid composite heat insulating coatings, for sample #1 λ = 0.086 W/(m·°C), for sample #2 λ = 0.091 W/(m·°C). We performed the measurement error calculation. The actual thermal conduction coefficient of the studied samples was higher than the declared one. The manufactures of liquid coatings might have used some "ideal" conditions when defining heat conductivity in the laboratory or the coefficient was obtained by means of theoretical solution of heat conduction problem in liquid composite insulating media. However, liquid insulating coatings are of great interest to builders, because they allow to warm objects of complex geometric shapes (valve chambers, complex assemblies, etc.), which makes them virtually irreplaceable. The proper accounting of heating qualities of paints will allow to avoid heat loss increase above the specified limits in insulated pipes with heat transfer materials or building structures, as well as protect them from possible thawing in the period of subzero weather.

1. Introduction

At present, the construction market has a wide range of different heat-insulating materials. Recently, some firms began to offer advanced ultra-thin insulating liquid composite coatings (hereinafter insulating paints) which thermal conductivity is λ = 0.001 ÷ 0.0015 W/(m·°C) for heat insulation of house fronts, and utilities. As a comparative example, they often provide the data that a layer of such paint with the thickness of 1 to 3 mm coated on engineering pipes can successfully replace the insulation of widely known mineral wool heat insulators with the thickness of a few centimeters [1, 2].

In the sphere of housing and communal services the application of heat insulating materials can result in substantial energy savings [3]. However, the heat transfer properties of the submitted data of heat-insulating coatings are not fully studied. The existing studies of various authors [4-7] on determination of thermal conductivity of the same type of heat-insulating paints often demonstrate a significant difference.

In this connection, it was decided to conduct the experimental determination of the thermal conductivity of some samples of insulating paints (as one of the main thermal characteristics of these coatings) in order to reveal their true value.

2. The existing methods of determination materials thermal conductivity

To review the methods of determination of materials thermal conductivity we should noted the work of such authors as Y.Y. Golovach (FSUE RI «Santechniki»), A.V. Shvetsov (Capstone Manufacturing), Y.F. Kolkhir (JSC «Predpriyatie Itil»)) [5]. The present method determines the thermal conductivity of insulation under strictly defined environmental conditions, which is not always possible to achieve. There are also several legal methods of determination of thermal conduction coefficient of various building materials [8, 9]. The relative error in determining the effective thermal conductivity and thermal resistance by method [9] does not exceed ± 3% if the test is carried out in full compliance with the requirements of the standard.

There is a method for determining the thermal conductivity using the "support wall" (the layer may be used as heat insulation) [10] consisting of two layers of material, one of which is with a known thermal conduction coefficient, placed on a heat source.

There is also a modern patented method for determination of the thermal conduction coefficient of ultra thin liquid heat insulating coatings [11].

The analysis of the usability conditions of the existing methods showed that the standard method [8] is designed to measure mainly the heat-conduction coefficient of bulk materials that does not meet the initial requirements. The method [5] is highly demanding to the accuracy of maintaining the temperature conditions. The method [11] is developed for section $\lambda = 0.01 \div 0.009$ W/ (m·°C).

Besides the works of Russian scientists in the sphere of the thermal conduction coefficient determination of insulating paints the similar experiences of foreign experts in the solution of similar problems has been analyzed [12-17].

In accordance with the performed analysis of the existing methods it was decided to use some samples of insulating paints of standard method [9] with the replacement of a heat meter with the layer of material with a known conductivity for the experimental determination of the thermal conduction coefficient. Such change is correct and does not contradict the theory of the study of thermal processes [10]. The method [9] meets all the requirements for conducting the experiment (the standard is not applied to materials and products with a thermal conductivity of more than 1.5 W/(m ·°C) and the relative error in determining the effective thermal conductivity and thermal resistance according to the method [9] does not exceed ± 3%).

3. Description of the experiment

For the experiment, the authors have reviewed the list of the most well-known manufacturers of liquid heat insulating coatings (mascoat, tsmceramic, thermalcoat, Isollat, Astratek, Alfatek, Teplokott, Corund etc.).

Two samples of the above brands of liquid heaters were selected for the study, which were named "Sample" (here in after Sample #1 and Sample #2). Several sample characteristics are given in the Table. 1.

Table 1. Some the stated characteristics of experimental samples.

Characteristic's name	Unit of measurement	Sample #1	Sample #2
Colour of coating		white	white
Heat conductivity	W/(m·°C)	0.0011	0.002 – 0.007
Density	kg/m^3	390	280
Vapor transmission	mg/(m·h·Pa)	0.012	0.012
Water permeability	kg/(h·m·0.1GA)	less than 30	less than 30

The manufacturers of heat insulating paints often do not provide full information on the physico-chemical composition of the heat insulating coatings produced by them, so the information on the composition of the heat insulating paints of the studied samples was obtained from official sites of the manufacturers. The paints include: glass ceramic microspheres, binders, dispersants, fillers, pigments, thinners. Typically, the suppliers do not provide the percentage of ingredients.

The measuring complex was developed for the experiment, it includes: The apparatus for testing samples (figure 1); the appliance "Terem-4.0" for measurement of thermocouples readings; "Chromel-Copel" thermocouples are made of wire with the thickness of δ = 0.2 mm; the plate of a material with a known heat conductivity (plexiglas, with the thickness δ = 3.2 mm, λ = 0.19 W/(m·°C), which is a replacement for heat meters.

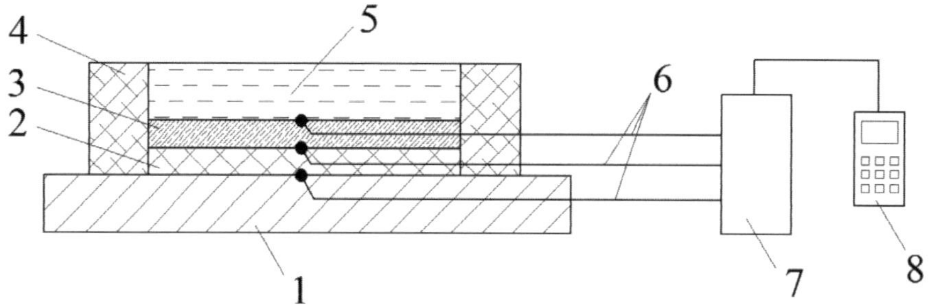

Figure 1. Schematic diagram of measuring complex:
1 - source of continuous heat flux; 2 - the layer of the material with known thickness and heat-conducting coefficient (plexiglas); 3 - the layer of heat insulating paint; 4 - insulator (polystyrene); 5 - "refrigerator" (a container with water); 6 - thermocouple between the layers; 7 - Switch; 8 - measuring device "Terem 4.0"

Paint in the layer 3 (figure 1) was applied evenly onto a copper plate of 0.5 mm thickness. The heat flux density q_i, W/m^2 depending on (1) is determined by calculation:

$$q_u = \frac{\lambda_{\text{layer}\,2}(t_1 - t_2)}{\delta_{\text{layer}2}}, \tag{1}$$

where $\lambda_{\text{layer 2}}$, $\delta_{\text{layer 2}}$ is a thermal conducting coefficient and the thickness of plexiglas layer (figure 1), t_1, t_2 are temperatures on the borders «the source of heat is a layer of plexiglas» and «the layer of plexiglas is a tested sample» accordingly (figure 1).

The thermal conductivity of copper plate with the thickness δ = 0.5 mm, equals λ = 384 W/(m·°C). The temperature and relative air humidity indoors during the experiment were equal to $t_в$ = 24 °C, and φ = 40%.

To stabilize the readings of the appliance during its "warming" and heat flux transfer into a stationary mode the control measurements of appliance reading dynamics were performed according to 3 sensors of thermocouples within 0.5 hours in increments of of 5 minutes measurement. At steady state the readings came to stationary mode in 20 minutes after the start of its operation. For defining the individual error of thermocouple sensors the temperatures measurement for each sensor immersed in a "Dewar" vessel filled with snow slush were performed before the experiments. The obtained temperature deviations from 0 °C temperature were also noted during the experiments.

The calibration was performed to test the adequacy of the developed appliance, measurement of thermal conduction of insulating paints. The plexiglas plate similar in size, thickness and the thermal conductivity to the plate in layer 2 was placed instead of layer 3 (figure 1) into the appliance. The calibration measurements were conducted. Due to the measurement results it was found that thermal conductivity of tested plexiglas plate was λ = 0.186 W/(m·°C). The error of the method of measuring the thermal conductivity was 2.1%.

The findings suggest that the error of this method does not exceed the error stated in the State Standard [12] (± 3%), which indicates the correctness of the chosen research scheme.

4. The analysis of the experiment results

Sample #1 and Sample #2 were tested at various temperature modes in different heat fluxes in purpose to analyze the dynamics of changes of thermal conduction coefficient depending on temperature of paint samples. The obtained results are presented in figure 2 and figure 3 correspondingly.

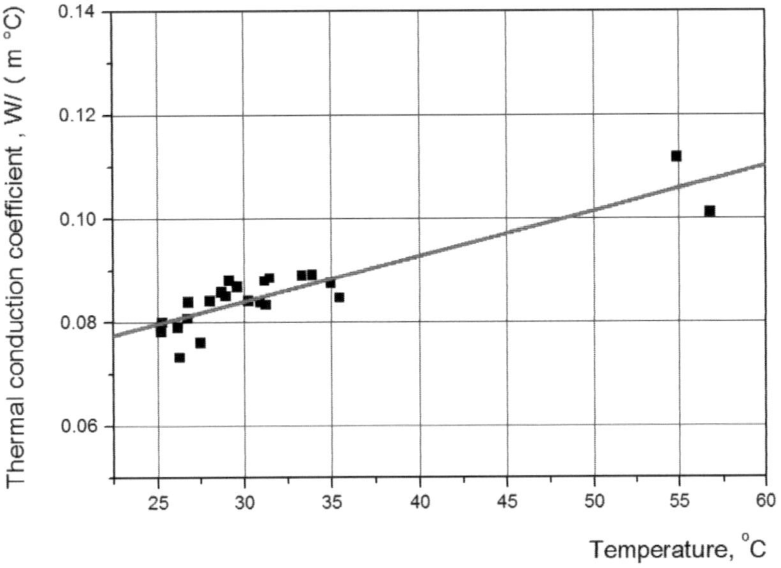

Figure 2. The results of the experiment on defining the paint thermal conductivity (sample #1)

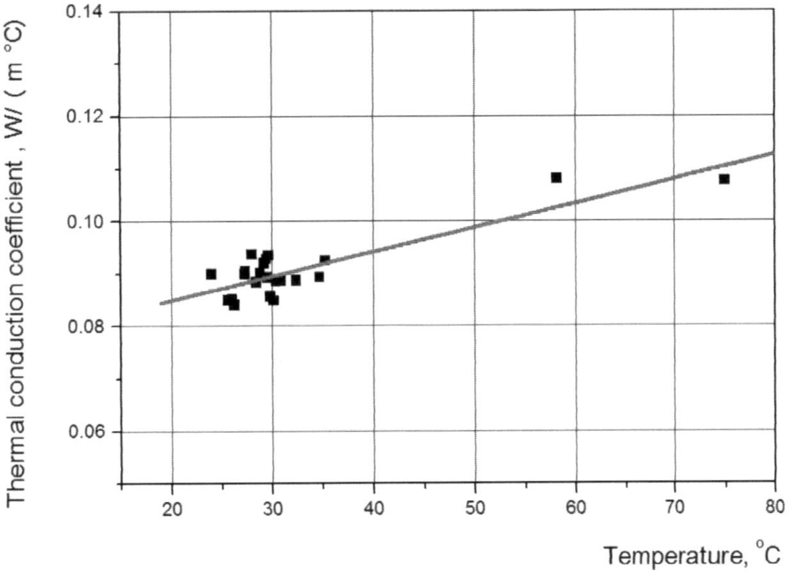

Figure 3. The results of the experiment on defining the paint thermal conductivity (sample #2)

The measurement error calculations was conducted by dependences (2-4)

$$\bar{A} = \frac{\sum_{i=1}^{n} A_i}{n}, \tag{2}$$

where A is a measured value, $\bar{A}$ is a medium value of the calculated value, $\Delta\bar{A}$ is an absolute error of medium value of the calculated value defined by the formula

$$\Delta\bar{A} = t_{y,n-1}\frac{\sqrt{\frac{\sum_{i=1}^{n}\left(A_i - \bar{A}\right)^2}{n-1}}}{\sqrt{n}}, \tag{3}$$

where $t_{y,n\text{-}1}$ is Student's coefficient (by $t_{95\%,23}$= 2.074).

The relative error of medium value of the measured value is calculated by dependences [17]

$$\varepsilon = \frac{\Delta\bar{A}}{\bar{A}}. \tag{4}$$

The measurement error made ε = 1.85%. The total error in the determination of thermal conductivity taking into account the error of the study method (3%), and the appliance error (1%) was 5.85%.

5. Conclusion

As a result of this work the thermal conducting coefficient of ultrathin liquid composite thermal insulating coatings has been experimentally determined. For «Sample #1» it makes λ = 0.086 W/(m·°C), for «Sample #2» it makes λ = 0.091 W/(m·°C).

Undoubtedly, the given liquid coatings can be attributed to the heat insulating materials. According to their thermal conducting characteristics they are equal, for example, to mineral wool (λ = 0.07 W/(m·°C) or foam glass (λ = 0.1 W/(m·°C). However, their real thermal conducting coefficient is higher than declared. The given difference is possible due to the fact that the manufacturers of liquid coatings used some "ideal" conditions, or the coefficient was obtained by the theoretical solution of the problem of thermal conduction in liquid composite insulating media at laboratory determining of the thermal conductivity

Despite this, these liquid insulating coatings are of great interest to builders, because they allow to warm objects of complex geometric shapes (valve chambers, complex assemblies, etc.), which in some cases makes them virtually irreplaceable.

The proper accounting of the thermal conducting characteristics of paints will prevent excess increase of heat losses in insulated pipes with thermal transfer materials or building structures, as well as protect them from possible thawing in the period of subzero weather.

This study does not claim to be the "definitive" and is probably aimed at the prerequisites for further study of thermal insulating properties of modern composite liquid thermal insulating coatings, because their thermal properties are not adequately studied to date.

Acknowledgments

We express our gratitude for provided measuring equipment to the "TGS" department, TSUAB (Tomsk), as well as to the building organizations of Tomsk for provided samples of materials.

References

[1] Dombrovsky L.A. 2005 Modeling of thermal radiation of polymeric coating containing hollow microsphere *Thermal physics high temperatures* **43 1** 1–11.

[2] German M.L. 2002 Mathematical model of calculating thermal properties of composite coating «ceramic microsphere – binder» *Engineering Journal of Physics* **75 6** 43-53.

[3] Komkov V.A., Timakhova N.S. 2010 Energy saving in housing and communal services *Moscow Infa-M Publ.* **320**.

[4] Shirinian V.T. 2007 The trip of liquid ceramic «ultra heat-insulating» coating through Russian heat supply networks *heating news* **9** 46-51.
[5] Golovach Yu.Yu., Shvetsov A.V., Kolkhir Yu.F. 2008 The method of the experiment and the calculation of the thermal conductivity for ultra-thin thermal insulation materials *Kazan* Available at: http://inn-t.com/teploprovodnost/index.html.
[6] Maneshev I.O., Pravnik Hu.I., Sadykov R.A., Safin I.A., Eremin S.A. 2013 Experimental determination of thermal conductivity coefficients and efficiency of ultrathin thermal insulation coatings *Izvestia KazGASY* **1 23** 135-142.
[7] Loginova N.A. 2010 Efficiency of thin-film thermal insulation coatings in centralized heat supply systems *Moscow* **133**.
[8] GOST 30290-94. State Standard 30290-94 1996 Building materials and products. The method of determining the thermal conductivity of the surface transducer *Moscow* **12**.
[9] GOST 7076-99. State Standard 7076-99. 2000 Building materials and products. Method for determination of thermal conductivity and thermal resistance under steady-state thermal conditions *Moscow* **20.**
[10] Mikheev M.A. 1973 Bases of thermal conductivity *Moscow graduate School* **309.**
[11] Pravnik Yu.I., Sadykov R.A. Ivanova R.V. 2013 A method for determining the thermal conductivity of ultrathin liquid insulation coating *Patent RF #* 2478936.
[12] D. Kuvshinov, M.R. Bown, J.M. MacInnes, R.W.K. Allen, R. Ge, L. Aldous, C. Hardacre, N. Doy, M.I. Newton, G. McHale. 2010 Thermal conductivity measurement of liquids in a microfluidic device *Springer-Verlag* **123** 122.
[13] Gustavson M, Nagai H, Okutani T. 2003 Thermal effusivity measurements of insulating liquids using microsized hot stripprobes *Rev Sci Instrum* **42** 48.
[14] Hammerschmidt U. 2010 A quasi-steady state technique to measurethe thermal conductivity *Int J Thermophys* **291** 312.
[15] Kuntner J, Kohl F, Jakoby B 2006 Simultaneous thermal conductivity and diffusivity sensing in liquids using a micromachined device *Sens Actuator* **10 1** 62-67.
[16] Xie H., Gu H, Fujii M, Zhang X. 2006 Short hot wire technique for measuring thermal conductivity and thermal diffusivity of various material *Meas Sci Technol* **12 2** 208-214.
[17] Zhang H, Zhao G, Ye H, Ge X, Cheng S 2005 An improved hotprobe for measuring thermal conductivity of liquids *Meas Sci Technol* **10 5** 430-435.

IOP Conf. Series: Materials Science and Engineering **116** (2016) 012015 doi:10.1088/1757-899X/116/1/012015

The evolution of phase transformation in Ni/Ni_3Al laminated composite under high temperature treatments

V Shmorgun, L Gurevich, A Bogdanov and M Trunov
Materials Science Chair, *Volgograd State Technical University*,
400005, Lenin Avenue 28, Volgograd, Russia

E-mail: mv@vstu.ru

Abstract. In this study the impact of isothermal annealing on the phase transformation rate in laminated Ni/Ni_2Al_3 composite was investigated. The method of nickel-aluminide coatings of the required chemical composition fabrication was proposed.

1. Introduction

One of the perspective methods of steel surface protection from high temperature oxidation is the formation of Ni/Ni_2Al_3 coatings [1, 2]. The Ni layer prevents the penetration of Al atoms in steel. The surface of Ni_2Al_3 layer is covered by Al_2O_3 and thus protects the covered material from oxidation. The lifetime of such coatings depends on the diffusion processes which according to the Ni-Al phase diagram lead to the formation of NiAl, Ni_5Al_3 and Ni_3Al intermetallics.

The operability of aluminide coatings is limited by 1100°C [3]. Above 1100°C Ni atoms start to diffuse in the coating which results in the responsible for protective coating main NiAl phase vanishing.

The purpose of the present study was to investigate the impact of heat treatment at 900-1000°C on the sequence and evolution of phase transformations in Ni-Ni_2Al_3 composite.

2. Materials and methods

The Ni-Ni_2Al_3 composite (figure 1) was obtained via method described in [4], namely explosion welded Ni-Al (10 and 0.7 mm thickness respectively) (figure 2a) laminated composite was heat treated at 600°C for 5 hours which contributed to the formation of two intermetallic interlayers in the bond interface between Ni and Al (Ni_2Al_3 adjacent to Ni layer and $NiAl_3$ adjacent to Al layer).

The subsequent to heating cooling of the composite resulted in the rupture of $NiAl_3$ interlayer [5] (figure 2 b).

Metallographic investigation of the composite was carried out using optical microscope Olympus BX-61. XRD investigation of the interlayers was performed via X-Ray diffractometer DRON-3 with CuKα radiation with Ni-filter at an operating voltage of 30 kV and current of 20 mA. EDX study was applied to determine the chemical composition of the interlayers. SEM Versa-3d (Check Republic) was used.

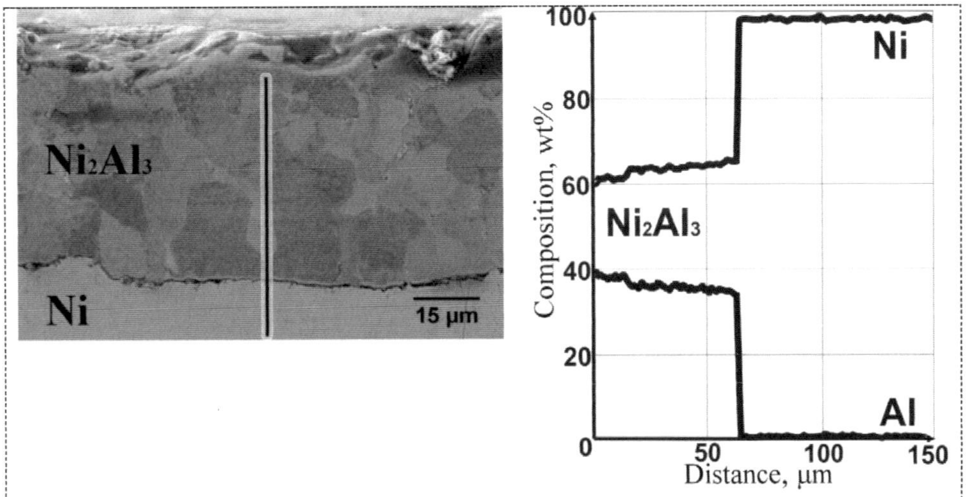

a) b)

Figure 1. SEM image of the Ni-Ni_2Al_3 composite (a) and element distribution (b) over the region indicated in *a*

3. Results and Discussion

Heat treatment of the Ni-Ni_2Al_3 (with Ni_2Al_3 thickness 50μm) composite at 1000°C for 2 hours promotes the formation of NiAl intermetallic between Ni layer and Ni_2Al_3. The width of the formed interlayer was measured to be 25 μm (figure 3 a and b).

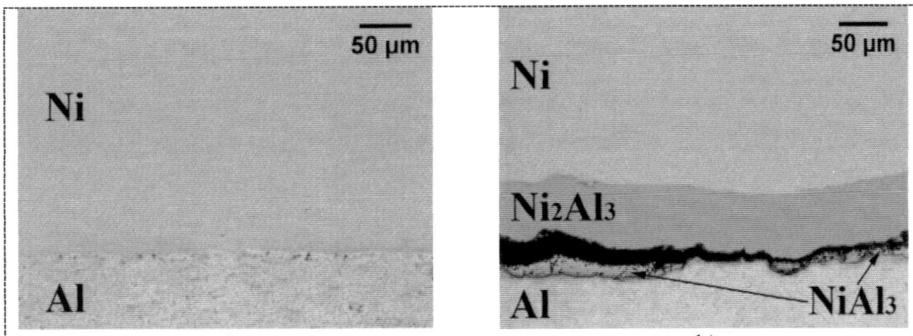

a) b)

Figure 2. Optical image of the Ni/Al interface after explosion welding (a) and after explosion welding and subsequent heat treatment at 600°C for 5 hours (b)

Heat treatment of the Ni-Ni_2Al_3 composite for 4 hours at 1000°C resulted in the uniform NiAl layer formation (figure 3 c and d). XRD patterns identified the presence of the protective Al_2O_3 layer on the surface of NiAl aluminide.

Heat treatment at 1000°C for 50 hr contributed to the NiAl phase transformation into Ni_3Al, while solid Ni(Al) solution was formed between Ni and Ni_3Al. The solution was composed of Ni austenite with alternating concentration (figure 3 e and f).

The increase in heat treatment time up to 100 hours promoted the solid solution layer thickness growth up to 15 μm, while the Ni_3Al layer thickness decreased (figure 4).

The reduction in the annealing temperature from 1000°C to 900°C has significantly decreased the rate of phase transformation. Thus the transformation of Ni_2Al_3 into NiAl and subsequently into Ni_3Al required 4 and 30 hours respectively at 1000°C, while the same transformation at 900°C proceeded 15 and 150 hours respectively.

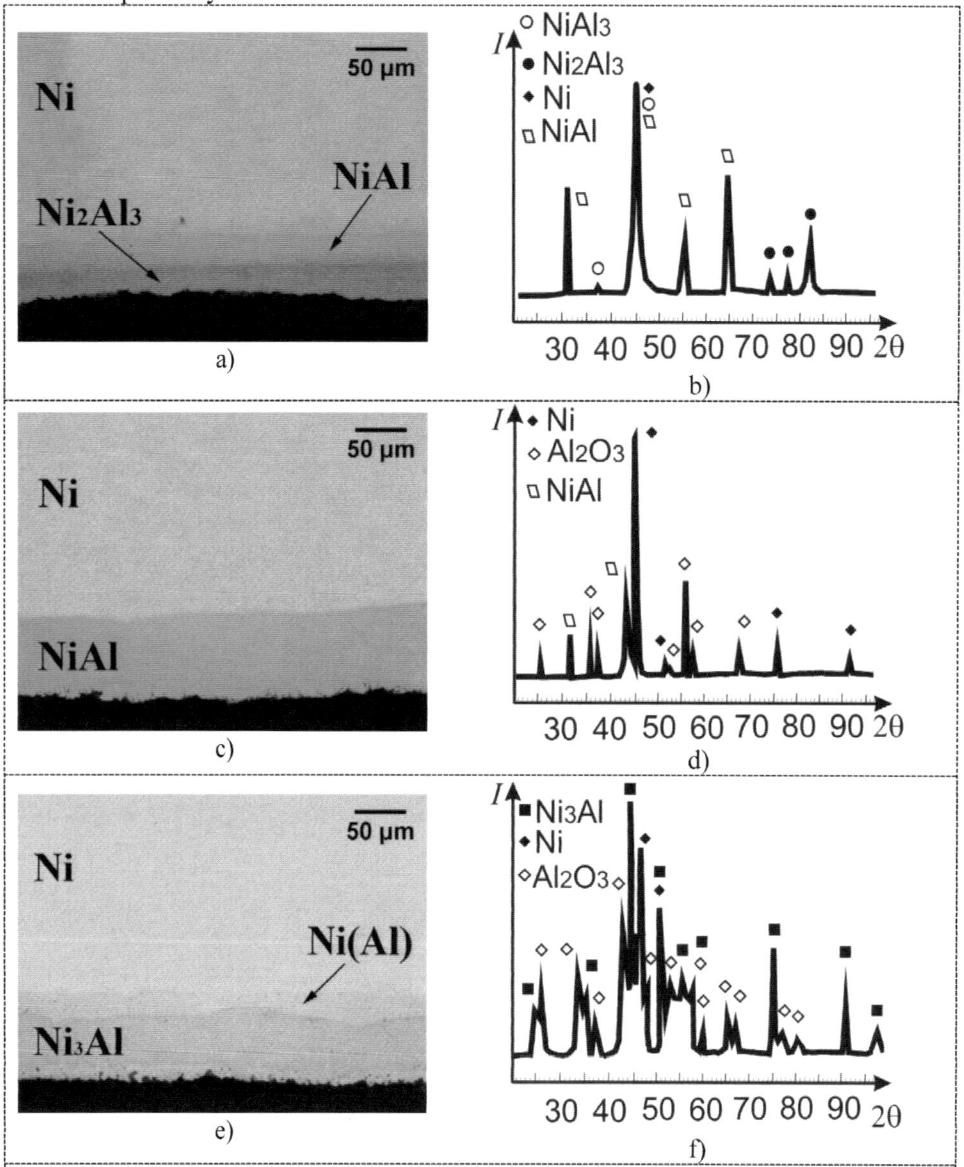

Figure 3. Optical image of the Ni/Al interface after explosion welding and after subsequent heat treatment at 600°C for (a) 2 hours (c) 10 hours (e) 50 hours and corresponding XRD patterns

The Ni_2Al_3 thickness - phase transformation duration curve and Ni_2Al_3 thickness – phase transformation temperature curve have inverse to each other shapes. Thus the increase in Ni_2Al_3 layer thickness value from 50 to 70 μm resulted in the rise of $Ni_2Al_3 \rightarrow$NiAl phase transformation duration from 4 to 8 hours, NiAl $\rightarrow Ni_3Al$ from 30 to 58 hours and $Ni_3Al \rightarrow$ Ni(Al) from 320 to 628 hours.
The obtained experimental results formed the basis of the nickel-aluminide coatings of the required chemical composition fabrication method. The method is as follows:

To calculate the initial Ni layer thickness, required to obtain NiAl (δ_{Ni}^{NiAl}), Ni_3Al ($\delta_{Ni}^{Ni_3Al}$) coating of the defined thickness consider reaction zone Ni/Ni_2Al_3 infinitesimal $dx \times dz$ element.

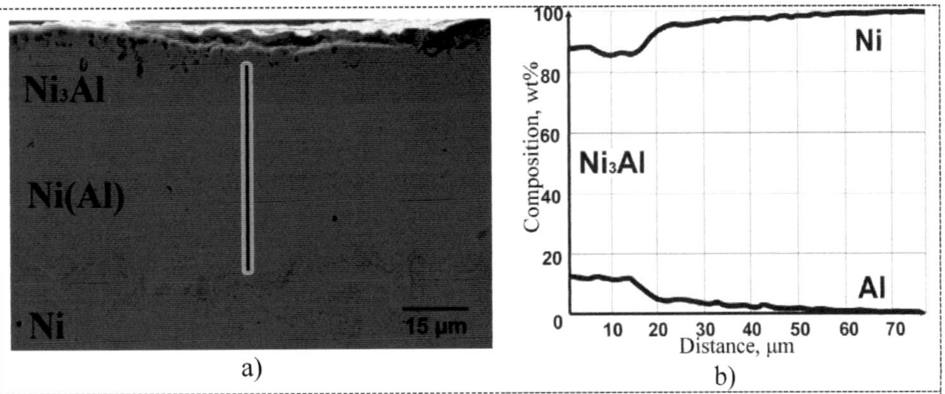

Figure 4. SEM image of the reaction zone (a) and element distribution (b) over the region indicated in *a*

According to the law of conservation of mass in order to obtain the intermetallic layer of the required stoichiometric composition the m_{Al}/m_{Ni} ration in the reaction zone has to be equal to mass fraction ratio v_{Al}/v_{Ni} in the considered intermetallic:

$$\frac{m_{Al}}{m_{Ni}} = \frac{v_{Al}^{NiAl(Ni_3Al)}}{v_{Ni}^{NiAl(Ni_3Al)}}. \tag{1}$$

The reaction zone contains Al in Ni_2Al_3 intermetallic and its mass is given by:

$$m_{Al}^{Ni_2Al_3} = V_{Ni_2Al_3} \cdot \rho_{Ni_2Al_3} \cdot v_{Al}^{Ni_2Al_3}, \tag{2}$$

where $V_{Ni_2Al_3} = \delta_{Ni_2Al_3} \cdot dx \cdot dz$ – denotes Ni_2Al_3 intermetallic volume; $\delta_{Ni_2Al_3}$ – Ni_2Al_3 interlayer thickness; $\rho_{Ni_2Al_3}$ – Ni_2Al_3 density (4787 kg/m^3); $v_{Al}^{Ni_2Al_3}$ – mass fraction of Al in Ni_2Al_3 intermetallic (40.81%) [6].

The total Ni mass in the reaction zone consists of the Ni in Ni_2Al_3 intermetallic ($m_{Ni}^{Ni_2Al_3}$) and Ni in the Ni layer (m_{Ni}):

$$m_{Ni}^{\Sigma} = m_{Ni}^{Ni_2Al_3} + m_{Ni}, \tag{3}$$

where

$$m_{Ni}^{Ni_2Al_3} = V_{Ni_2Al_3} \cdot \rho_{Ni_2Al_3} \cdot v_{Ni}^{Ni_2Al_3}, \tag{4}$$

$$m_{Ni} = \delta_{Ni} \cdot dx \cdot dz \cdot \rho_{Ni}, \tag{5}$$

$v_{Ni}^{Ni_2Al_3}$ – mass fraction of Ni in Ni_2Al_3 intermetallic; δ_{Ni} – Ni layer thickness; ρ_{Ni} – Ni density.

Applying (2) and (3) to (1) and then solving (1) with respect to δ_{Ni} gives the formula for aluminides layer thickness value and Ni layer thickness value. The values are required for the total transformation of Ni/Ni_2Al_3 into NiAl or Ni_3Al:

$$\delta_{Ni}^{NiAl} = \frac{\delta_{Ni_2Al_3} \cdot \rho_{Ni_2Al_3} \cdot \left(v_{Ni}^{NiAl} \cdot v_{Al}^{Ni_2Al_3} - v_{Al}^{NiAl} \cdot v_{Ni}^{Ni_2Al_3}\right)}{v_{Al}^{NiAl} \cdot \rho_{Ni}}, \tag{6}$$

$$\delta_{Ni}^{Ni_3Al} = \frac{\delta_{Ni_2Al_3} \cdot \rho_{Ni_2Al_3} \cdot \left(v_{Ni}^{Ni_3Al} \cdot v_{Al}^{Ni_2Al_3} - v_{Al}^{Ni_3Al} \cdot v_{Ni}^{Ni_2Al_3}\right)}{v_{Al}^{Ni_3Al} \cdot \rho_{Ni}}, \tag{7}$$

where $\nu_{Al}^{Ni_3Al}$ and $\nu_{Ni}^{Ni_3Al}$ – mass fraction of Al (13.28%) and Ni in Ni_3Al intermetallic; ν_{Al}^{NiAl} and ν_{Ni}^{NiAl} – mass fraction of Al (31.49%) and Ni in NiAl intermetallic.

Taking (6) and (7) into consideration it is possible to clearly define δ_{Ni}^{NiAl} and $\delta_{Ni}^{Ni_3Al}$ from Ni_2Al_3 intermetallic thickness value, which depends on the time-temperature conditions of the diffusion annealing [4] and is given by:

$$\delta_{Ni_2Al_3}^2 = 0.38\cdot 10^{10}\exp(-\frac{171700}{RT})\cdot\left[\tau - 0.786\cdot 10^{-6}\exp(\frac{118060}{RT})\right], \quad (8)$$

where τ – annealing time; R – gas constant.

Figure 5 presents the results of Ni layer thickness value calculation required to obtain NiAl and Ni_3Al intermetallics.

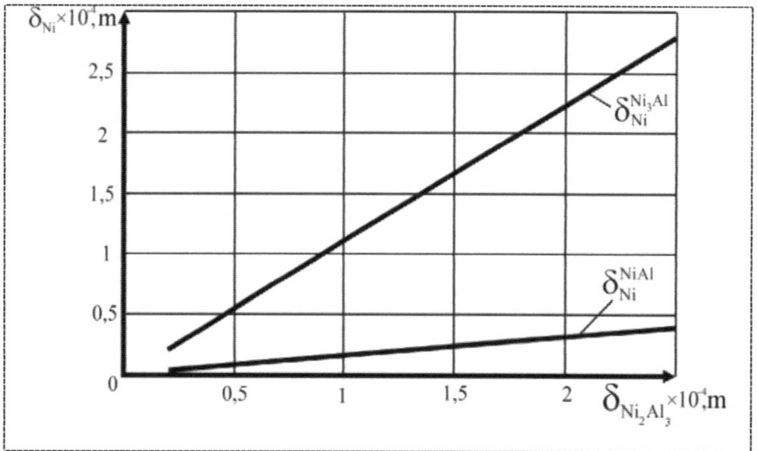

Figure 5. The relation between the initial Ni and Ni_2Al_3 layers thickness value and corresponding thickness values of NiAl and Ni_3Al layers which can be formed after annealing

4. Conclusion

1. Heat treatment at 900°C or 1000°C of the Ni/Ni_2Al_3 laminated composite result in the following order of phase transformations Ni/Ni_2Al_3 → Ni/NiAl/Ni_2Al_3 → Ni/NiAl → Ni/Ni_3Al → Ni(Al). The phase transformation time depends on the heat treatment temperature and the initial Ni_2Al_3 thickness value.
2. The increase in the thickness value of each formed layer contributes to the Ni layer thickness value change, while the intensity of the intermetallic layer growth and transformation is defined by temperature and time of the heat treatments.

Acknowledgements
The study was supported by Russian Science Foundation grant No 14-19-00418.

References

[1] Xiang Z D, Zeng D, Zhu C Y, Wu D. J. and Datta P K 2012 *Journal of Materials Science* **47(1)** 257-266
[2] Xiang Z D, Zeng D, Zhu C Y, Rose S R and Datta P K 2011 *Corrosion Science* **53(1)** 496-502
[3] Poklad V, Shkretov Yu and Abraimov N 2010 *Dvigatel* **4(70)** 4-8

[4] Shmorgun V G, Trykov Y P, Slautin O V, Metelkin V V and Bogdanov A I 2009 *Russian Journal of Non-Ferrous Metals* **50 (3)** 286-289
[5] Shmorgun V G, Bogdanov A I and Trykov Y P 2013 *Deformatsiya i razrusheniye materialov* **9** 43-46
[6] Deevi S C, Sikka V K, Swindeman C J and Seals R D 1997 *Journal of thermal spray technology* **6(3)** 335-344

Formation and Characterization of Crystalline Hydroxyapatite Coating with the (002) Texture

I Grubova[1], T Priamushko[1], E Chudinova[1], M Surmeneva[1], O Korneva[2], M Epple[3], O Prymak[3], I Shulepov[2] and R Surmenev[1, 4, *]

[1] Department of Theoretical and Experimental Physics, National Research Tomsk Polytechnic University, 634050 Tomsk, Russia
[2] Material Properties Measurements Centre, National Research Tomsk Polytechnic University, 634050 Tomsk, Russia
[3] Inorganic Chemistry and Center for Nanointegration Duisburg-Essen (CeNIDE), University of Duisburg-Essen, 45117 Essen, Germany
[4] Fraunhofer Institute for Interfacial Engineering and Biotechnology (IGB), 70569 Stuttgart, Germany

[*]E-mail: rsurmenev@gmail.com

Abstract. This study reports the effect of titanium (Ti) microstructure on the mechanical properties and surface wettability of thin (<800 nm) hydroxyapatite (HA) coating deposited via radio-frequency (RF) magnetron sputtering. It was revealed that the sand-blasting (SB) and acid etching (AE) of Ti prior deposition led to a wide range of surface roughness in nano/micro scale. After nanostructured HA coating deposition such physico-mechanical characteristics as nanohardness H, Young's modulus E, H/E ratio and H^3/E^2 were significantly improved. Moreover, HA coatings exhibited improved wear resistance, lower friction coefficient and ability of the coating to wetting.

1. Introduction

Permanent implants based on Ti and its alloys are clinically used this time due to their good mechanical properties and a high corrosion resistance [1-3]. Unfortunately, a wide application of such materials is limited due to insufficient wear resistance, high friction coefficient and lack of osteoconductivity. Recently, there is increasing interest to the investigations of different ways of surface modification to enhance their physico-chemical, mechanical and biocompatible properties [4].

Crystalline HA thin films is the best choice for bone repair and replacement [5-7]. However, a low ductility of HA limits its load bearing applications. In order to fabricate high quality bone implants with rapid healing the combination of good mechanical properties of Ti based metals with excellent biocompatibility and bioactivity of HA have been applied. Nevertheless, the long-term stability of HA coating in body environment is still a subject for investigations.

In order to maintain long-term stability of Ti implants with HA coating, several requirements are usually should be met. First, HA coating itself should have good mechanical properties (high hardness, low friction etc.); second, the bonding strength of the interface between HA coating and Ti substrate should be high enough to sustain possible fatigue stress during surgical operation or after implantation. In this study, two ways have been studied to overcome the above mentioned challenges. One is to use

RF magnetron sputtering, which allows to prepare dense, pore-free films with high adhesion strength [8]. The other is to increase the bonding strength via several surface pretreatment methods, such as SB and AE prior a coating deposition procedure.

Table 1. Effect of the titanium surface treatment procedure on the surface roughness, local physico-mechanical characteristics and wettability

Sample type	R_a (μm)	H (GPa)	E (GPa)	H/E	H^3/E^2 (GPa)	h_c (nm)	θ_{adv} (°)	$\Delta\theta$ (°)
SB	1.3±0.4	3.9±0.3	87±14	0.044	0.008	126.5±9.6	63.7±0.4	28.18
SB+AE	1.0±0.1	3.1±0.2	81±12	0.038	0.005	130.1±4.5	69.4±2.7	31.03
SB+AE+HA	0.8±0.1	15.2±0.7	147±16	0.101	0.164	71.1±2.9	87.1±0.2	36.55

2. Materials and methods

The commercially pure Ti substrates were used in the study. The samples were prepared via SB with Al_2O_3 particles (50 μm) for 10 s at the pressure P=1 MPa Pressure was measured using a pressure gauge via a coaxial compressor Fini SUPER TIGER 245. Then the samples were AE using 1 ml HF + 2 ml HNO_3 + 2.5 ml H_2O solution for 10 s. Finally, after AE, the samples were thoroughly cleaned with acetone in an ultrasonic bath (Eurosonic Micro, Euronda SPA, Italy) for 10 min followed by ethanol and deionised water. The coatings were deposited using a commercially available apparatus with an RF generator COMDEL (13.56 MHz). The precursor-powder of HA ($Ca_{10}(PO_4)_6(OH)_2$) was synthesized using a mechanochemical activation method. Preparation and investigation of the target was described elsewhere [8, 9].

Surface morphology and chemical composition of the biocomposites were investigated via scanning electron microscopy (SEM) using a Quanta 400 instrument equipped with an energy-dispersive X-ray analysis (EDX analysis system Genesis 4000, SUTW-Si (Li) detector) operated under high vacuum. The phase composition of the coating was identified by X-ray diffraction (Bruker D8 Advance XRD). Lattice parameters, crystallite size and the texture coefficient (TC_{hkl}) were calculated. To calculate the average crystallite size and the lattice parameters, Rietveld refinement (using the Le Bail method) with the program package TOPAS 4.2 from Bruker was performed. The calculations of TC_{hkl} were performed using the following HA reflexes: 25.8 ° (002), 31.7 ° (211) and 32.9 ° (300) according to the study [10]. Wettability and contact angle hysteresis of the samples were studied using Easy Drop Instrument (Kruess, Germany) with the use of deionized water.

The surface roughness of the samples was investigated using a Dektak 6M Stylus mechanical profilometer. Mechanical properties of the samples were studied via a Nano Hardness Tester «NHT-S-AX-000X» using a Berkovich indenter at the load 2.5 mN. Young's modulus (E) and nanohardness (H) were determined in accordance with the Oliver–Pharr method [11]. Wear resistance was evaluated using the plastic resistance parameter (H/E) as well as resistance of the material to plastic deformation (H^3/E^2 factor). Adhesion of the coatings was investigated via Micro-Scratch Tester "MST-S-AX-0000". Optical ellipsometry (Ellipse 1891-S AG) was used to determine the thickness of the deposited coatings [12].

3. Experimental results and discussion

Typical SEM-images of the investigated surfaces of treated Ti and thin (500-800 nm) HA coating are presented in figure 1. The roughness parameters after abrasive blasting were as follows R_a=1.3±0.4 μm. The chemical etching of Ti surface was selective and led to the smoothing of the initial microscale topography, which in turn led to decrease of the roughness to R_a=1.0±0.1 μm (table 1).

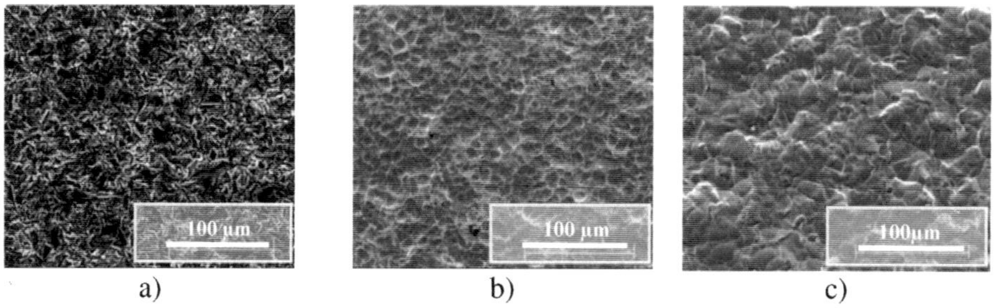

Figure 1. SEM-images of the SB Ti surface (a), SB+AE Ti surface (b) and HA coating surface (c).

SEM micrographs of HA coating are shown in Fig. 1c. The samples were rotated in the deposition chamber. The resulting coatings are deposited without any cracks and repeat the relief of the initial Ti substrate (figure 1c). It can be seen a homogeneous coating with the unclear grain boundaries. Deposition of the HA coating resulted in surface smoothing in a nanometer range (R_a=0.8±0.1 μm).

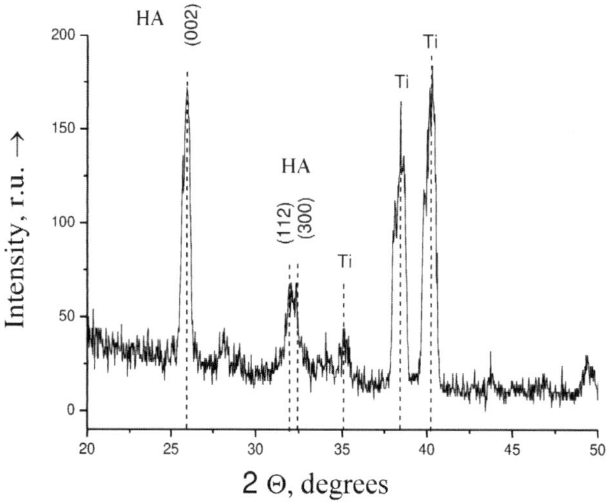

Figure 2. XRD-patterns of the HA-coated Ti substrate.

EDX analysis revealed that the Ca/P and O/P ratios of the coatings were 1.79 and 3.36, respectively, which were close to that reported for CaP coatings deposited via RF magnetron sputtering [8].

It can be seen from the figure 2 that the HA coating is crystalline with (002), (211) and (300) Bragg peaks of HA and several peaks corresponding to the substrate. Other phases such as CaO and (α, β) - TCP were not observed. An average crystallite size estimated from the XRD patterns was 96±2 nm. The calculations of TC_{hkl} revealed the predominant (002) growth texture of the HA coating (TC_{002}=1.40±0.01, TC_{112}=0.80±0.05, TC_{300}=0.75±0.05), which is common phenomenon for RF magnetron sputtering [13, 8-9]. This reveals the formation of HA grains with the c-axis preferentially aligned in the direction perpendicular to the substrate. The determined unit cell parameters were a=b=0.9430 nm and c=0.6893 nm, which were in good agreement with that of HA standard (ICDD, #09-432, a=b=0.9418 nm, c=0.6884 nm (space group $P6_3/m$).

The typical loading/unloading curves performed at the load of 2.5 mN recorded for uncoated Ti substrate and the HA coating are shown in Fig. 3a. The penetration depth of the indentor decreased significantly in the case of the deposited coating compared with that of the treated Ti. This was expected as the average coating nanohardness is superior to that of the uncoated Ti substrate. The nanohardness of the HA films was 15.2±0.7 GPa while Young's modulus was 147±16 GPa. On the other hand, the nanohardness of the treated Ti substrate prior deposition was 3.1±0.2 GPa while Young's modulus was 81±12 GPa (figure 3a, table 1). Based on the data obtained before for pure HA coatings deposited on the surface of untreated Ti (nanohardness ~10 GPa [9]), we can assume that substrate surface microstructure affected the mechanical properties of HA films. The HA coated samples exhibited better mechanical resistance. The maximum penetration depth h_c was 130.10±4.47 nm before deposition and decreased to 71.11±2.87 nm after HA coating fabrication.

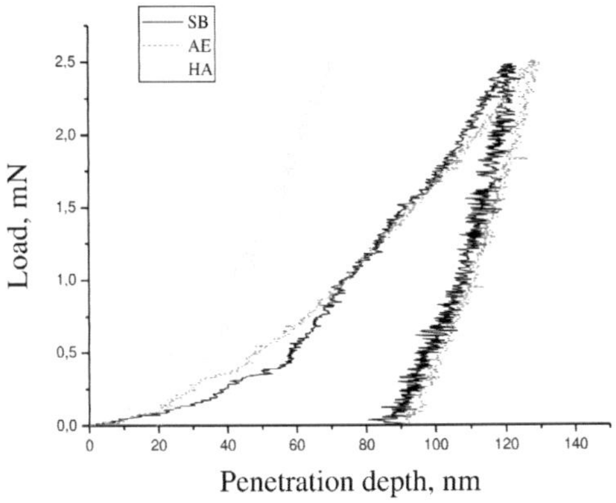

a)

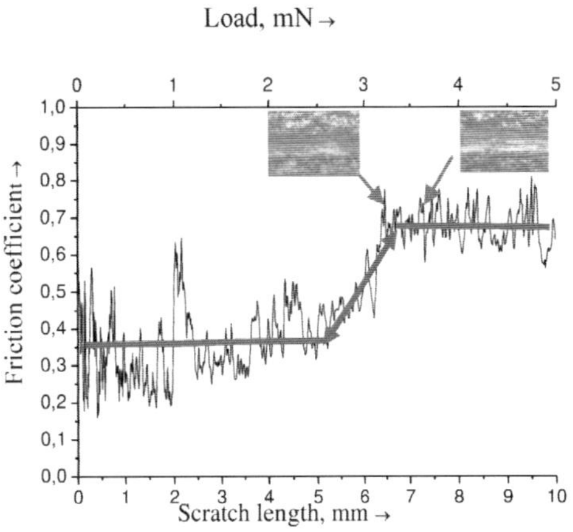

b)

Figure 3. Loading/unloading curves (a) and results of the HA coating scratch test (b).

The coating wear resistance and adhesion can be estimated taking into account *H* and *E* values. The ratio between *H* and *E* is called elastic strain to failure and it was used for the first time as a materials ranking parameter by Oberle [14, 15]. The *H*/*E* ratio of the composite increased from 0.038 up to 0.101 after the HA coating deposition, which indicated the improvement of the wear resistance. The similar trend was observed in the case of the H^3/E^2 factor (table 1).

Figure 3b shows the typical scratch test diagram of the friction coefficient change as a function of the scratch length for the HA-coated sample. A clear jump of the friction coefficient from ~0.4 to ~0.7 is visible (figure 3b). We conclude that the HA coating resists against failure better at the loading than uncoated Ti substrate. For the deposited HA coatings, the evolution of the coating failure can be divided into three stages. At low loads (under 2.6 N), cracking on the coating's trackside was evident. As the load gradually increased, delamination at the trackside occured. Eventually, the coating was delaminated from the substrate along the scratch path when the load increased up to ~3.14 N.

Contact angle hysteresis $\Delta\theta$ was used to study the properties of the HA coating and uncoated Ti [9]. Analysis of the results obtained for the samples after SB and AE showed an increase in the hysteresis (table 1). The HA coatings were hydrophobic ($\theta_{\text{advancing}}$ = 87.1 ± 0.2°), while receding contact angle had a tendency to decrease which resulted in increase in hysteresis up to $\Delta\theta$ = 36.55°. Thus, dynamic contact angle measurements indicated the hydrophilic properties of the deposited HA coating: a high hysteresis led to better wetting of the coating.

4. Conclusion

The HA coatings with the (002) texture were deposited on the surface of SB and AE Ti substrates via RF magnetron sputtering. Nanoindentation results indicated an intensive surface hardening after nanocrystalline HA coating deposition. Nanohardness *H* and Young's modulus *E* of the coatings prepared on Ti at the penetration depth of h_c =71.11±2.87 nm are 15.2±0.7 and 147±16 GPa, respectively. The values of H/E and H^3/E^2 for the HA coating (0.101 and 0.164 GPa, respectively) were significantly higher than that of the uncoated substrate after SB+AE treatment (0.038 and 0.005 GPa). Scratch test results revealed that the deposited HA coatings exhibited improved wear resistance and lower friction coefficient. Dynamic contact angle measurements revealed the hydrophilic properties of the deposited HA coating due to a higher hysteresis compared with uncoated Ti surface. In this respect, we assume that HA coating can decrease stress-shielding effect and improve the fixation of an implant in bone.

Acknowledgments
The authors would like to express their gratitude to Mr. T.M. Mukhametkaliyev for the help with the fabrication of thin HA films. The study was supported by the Russian Fund for Basic Research (project numbers 15-08-08652 and 14-08-31027 mol_a), the Russian President's Stipend SP-6664.2013.4, grant MK-485.2014 and the State order NAUKA #11.1359.2014/K.

References
[1] Surmenev R A, Surmeneva M A and Ivanova A A 2014 Significance of calcium phosphate coatings for the enhancement of new bone osteogenesis – a review *Acta Biomater.* **10** 557-579
[2] Li Y, Zou S, Wang D, Feng G, Bao C and Hu J 2010 The effect of hydrofluoric acid treatment on titanium implant osseointegration in ovariectomized rats *Biomater.* **31** 3266–3273
[3] Khan M R, Donos N, Salih V and Brett P M 2012 The enhanced modulation of key bone matrix components by modified Titanium implant surfaces *Bone.* **50** 1-8
[4] Hornberger H, Virtanen S and Boccaccini A R 2012 Biomedical coatings on magnesium alloys – A review *Acta Biomater.* **8** 2442-2455
[5] Sun L, Berndt C C and Grey C P 2003 Phase, structural and microstructural investigations of plasma sprayed hydroxyapatite coatings *Mater. Sci. Eng.* A **360** 70-84
[6] Lynn A K and DuQuesnay D L 2002 Hydroxyapatite-coated Ti-6Al-4V Part 1: the effect of coating thickness on mechanical fatigue behavior *Biomater.* **9** 1937–1946
[7] Ducheyne P, Beight J, Cuckler J, Evans B and Radin S 1990 Effect of calcium phosphate coating characteristics on early post-operative bone tissue ingrowth *Biomater.* **8** 531–540
[8] Surmenev R A, Surmeneva M A, Evdokimov K E, Pichugin V F, Peitsch T and Epple M 2011 The influence of the deposition parameters on the properties of an rf-magnetron-deposited nanostructured calcium phosphate coating and a possible growth mechanism *Surf. Coat. Technol.* **205** 3600-3606
[9] Pichugin V F, Surmenev R A, Shesterikov E V, Ryabtseva M A, Eshenko E V, Tverdokhlebov S I, Prymak O and Epple M 2008 The preparation of calcium phosphate coatings on titanium and nickel-titanium by rf-magnetron sputtered deposition: composition, structure and micromechanical properties *Surf. Coat. Technol.* **202** 3913-3920
[10] Perry A J 1986 The structure and colour of some nitride coatings *Thin Solid Films.* **135** 73–85
[11] Oliver W C and Pharr G M 1992 An improved technique for determining hardness and elastic modulus using load and displacement sensing indentation experiments *J. Mater. Res.* **7** 1564–1583
[12] Bu-Abbud G H, Bashara N M and Woollam J A 1986 Variable wavelength, variable angle ellipsometry including a sensitivities correlation test *Thin Solid Films.* **138** 27-41
[13] Sima L E, Stan G E, Morosanu C O, Melinescu A, Ianculescu A, Melinte R, Neamtu J and Petrescu S M 2010 Differentiation of mesenchymal stem cells onto highly adherent radio frequency-sputtered carbonated hydroxylapatite thin films *J. Biomed. Mater. Res.* **95** A 1203-1214
[14] Leyland A and Matthews A 2000 On the significance of the H/E ratio in wear control: a nanocomposite coating approach to optimised tribological behavior *Wear* **246** 1-11
[15] Oberle T L 1951 Properties influencing wear of metals *J. Metals.* **3** 438–439

Complex technology of production of three-ply titanium composite sheets

V G Shmorgun[1], A I Bogdanov[1], A O Taube[1]* and A G Serov[1]

[1]Department of Material Science and Composite Materials, Volgograd State Technical University, Volgograd, Russia

*E-mail: alekstaube@yandex.ru

Abstract. The mechanical properties of the explosion welded three-ply compositions of titanium alloys in dependence on the conditions of the subsequent rolling and heat treatment were studied.

1. Introduction

A well-known feature of high-strength titanium alloys and hardened steels is a low ductility and high propensity to brittle fracture. So alloy VT5 in comparison with the technical titanium BT1-0 has almost twice as high tensile strength and so much the less stretch ratio at five-fold decrease of the homogeneous deformation, which limits its application to critical structures [1, 2].

In [3 - 6] it is shown, that significantly improve of the ductility of high strength steels can be made by creating with the help of thermal operations surface decarbonized layers, and double-sided facing of steel 30HGSA with layers of steel 12H18N10T with a total thickness of 30% of the base layer increases the elongation of a three-ply composition to 7 - 8% in comparison with 3 - 4% of the steel 30HGSA, homogeneous deformation – from 0-2 up to 7% by lowering the tensile strength by 16%. And the improvement of properties of high strength steels is achieved in case of strong connection of layers, which is easily realized by using the explosion welding layered compositions [7].

Explosion welding in combination with the subsequent rolling is advisable to use to enhance the resistance to brittle fracture of high-strength titanium alloys [7]. This technology allows not only to obtain compositions with high-strength layer interconnection, but also to avoid the use of expensive vacuum or inert atmosphere rolling equipment.

2. Materials and methods

We used commercial titanium VT1-0 and titanium alloys VT5 and VT14. Their initial properties before explosion welding are shown in table 1. VT5 alloy was cladded on both sides with titanium VT1-0 thickness of 1 mm side, VT14 alloy – with titanium VT 1-0 thickness of 1, 1.5 and 2 mm.

Operational sequence was:

1. The simultaneous explosion welding of three-ply compositions VT1-0+VT5+VT1-0 (CM1) and VT1-0+VT14+VT1-0 (CM2) in the optimal conditions (speed of the contact point of 1800-2000 m/s, impact velocity 480-520 m/s) to ensure strength uniformity of the welded joints to the least durable alloy VT1-0.

Table 1. The mechanical properties of the alloys VT5, VT14 and titanium VT1-0

Material	Tensile strength, MPa	Stretch ratio, %	Contraction ratio, %	Note
VT5	800 700	12.0 15.0	- -	After rolling, tempering 800 ^{0}C
VT14	1070	1.9	3.1	Quenching 920 ^{0}C, aging treatment 480 ^{0}C, 16 h
VT1-0	480	25.0-40.0	45.0-55.0	Annealed

2. Hot rolling of the three-ply blanks of CM1 to the sheet thickness of 4 mm and CM2 to the sheet thickness of 2.5 mm at a three thickness ratios at the following order: rolling start temperature 850 – 900 ^{0}C, rolling closure 650 – 750 ^{0}C, number of passes - 4. For comparative studies simultaneously with the three-ply blanks the single-layer sheets of alloys VT5 and VT14 were rolled to the same thickness.
3. Tempering of the alloy VT5 and CM1 for 1 hour at 400, 600, 700, 800, 900 ^{0}C with the air cooling. Alloy VT14 and CM2 were tempered at 600 ^{0}C for 0.5 hours, quenched at 920 ^{0}C with water cooling, and the same quenching, followed by aging at 400 ^{0}C for 16 hours.
4. Cutting from one- and three-ply heat-treated sheets the standard samples and their subsequent mechanical testing. In addition, the notched samples and samples with pre-formed cracks from the alloy VT14 and CM2 were produced and tested for impact strength with the registration of the fracture.

3. Results and Discussion

To determine the influence of the direction of rolling on the mechanical properties of the composite material rolling of the welded samples was made across and along the direction of welding. Since the boundary between the layers in the titanium composition is irregular wavy surface, warped waves in the process of tensile tests are stress concentrators. Rolling across the direction of welding provides a lower variation in thickness of cladding layers, so the influence of the waves irregularity on the mechanical properties is smaller than when rolling in a direction similar to the direction of welding (table 2).
Mechanical tests shown that at 15 percent of the total thickness of the cladding layers strength of CM1 in the directions along and across the rolled is reduced by 1.5 - 2%, and the total elongation is increased by 15 - 20% (Table 2). Besides cladding of VT5 by plastic titanium alloy VT1-0 led to a twofold decrease of variation in individual strength values along the rolling direction and two - three times decrease in the across rolling direction. At about the same extent variation of values of total elongation and uniform strain decreased.
Similar results were obtained in the test of sheet CM2. As seen from the graphs of Figure 1, the increase in the total thickness of the cladding to 40% of the thickness of the base layer decreases strength of CM2 in comparison with the alloy VT14 after quenching and aging at 7.5%, after tempering - at 13% (Figure 1a) and increases total elongation and contraction after thermal operations in 2 - 3 times (Figure 1 b, c). The specific work of fracture of samples with pre-applied crack under impact bending increased with an increase of the thickness of the cladding, in all cases, roughly equally: the δ_{pl}/δ_0 = 25% - 50%, and δ_{pl}/δ_0 = 40% - 100% (Figure 1d).

Table 2. The results of mechanical tests of unclad and cladded sheets of alloy VT5 depending on the rolling direction and tempering temperature

Cutting directions	Tempering temperature, ^0C	1-layer sheets			Three-ply sheets		
		Tensile strength, MPa	Uniform strain,%	Total elongation, %	Tensile strength, MPa	Uniform strain,%	Total elongation, %
Along rolling	-	750-830	0-5	10-20	750-790	2-4	15-20
	400	740-830	1-5	9-21	760-790	3-5	15-22
	600	740-810	1-7	10-21	720-760	4-7	15-20
	700	630-710	2-9	9-22	660-700	5-9	16-22
	800	620-750	2-8	8-20	660-690	5-9	15-24
	900	610-740	2-8	7-18	590-680	5-8	15-22
Across rolling	-	650-800	0-5	7-19	710-760	1-5	10-18
	400	670-810	1-5	8-19	710-750	2-5	10-19
	600	610-750	1-5	9-20	650-690	2-6	12-20
	700	600-750	2-7	10-21	630-650	4-8	14-22
	800	590-690	2-7	9-22	600-650	4-8	14-25
	900	580-670	2-5	5-18	600-650	4-7	10-21

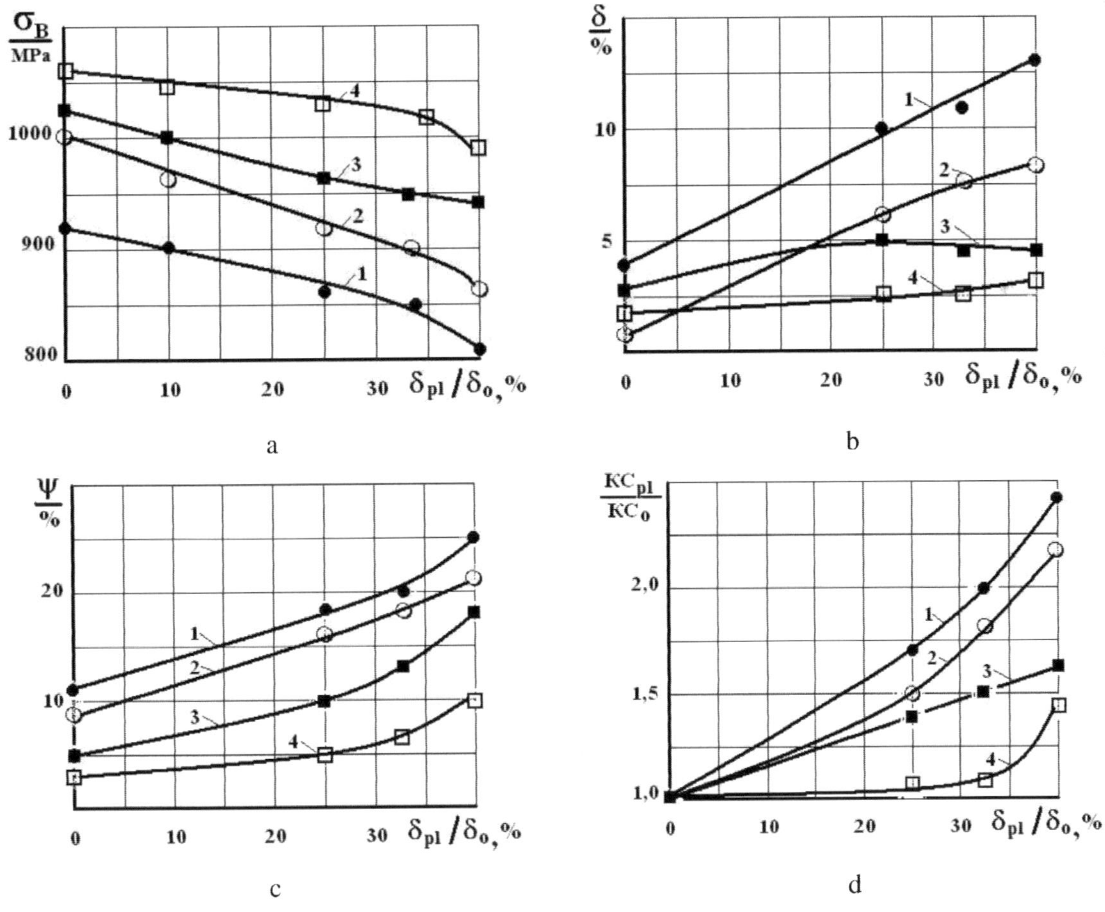

Figure 1. Changes of the mechanical properties of the alloy VT14 with the thickness of cladding layers: a - tensile strength; b - total elongation; c - percentage reduction; d - a relative reduction of the

specific work of fracture of samples with a crack under shock loading in comparison with not cladded. Curves legend: 1 – Tempering at 600 ^{0}C; 2 - in the state after rolling; 3 – quenching at 920 ^{0}C with water cooling; 4 - quenching and aging at 400 ^{0}C for 16 hours. δ_{pl} - total thickness of the layers of titanium VT1-0, δ_o - the thickness of the alloy VT14.

4. Conclusions

The complex technology of production by explosion welding and subsequent hot rolling of three-ply compositions VT1-0+VT5+VT1-0 and VT1-0+VT14+VT1-0 was developed. Compositions have the interconnection strength at the level of the strength of titanium VT1-0, which have in comparison with alloys VT5 and VT14 2 - 3 times less variation in individual strength values, total elongation and uniform strain with the considerable reduce of the tendency to brittle fracture.

To reduce variations in thickness of explosion welded cladding layers of three-ply the compositions of high-strength and ductile titanium alloys, following hot rolling must be done in the cross direction to the direction of the welding process.

Acknowledgments

The investigation was performed by a grant from the Russian Science Foundation (project № 14-29-00158).

References

[1] Glazunov S G, Moiseev V M 1974 *Structural titanium alloys*. (Moscow, Metallurgy)

[2] Kolachev B A, Livanov V A, Elagin V I 1981 *Metallurgy and heat treatment of non-ferrous metals and alloys* (Moscow, Metallurgy)

[3] Ioffe A F, Kirpicheva M V, Levitskaya M A 1924 *Deformation and strength of crystals* (JRFHO)

[4] Pashkov P O 1950 *The ductility and fracture of metals* (St. Petersburg, Sudpromgiz)

[5] Friedman Y 1946 *Mechanical properties of metals* (Moscow, Oborongiz)

[6] Katihin V D, Kofman A P, Javor A A 1965 *Cladding as a means of reducing the tendency of hardened steels to brittle fracture* (Volgograd)

[7] Cossack N N, Sedykh V S, Trykov Y P, Ulitin A I 1974 The mechanical properties of titanium multilayer composition after explosion welding and hot rolling *Physics and Chemistry of Materials Processing* **1** 215-220

Investigation of heat-resistant layered coating of Al-Cr-Ni

V G Shmorgun[1], Y P Trykov[1], A I Bogdanov[1] and A O Taube[1]*
[1]Department of Material Science and Composite Materials, Volgograd State Technical University, Volgograd, Russia

*E-mail: alekstaube@yandex.ru

Abstract. The paper shows the transformation of the structure and phase composition of the layered coating system Al-Cr-Ni, obtained by the heat treatment of multilayered composite H20N80+AD1, welded by explosion, in the time range 1-300 hours. The cyclic heat resistance of the coating at 1150 ° C is studied.

1. Introduction

The necessity of increasing the operational reliability and service life of high-tech products of power engineering functioning in extreme conditions - at high and ultra-high temperatures, in corrosive environments and erosive action puts the task of developing fundamentally new materials. This task can be solved in two directions - improvement of existing structural materials [1,2] by their doping, optimizing the structure, creating compositions of complex composition and the formation of functional protective coatings on the surface of materials [3,4].

Recently, the second direction is being developed the most rapidly, ensuring reliable functioning of the elements of gas turbines and rocket engines for various applications, manufactured as; the composition of the heatproof alloy and heat-resistant protective coating.

Among the heat-resistant coatings which protect the materials in oxidative environments at temperatures up to 1200 °C, there are very effective coatings of nickel aluminide of which high barrier properties are based on the ability of the surface layer of the coating to oxidize with forming a protective film of aluminum oxide Al_2O_3.

An analysis of studies [5-8] showed that along with the widely used methods (SHS, spraying, mechanosynthesis etc.) combined processing techniques become increasingly important, allowing to obtain layered coating, such as "nickel + nickel aluminide." In such coatings each layer has a certain functional load: nickel layer prevents penetration of the aluminum atoms to the bottom layer and provides high adhesion of the coating to the substrate, and the nickel aluminide base protects against oxidation by forming an oxide film Al_2O_3.

It is known that it is reasonable to add chromium to the binary system Al-Ni as an alloying element, since it reduces the necessary for the formation of Al_2O_3 aluminum concentration from 40 to 10 at.%, and also promotes the formation of an additional protective Cr_2O_3 oxide film [9].

So the aim of this work was to study the heat resistance of the coating of Al-Cr-Ni, obtained by complex technology [10], as well as the transformation of its structure and phase composition.

2. Materials and methods

Studies were conducted on the samples: a substrate of a nickel alloy H20N80 + coating (100-200 microns thick) obtained from an explosion welded bimetallic blanks of composition H20N80 +

aluminum AD1 (2 + 2 mm) by using a diffusion annealing. Temperature effect on the coating was carried out by heating the samples to 1150 ° C with holding 1-300 hours. The chemical composition of the structural components of the diffusion zone determined by energy-dispersive analysis in a scanning electron microscope, a double-beam system Versa 3D.

The phase composition was performed on diffractometer "DRON-3" using the emission of copper anode with a nickel filter, X-ray pictures are taken in a range of angles 2Θ from 20 to 110 degrees.

The study of heat resistance was carried out in according to GOST 9.312-89 "Unified system of corrosion and aging protection. Protecting covers. Methods for determination of heat resistance "and GOST 6130-71"Metals. Methods for determination of heat resistance" by the gravimetric method by the weight reduce of the sample.

3. Results and Discussion

Microstructure and distribution of chemical elements over the thickness of the coating before a heat treatment are shown in figure 1. The analysis of presented in figure 1b distribution has allowed to establish the following: the concentrations of Al and Ni in a surface layer are respectively 62 and 30 wt. %. At the approach to the H20N80 aluminum content reduces to 42 wt. %, and nickel increases to 48 wt. %. According to the phase diagram of the Al-Cr-Ni, given in [11], the chemical composition corresponds to the phase composition $NiAl_3$ + Al (Cr) + Cr (Al).

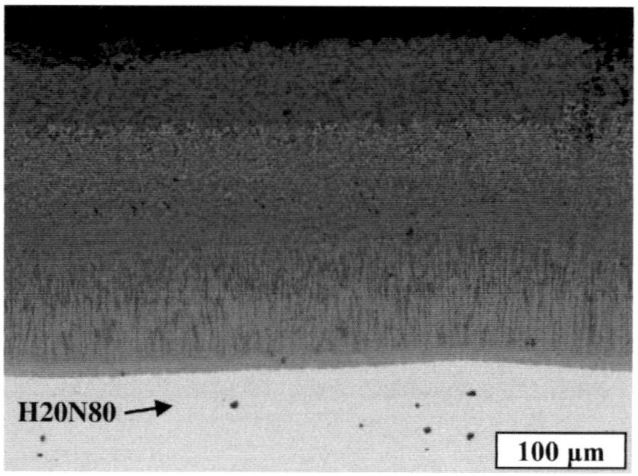

a

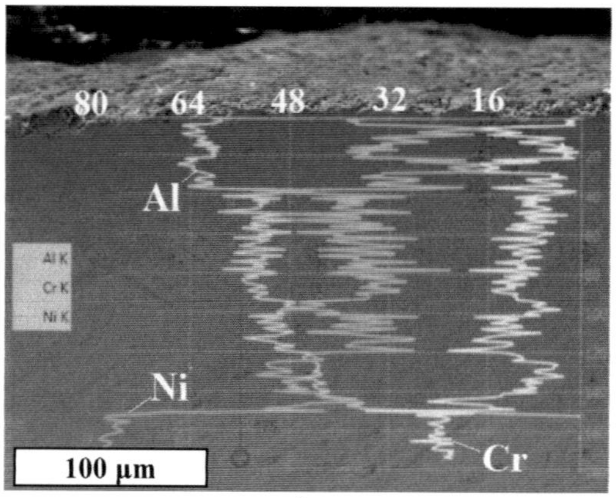

b

Figure 1. Microstructure (a) and SEM image (b) of coating

During heat treatment with increasing of holding time from 1 to 50 hours at 1150 °C there is an aligning of the Al and Ni content in the thickness of the coating (figure 2). The content of Ni in the surface layer increases from 30 to 73 wt.%, Al - decreases from 62 to 18 wt.%, the Cr content is almost constant and varies from 7 to 8 wt.%. The composition and structure of the inhomogeneous zone between H20N80 and coating are given in [12].

Reduction of the Al content in the surface layer leads to the transformation of its phase composition: $NiAl_3$+Al(Cr)+Cr(Al) → $NiAl_3$+Ni_2Al_3+Cr(Al) → Cr(Al)+Ni_5Al_3+Ni_2Al_3.

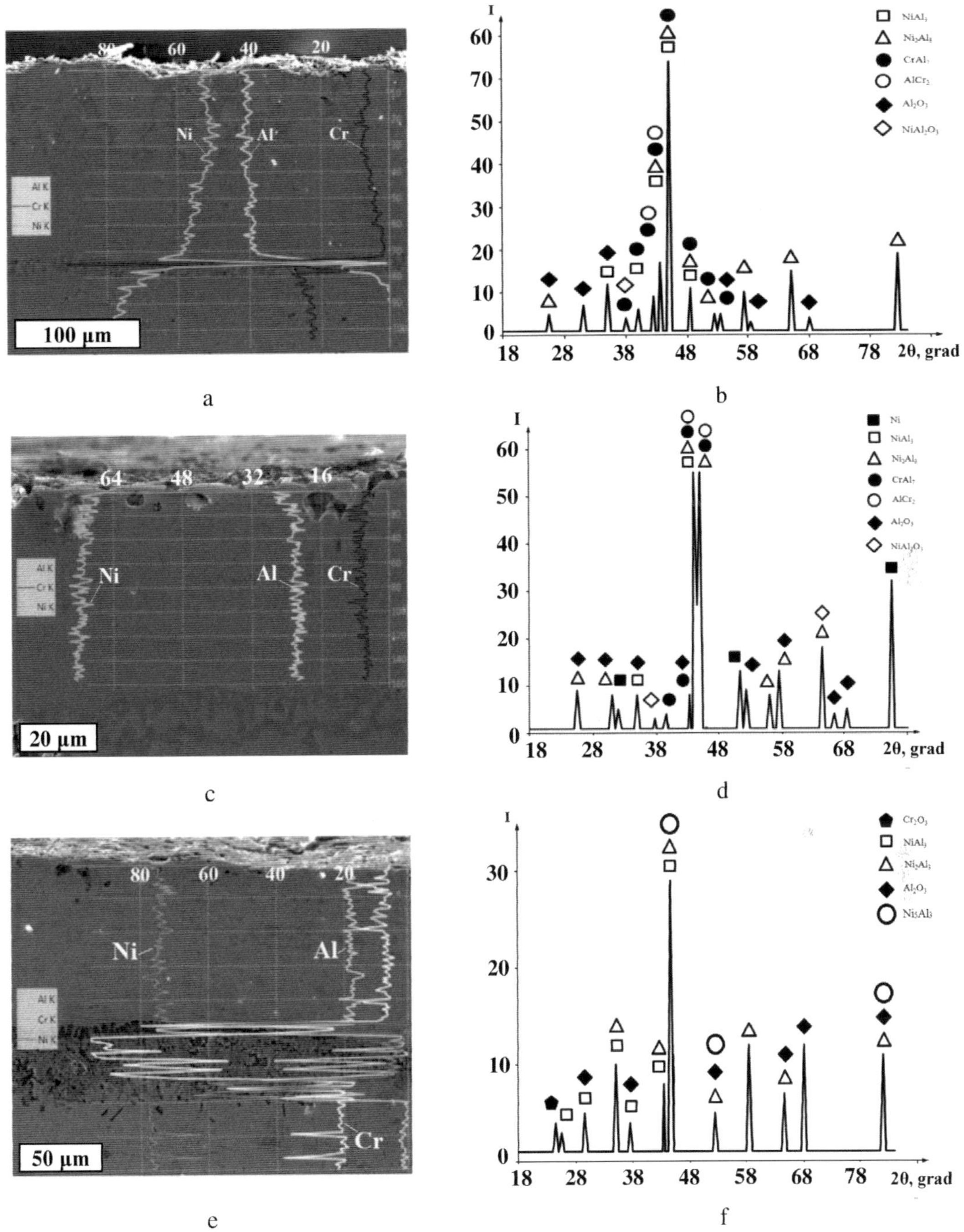

Figure 2. SEM images and phase composition of the coating after heat treatment 1150 °C 1 (a, b), 10 (c, d) and 50 hours (e, f)

At holding time of 100, 200 and 300 hours the structure and distribution of chemical elements over the thickness of the coating is not substantially changed (figure 3). Al concentration in the coating is at a level of 18 wt.% and Ni – 70-74 wt.%.

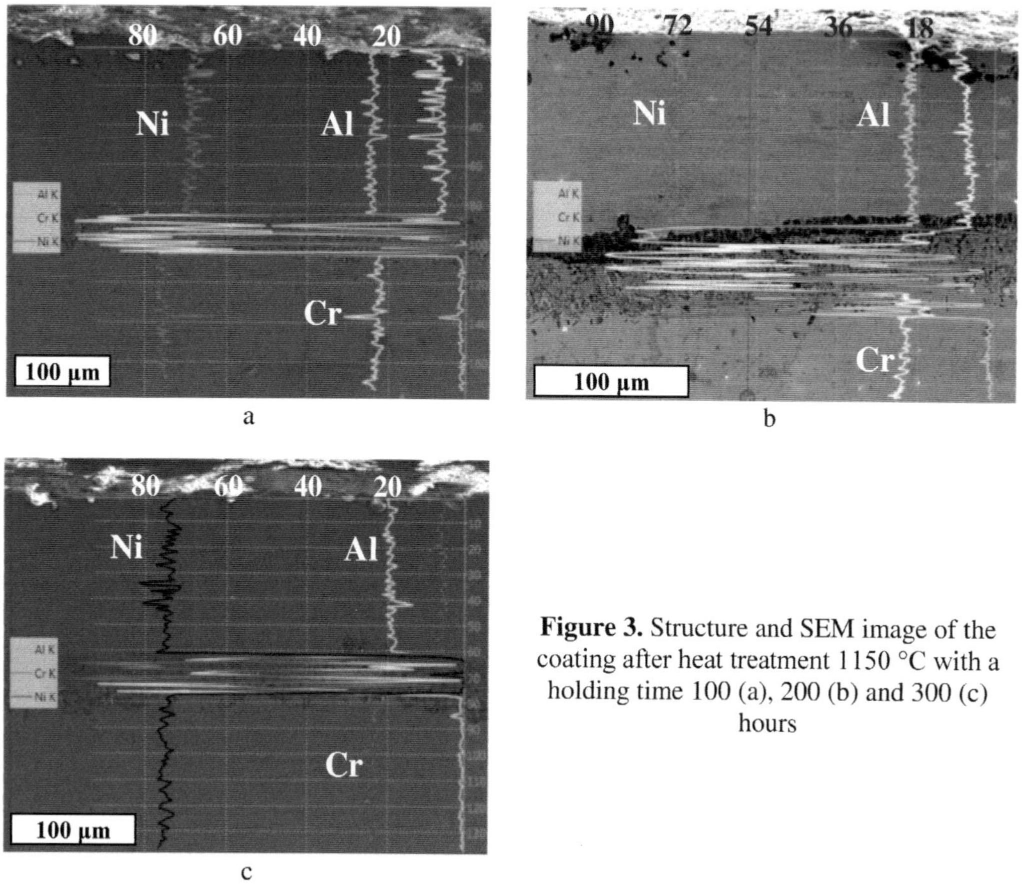

Figure 3. Structure and SEM image of the coating after heat treatment 1150 °C with a holding time 100 (a), 200 (b) and 300 (c) hours

The results of the research of the cyclic oxidation resistance of alloy H20N80 samples and samples with intermetallic coating at 1150 °C are shown in figure 4. The figure 4 shows that the weight loss of the alloy H20N80 is more, and thus the formation of oxides is faster.

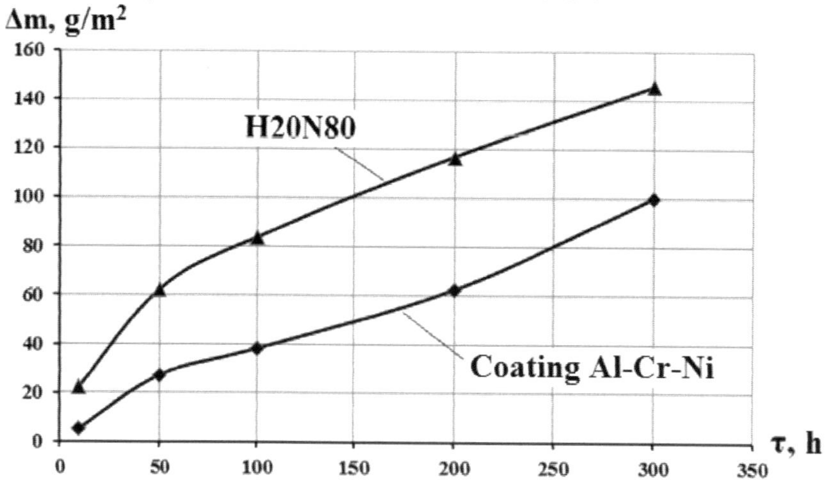

Figure 4. The mass loss of samples over time at a temperature of 1150 °C

According to the GOST 9.312-89 corrosion durability of the protective coating is estimated by changing of the contents of the components which determine its heat resistance, during the test period. In case of this coating, these components are Al and Cr, changing contents of which in the time range of 1-300 h is shown in Figure 5. It is shown that in the first stage (1-10 hours) there is a sudden decrease of aluminum content from 40 to 24% which is probably due to the participation of aluminum in the diffusion processes in the interaction with the substrate and the formation of H20N80 oxide film. For 50 hours decreasing of Al content slows down and then remains practically at the same level (~ 18%). The chromium content gradually decreases from 10 to 6%.
Thus, in the studied time interval required contents of aluminum and chromium which determine the heat resistance of the test coating (due to the formation of oxides) remain, enabling the maintenance of efficiency of the coating.

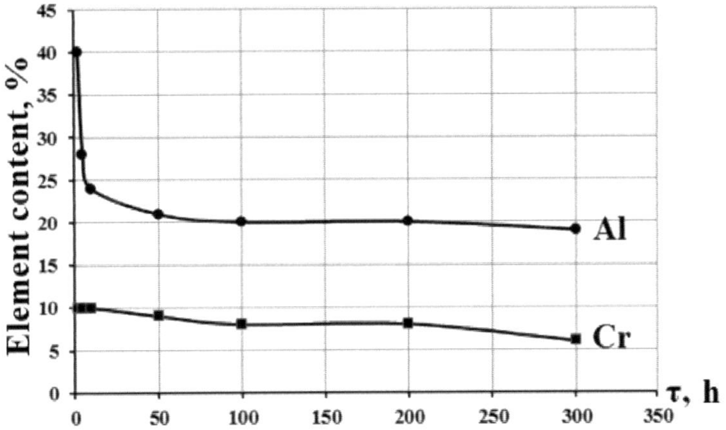

Figure 5. Al and Cr content in the coating at 1150 °C

4. Conclusions

1. Thermal effect on the coating of Al-Cr-Ni at 1150 ° C for 50 hours leads to the transformation of its phase composition $NiAl_3$+Al(Cr)+Cr(Al) → $NiAl_3$+Ni_2Al_3+Cr(Al) → Cr(Al)+Ni_5Al_3+Ni_2Al_3 by reducing the Al content in the surface layer and increasing the content of Ni. At a longer holding time (300 hours) concentrations of aluminum and chromium (which determine the heat resistance of the coating due to the formation of oxides) are practically unchanged which enables the maintenance of its efficiency.
2. Comparative studies of the cyclic oxidation resistance of alloy samples H20N80 and samples with intermetallic coating at 1150 °C showed a lower weight loss in a layered coating and, accordingly, its higher heat resistance.

Acknowledgments
The investigation was carried out according to the assignment of Russian Ministry of Education No 11.1865.2014/K.

References
[1] Colin J 2002 *Journal of Materials Engineering and Performance* **5** 487-493
[2] Kablov E N 2010 Foundry structural alloys based on nickel aluminide *Engine* **4** 22-26
[3] Kablov E N, Muboyadzhyan S A 2001 *Gas turbine technology* **3** 30-35
[4] Desai V 2006 *Journal of Materials Engineering and Performance* **58** 15-22
[5] Kissel V M 2009Aerospace tehnology **10** 50-53
[6] Culha 2008 *Journal of Materials Processing Technology* **204** 221-230
[7] Zadorozhnyy V 2011 *Journal of Alloys and Compounds* **509S** 507-510

[8] Morsi K, Shinde S, Olevsky E A 2006 *Journal of Materials Science* **41** 5699-06
[9] Kolomytsev P T 1991 *High-temperature protective coatings for nickel alloys* (Moscow: Metallurgy) 237
[10] Shmorgun V G, Bogdanov A I, Taube A O 2014 *Proceedings of the universities. Ferrous metallurgy* **5** 64-70
[11] Shmorgun V G, Taube A O , Krohalev A V 2015 *Applied Mechanics and Materials* **698** 430-436
[12] Shmorgun V G, Trykov Y P, Bogdanov A I, Taube A O, Serov A G 2015 *News VSTU. Problems of Materials Science, Welding and strength in engineering* **5** 17-23

Investigation of adsorption properties of alumina produced by vacuum spray method

K Khrustaleva*, G Voronova
Department of nanomaterials and nanotechnologies, Tomsk Polytechnic University, 30 Lenina Avenue, Tomsk, 634050, Russia

*E-mail: K.Khrustaleva@gmail.com

Abstract. In this paper nanopowders obtained by vacuum spray method were studied. The phase composition of the obtained powders is $\gamma - Al_2O_3$. Obtained nanopowders have specific surface area about 200 m^2/g. Adsorptive properties of these powders were studied by static adsorption from solutions. The anionic dye eosin was selected as adsorbate. It has been found that the powders obtained by vacuum spray method have significant capacity to adsorb eosin.

1.Introduction

Nowadays nanopowders of aluminium oxide and system based on it are widespread in various branches of technology. These nanopowders are used as adsorbents, catalysts and a raw for nanoceramics. A structure of these nanopowders provides their unique mechanical and physical-chemical properties, which appear as individually as when interacting with other substances.

There are a lot of methods for producing alumina ($\gamma - Al_2O_3$), method of dispersing and hydrothermal treatment of the alumina trihydrate for example. It should be noted that his method is a time-consuming and energy-intensive, because it consists of many stages [1]. Furthermore nanopowders of alumina are produced by combustion of large particles of alumina with a binder component. At the high temperature the binder burns down and it leads to obtaining of porous active alumina [2]. But this method has disadvantages too; the essential is nonreproducibility of results.

In this paper the vacuum spray method was used for obtaining of alumina. This technique is a three stage process: preparing a suspension, spraying the suspension on a vacuum spray drying system NanoSprayDryerB-90 and heat treatment of the samples.

In contrast to existing methods of producing active alumina, this method is simple to implement, provides reproducible results and is promising for use in many areas of production, for producing sorbents for example.

Nowadays humanity has a problem of rational use of water resources. There are new adsorption technologies for water treatment against various contaminants: oil products, phenol substances, microbiological objects. Adsorption of microbiological objects which have negative charge in aqueous media occurs most effectively at the surfaces, which are characterized by a positive charge as alumina surface. Therefore the searching and development of this type of adsorbents for water treatment technologies is actual scientific problem now [3].

Alumina produced by NanoSprayDryerB-90 has significant specific surface and positive zeta potential in aqueous solutions [4], so such an alumina is perfect material for sorbents. In this work our study was focused on producing of $\gamma - Al_2O_3$by vacuum spray method and investigation of its adsorption properties.

2. Experimental

2.1. Sample preparation

Vacuum spray drying method (NanoSprayDryerB-90, NSD-method). A 0.5 M aqueous solution of 9-aqueous aluminum nitrate was prepared, after preparation solution was heated at 70 °C for one day. Obtaining of the $Al(OH)_3$ suspension was produced from anaqueous solution by chemical deposition. Chemical deposition was carried out using aqueous 25% solution of NH_4OH.The resulting $Al(OH)_3$suspension was washed by distilled water to remove excess of ammonia. Then the suspension was fed to Nano Spray Dryer B-90 (Switzerland). The following parameters were used in the experiment: air flow rate 140 l/min, the relative intensity of dispersion is 60 – 63%, T = 60 – 80°C, P = 120 Pa, nozzle size 7 nm. The product was divided into pieces, and these pieces were annealed at temperatures 550 ° C, 600 ° C and 700 ° C for 4 hours respectively.

Chemical deposition (CD-method). Also for comparison samples of alumina were prepared by chemical deposition without vacuum spraying. Produced precipitate was filtered and washed with distilled water. After all operations precipitate was dried for 3 days at 60 °C, divided into three pieces and annealed at temperatures of 550 ° C, 600 ° C, 700 ° C for 4 hours respectively.

2.2. Adsorption tests

Static sorption method from the aqua solutions was performed for studying the adsorption properties of alumina. Adsorption was carried out at room temperature. Anionic dye eosin ($C_{20}H_6O_5Br_4K_2$) with maximum of light adsorption at 490 nm was selected as model adsorbate.The initial and the equilibrium concentration of adsorbate were determined photometrically by spectrophotometer (Pd Spectrophotometer, Japan) [5].Initial concentration of eosin in aqueous solutionsis presented in Table 1.

Table 1. Initial concentration of eosin in aqueous solutions

Number of solution	Eosin concentration(mg/l)
1	4
2	8
3	12

Alumina powders were placed in eosin solutions#1 – 3 (Table 1), were sonicated for 3 min and transmittance of solutions was measured after 1, 2, 3 and 4 hours. Before transmittances measuring the deposition of alumina powder from solutions was carried out by centrifugation at 15000 rpm by Centrifuge CM-50.

3. Results and discussion

3.1 XRD and TEM analyses

X-ray analysis was carried out for determining the phase composition and the region of coherent scattering of samples; results are shown in Table 2.Phase composition of ceramics supports was studied using XRD - 7000S Shimadzu diffractometer with CuKα irradiation (λ = 1.5406 A°). Analysis of phase composition was performed according to ASTM database.

According to the data of Table 2, all the samples content only crystalline phase in form of γ - Al_2O_3. Significant differences in the region of coherent scattering of samples prepared as NSD and CD have not been found.

Samples have a considerable surface area (Table 2) up to the largest one – 216 m^2 / g for the sample #1. Specific surface area slightly decreased with the increasing of temperature treatment for all the samples.

Table 2. Parameters and characteristics of the samples

Number of samples	**Method of producing**	**Temperature of annealing (^0C)**	**Size of CSR[a](nm)**	**Phase composition**	**Specific surface (m^2/g)**
1	NSD[b]	550	9,48	$\gamma - Al_2O_3$	216
2	CD[c]	550	9,49	$\gamma - Al_2O_3$	204
3	NSD	600	9,49	$\gamma - Al_2O_3$	177
4	CD	600	9,49	$\gamma - Al_2O_3$	200
5	NSD	700	9,19	$\gamma - Al_2O_3$	161
6	CD	700	9,46	$\gamma - Al_2O_3$	158

[a]Coherent Scattering Region
[b]Nano Spray Dryer
[c]Chemical Deposition

Investigation of the samples structure was carried out by transmission electron microscopy (JEM 2100, JEOL) and scanning electron microscopy (NeoScope JCM-6000). Bright field images of the sample obtained by the NSD shown, that the particles have irregular shape, close to the distorted spherical (Figures 1, 2). The medium size of particles was ~ 110 nm. The particles had a uniform grain structure and were partly agglomerated and consisted from the smaller ones. Particles had nanoscale unit at least on one dimension (Figures 1, 2).

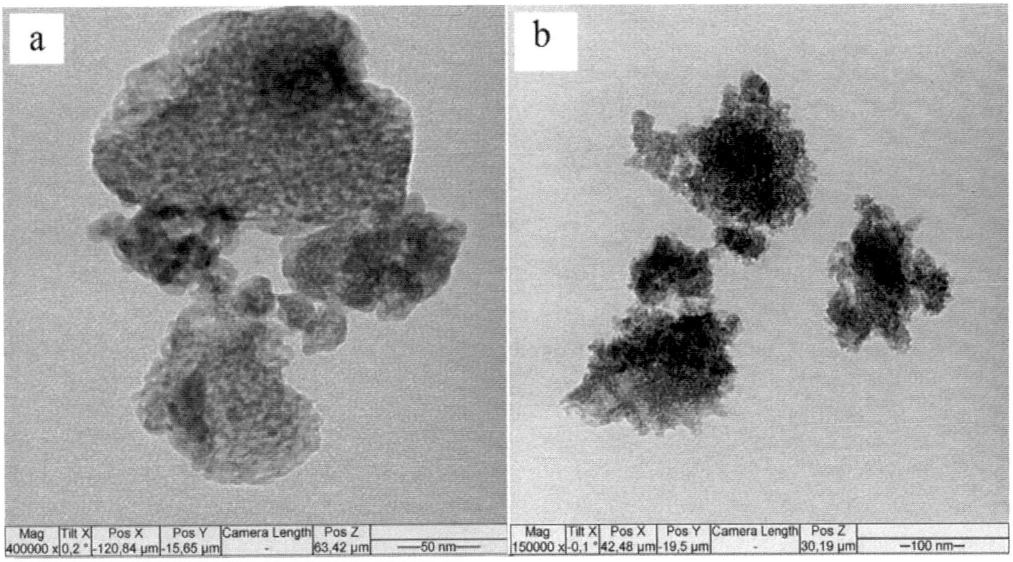

Figure 1. TEM microphotographs of sample #3(a), sample #4 (b)

Presumably, the presence of the granular structure shows that the particles consist of smaller crystallites, which is consistent with XRD (CSR is~ 9 nm). The gaps between crystallites presented the porous structure of the samples.

The powders produced by CD had different morphology. These particles had a jagged irregular shape (Figures1, 2).

Figure 2. SEM microphotographs of samples: #1 (a), #2 (b), #3 (c), #4 (d)

According to the Figure 2 particles obtained by CD had a smoother surface and a higher agglomeration.

3.2. Sorption properties of samples

All synthesized samples of alumina adsorbed eosin from the aqueous solution. Anionic dye eosin can absorb on positively charged surface, so all the obtained samples of alumina are characterized by a positively charged surface.

Fig. 3 shows that the samples obtained by NSD adsorbed a comparable to CD samples amount of eosin. The samples #3 and #5 obtained by spray dryer and annealed at 600 and 700 °C adsorbed a larger amount of eosin (with initial C 4 mg/l) than similar samples prepared by chemical precipitation (Figure 3, a).

Figure 3, b shows that the sample #6 prepared by CD have the bigger value of adsorption at initial concentration of eosin 8 mg / l. At the initial concentration of eosin 12 mg/l the higher adsorption demonstrate samples #1 (NSD) and #6 (CD).

All samples are characterized by large capacity for adsorption of eosin due to a positively charged surface.

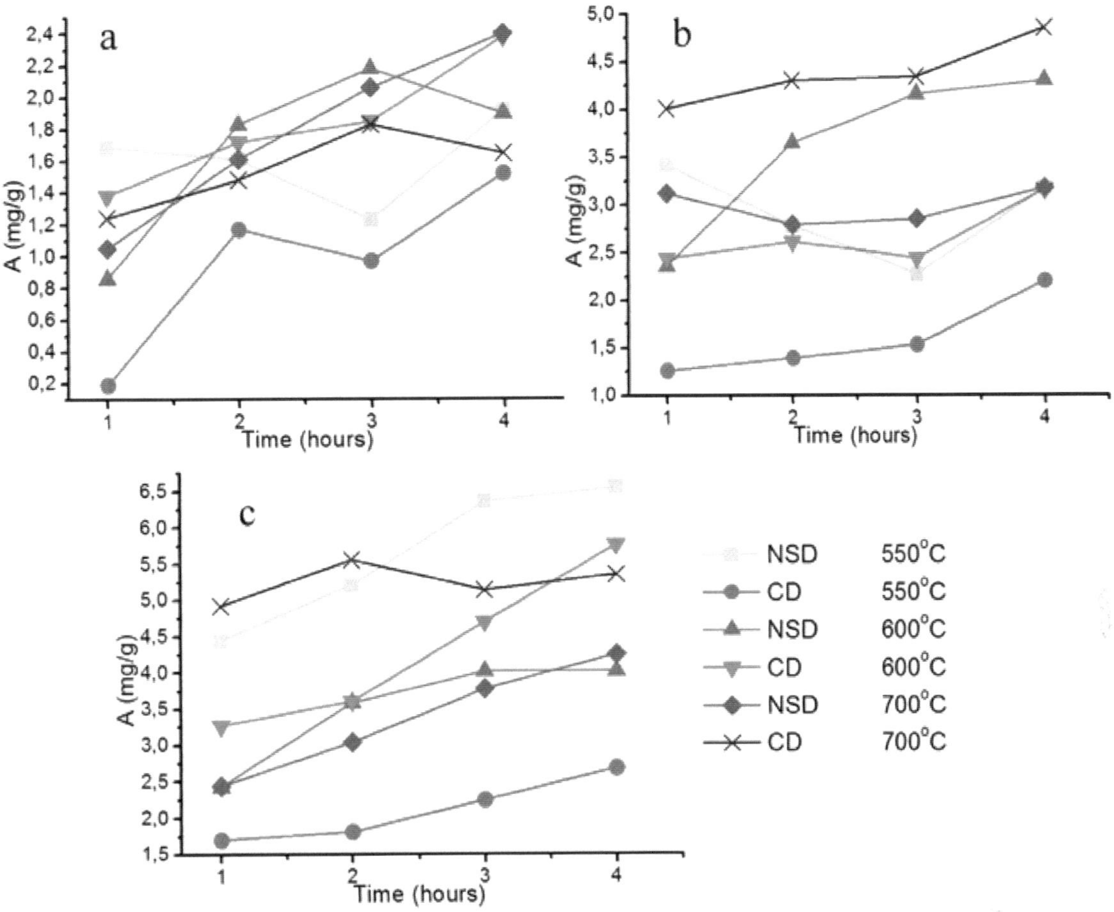

Figure 3. Adsorption of eosin (a – initial concentration of eosin is 4 mg / l; b - 8 mg / l; c – 12 mg / l)

4. Conclusion

The nanosized alumina powders were obtained by NSD method, it should be mentioned that all the samples content only γ - Al_2O_3 phase. The powders had large specific surface area ~ 200 m^2 / g. All samples are characterized by large capacity for adsorption of eosin due to a positively charged surface. So it can be used in adsorption of microbiological objects which have negative charge in aqueous media.

References

[1] Balashov V, et al 2011 *The method for producing the active alumina.* Patent RF, no. 2 473 468
[2] De Maio D 1972 *Ethylene oxide catalyst.* Patent USA, no. 3664970 B01J 23/50, B01J 23/66
[3] Lozhkomoev A, Glazkova E and Svarovskaya N 2013 The adsorption of microorganisms on fibrous sorbent with pseudoboehmite particles *J. Modern problem of science and education* **2**
[4] Bakina O, Svarovskaya N and Glazkova E 2013 Behavior of aluminum nitride composite nanopowders in aqueous media *J. Fundamental research* **4** 862
[5] Galanov A and Urmazova T 2010 Investigation of the adsorption mechanism of anticancer drugs on the iron carbide nanoparticles *J. Bulletin of Tomsk Polytechnic University.* **3** 29

IOP Conf. Series: Materials Science and Engineering **116** (2016) 012020 doi:10.1088/1757-899X/116/1/012020

High-strength fibrous concrete of Russian Far East natural materials

R Fediuk
Far Eastern Federal University,
20, Ajax St.,Vladivostok, Russky Island, Russia 690000

E-mail: roman44@yandex.ru

Abstract. Fiber-reinforced concrete is designed on composite binder. At 1.6% of reinforcing steel anchoring fiber maximum physical and mechanical properties (R_{compr} = 100.9 MPa) can be obtained. It was found that the combined effect of mechanical and chemical activation (the presence of limestone particles) increases the pozzolanic activity of acidic ashes. It has a catalytic effect on the reaction activity of the surface of ash and sand during machining in vario-planetary mill. Furthermore, the addition of limestone increases the alkalinity of the concrete, which leads to the formation of greater hydration products of cement per unit of time. Theoretical and experimental results can be recommended for expanded implementation of the construction in various regions of the Russian Federation, taking into account the availability of raw materials.

1. Introduction

In the production of concrete with improved properties it is necessary to pay special attention to the selection of aggregates (quality, size, surface configuration) which play an important role in the preparation of concrete with high physical-mechanical properties [1,2]. Increasing the size of aggregates entails correspondingly increase of stress concentrations in the concrete. Small aggregates also provide more homogeneous structure. The larger aggregate, the more concrete is prone to cracking, but at the same time a more rigid framework is formed, which is important to reduce the strain of the concrete under load.

In Primorye, it is advisable to use local raw materials, which are characterized by high activity silica-containing phases of man-made sand (in particular, elimination of crushed granite), obtained from crushing rocks.

Fine-grained structure, in addition to high homogeneity is also characterized by reduction of specific stress in the contact area and an increase of the integrated strength of adhesion between cement stone and aggregate.

Structure-forming role of filler is most developed with increasing surface of interaction, these conditions are realized in the fine-grained concrete with screenings of crushing of granite rubble of field Wrangel based on composite binders that due to the high surface intensify and accelerate the process of structure formation of concrete strength and seal structure.

To study the effect of composition on the binding properties of fine-grained concrete samples of concrete slump were produced in the range of 10-12 cm in the developed binder, which were further investigated for various technological parameters, standardized during construction.

2. Selection of the composition of high-strength fine-grained concrete

The author's analysis of the literature proves the possibility of producing dispersion-reinforced cellular concrete construction with high-performance. However, it is necessary to take into account that the implementation of the potential of fiber-reinforced concrete is only possible at creating the optimum structure of the material, the formation of which is determined by the following parameters:

-type and the quality of raw materials [3];

-technology of preparation of concrete mixtures [4];

- quantitative relationship between the components of fiber-reinforced concrete mixture.

The main factors determining the efficiency of designing of concrete and its construction and technical features are:

- obtaining the desired properties (strength, workability, and etc.) at the lowest possible consumption of costly and energy-intensive components;
- reducing the complexity of operations related to the selection of optimum compositions;
- verification of quality of control samples and testing at the factory.

The properties of fine-grained concrete are characterized by the same factors as the conventional heavy concrete, but the lack of it coarse aggregate leads to increased water demand of concrete, and in equally full-strength concrete cement consumption increases by 20 ... 40%. To reduce the consumption of cement it is advisable to use high-quality fillers, plasticizers, mixtures, to produce a good seal.

In the manufacture of fine-grained concrete it is advisable to use natural and enriched sand, crushed and granulated slag and ash, waste industry, including screenings of crushed granite. The properties of fine-grained concrete are highly dependent on the properties of the binder component in the concrete and aggregate used, grain structure, surface quality and water demand which determine the basic physical and mechanical properties of concrete. When you receive a fine concrete aggregate of smaller size and higher surface area compared to conventional heavy concrete is used, the presence of these factors leads to an increase of cement consumption, which negatively affects the physical and mechanical characteristics of fine-grained concrete. With increasing flow rate cement concrete shrinkage also increases significantly, fracture toughness decreases, its creep increases also.

Calculation of the composition is produced in accordance with the recommendations on the selection of compounds and heavy fine-grained concrete (state standard 27006-86).

Baseline data for the design of the composition:

- the desired strength - 100 MPa;
- the mobility of the concrete mix slump - 10-12 cm;
- activity of the composite binder - 77.3 MPa;
- screening of crushed granite with water demand 7% and the true density of 2.6 kg / l and a bulk density of 1.37 kg / l;
- sand with true density of 2.63 kg / l.

$$W/C = \frac{A_1 R_{CB}}{\mathrm{R_c} - A_1 \cdot 0.5 R_{CB}},$$

coefficient A_1 is accepted by the reference literature [5].

$$W/C = \frac{0.43 \cdot 77.3}{100 - 0.43 \cdot 0.5 \cdot 77.3} = 0.40$$

According to the schedule in reference [5] water flow turned to be 220 l / m^3.

$$\tilde{N}\hat{A} = \frac{220}{0.4} = 550 \hat{e}g / m^3$$

Emptiness of dropping out of granite rubble:

$$E_{gr} = 1 - \frac{1.37}{2.6} = 0.47$$

According to the handbook [5] extendable coefficient α = 1.5.
The amount of filler:

$$F = \frac{1000}{0.47 \cdot 1.5/1.37 + 1/2.6} = 1623\hat{e}g/m^3$$

Thus, an exemplary structure of the concrete mixture: composite binder - 550 kg/m³, a filler - 1623 kg/m³, water - 220 l/m³. The optimal composition will be selected by experimental verification.

3. Results

Research of physic-mechanical properties of fine-grained concrete showed that the use of the composite binder prepared by co-grinding of cement, fly ash and limestone, and hyperplasticizer allows to improve the characteristics of concrete, as compared to similar compositions made with other binding materials [6,7]. This fact is explained by more dense structure of cement stone of developed composite binder and lesser porosity due to the smaller amount of water in the concrete (Table 1).

Table 1. Physical and mechanical properties of fine-grained concrete depending on the composition of the binder

№ composition	Material consumption per 1 m³							Slump	The compressive strength R, MPa	Prism strength R, MPa	Elastic modulus
	Binder, кg			Hyperplasticizer	Screenings crushing of granite, kg	Sand, kg	Water, l				
	Cement	Fly ash	Limestone								
1	550	-	-				220		107.5	86.3	61.2
2	278	234	36				330		83.7	59.5	43.8
3	265	246	39	1.2	1000	623	334	10-12	84.2	60.3	44.5
4	252	257	46				337		76.3	55.2	40.9
5	239	268	48				343		75.2	55.0	40.8
6	225	278	52				348		75.0	54.9	40.8

The best physical and mechanical properties of the concrete compositions are '1 and '2: (they include cement - 48-51%, acidic ash - 43-45%, limestone - 4-9%).
It should be noted that with increasing amounts of ash and reducing amount of cement to ensure equal mobility formulations (slump = 10-12), it is necessary to increase the amount of mixture put into the concrete mixing water.
Despite a number of advantages compared, for example, with conventional heavy concrete, fine concrete has high shrinkage and modulus reduced by 20-25%.
Dispersion reinforcing of fine concrete with polymer and metal fibers opens great opportunities to improve the performance of the concrete.
In order to obtain high-density fiber-reinforced concrete the effect of the introduction of the concrete matrix of reinforcing fibers was studied. Composition number 2 from Table. 1 was adopted as a basis for the concrete matrix.
To determine the optimal percentage of reinforcement of fine steel fiber concrete samples of the same composition with different steel fiber content (Table. 2) were formed.

Table 2. The dependence of the strength of the fine steel fiber concrete reinforcement ratio

№	Material consumption per 1 m^3				Reinforcement,	R_{compr},
composition	Binder	Water	Filler	Fiber	%	MPa
2-1	550	330	1623	-	0	94.2
2-2	550	330	1623	23.97	1	96.1
2-3	550	330	1623	28.76	1.2	97.3
2-4	550	330	1623	33.56	1.4	99.8
2-5	550	330	1623	38.35	1.6	100.9
2-6	550	330	1623	43.15	1.8	99.5
2-7	550	330	1623	47.94	2	99.6

It was found that at the reinforcement of 1.4-1.6% by volume it is possible to obtain the maximum physical and mechanical properties. A further increase in the percentage of reinforcement is not advisable as it causes a decrease in strength and performance of steel fiber concrete.
For further optimization of structure, prevention of crack initiation, increase of strength and durability of the structure basalt fiber domestic production (Table. 3) was used.

Table 3. The dependence of the strength of the fine basalt fiber concrete reinforcement ratio

№	Material consumption per 1 m^3				Reinforcement,	R_{compr},
composition	Binder	Water	Filler	Fiber	%	MPa
2-1	550	330	1623	-	0	94.2
2-2	550	330	1623	1	0.0004	95.6
2-3	550	330	1623	1.2	0.0005	95.9
2-4	550	330	1623	1.4	0.0006	96.3
2-5	550	330	1623	1.6	0.0007	97.1
2-6	550	330	1623	1.8	0.0008	98.2
2-7	550	330	1623	2.0	0.0009	98.1
2-8	550	330	1623	2.2	0.001	98.2

Comparing the experimental data on the tables 2 and 3 for further investigations we take steel fiber concrete according to number 5 (1.6% reinforcement).

4. Conclusion

Based on the abovementioned, it is obvious that it is possible to achieve high physical-mechanical properties of concrete, using local materials of the Russian Far East south. These features are implemented in the fine-grained concrete with screenings of crushing of granite rubble field Wrangel based on composite binders that due to the high surface intensify and accelerate the process of structure formation of a set of concrete strength and seal the structure.

References

[1] D-Y Yoo and Y-S Yoon 2015 *Engineering Structures* **102** 409
[2] B Singh and K Jain 2014 *ACI Structural Journal* **111** 1191
[3] T Bhuvaneswari and D Elavarasi 2014 *Int. J of Applied Engineering Research* **9** 3335
[4] E S Khalifa 2012 *Ain Shams Engineering J* **3** 359
[5] Yu M Bazhenov 2007 *Concrete technology* (Moscow: ASV) 501 p.
[6] S Balaji and G S Thirugnanam 2013 *J of Structural Engineering (India)* **40** 262
[7] S A AlTaan, A A Mohammed and A A Al-Jaffal 2012 *Asian J of Appl. Sci.* **5** 485

Composite binders for concrete with reduced permeability

R Fediuk, A Yushin
Far Eastern Federal University,
20, Ajax St.,Vladivostok, Russky Island, Russia 690000

E-mail: roman44@yandex.ru

Abstract. Composite binder consisting of cement (55%), acid fly ash (40%) and limestone (5%) has been designed. It is obtained by co-milling to a specific surface of 550 kg/m², it has an activity of 77.3 MPa and can produce a more dense cement stone structure. Integrated study revealed that the concrete on the composite binder basis provides an effective diffusion coefficient D. So we can conclude that the concrete layer protects buildings from toxic effects of expanded polystyrene. Low water absorption of the material (2.5% by weight) is due to the structure of its cement stone pore space. Besides lime powder prevents the penetration of moisture, reduces water saturation of the coverage that has a positive effect on useful life period. It also explains rather low water vapor permeability of the material – 0.021 mg/(m-hour-Pa).

1. Introduction.

In modern construction concrete and reinforced concrete are ones of the most commonly used construction materials, foam materials are most commonly used heat insulators.

At the same time, it is well known that the materials from foamed polymers have ecological and fire hazard, and, in the absence of adequate protection, also have short time of exploitation.

It is obvious, that influence of interior and exterior climatic conditions on polystyrene foam should be avoided. This problem can be solved by the internal arrangement of insulation between two structural concrete (reinforced concrete) layers [1,2].

However, concrete is cellular material and the evaporation of styrene from polystyrene occurs continuously, and when the temperature rises, this process significantly accelerates. So, only the infusion of special admixtures, which treat concrete pores and strengthen its structure, is able to neutralize the environmental and fire hazard of polystyrene, as well as to increase its durability by preventing the penetration of atmospheric moisture (which is very important for the monsoon climate of the south of the Russian Far East).

Thus, the development of high gas, water and steam impermeability concrete, able to protect both cellular polystyrene from moisture, and a building from styrene vapors is an urgent task.

2. Design of composite binding material

To achieve this goal the composite binders obtained by co-grinding cement, hyperplasticizer, ash and limestone in the vario-planetary mill have been designed.

Fly ash of the Primorye Territory's largest thermal power plants: Vladivostok TPP and Artem TPP, Primorye SRPS and Partisansk SRPS was used as a component of the composite binder. An important factor is the capacity of dry separate ash extraction that is currently implemented at these power plants. The use of man-made products in the manufacture of building materials contributes to the solution of the following main objectives: saving energy and raw resources; waste management; improvement of the environmental situation in the regions. It should be noted that revenues in the seaside ash dumps are one of the largest in the Far Eastern Federal District. In particular, the annual flow of ash in the ash dumps of the Primorsky Territory is from 2.5 to 3.0 mn tons per year, Khabarovsk Territory is up to 1.0 mn tons. Thermal power plants fly ash is an effective material for the production of fine and active mineral admixtures. These waste substances do not require special preparation when infused into the concrete.

Mineral powder from limestone Dinnogorsk deposit (Primorye) of "Spasskcement" production (MP-1) was employed.

The obtained composite powder had a specific surface area equal to 600 m^2 / kg (determinations were performed on surface meter PSH-11), particle size is 0.15-500 micron, the average particle diameter is shifted to 0.65-11.2 mm, density is 930 kg / m^3 (Figure 1).

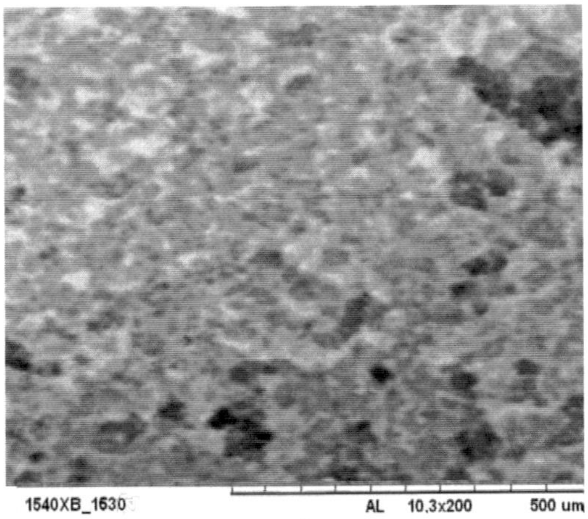

a)

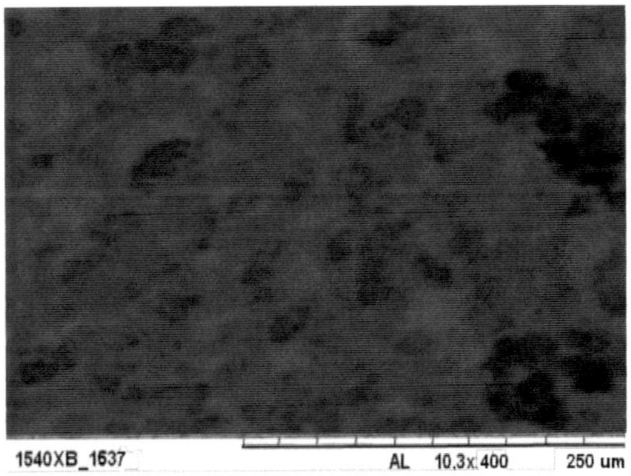

b)

Figure 1. A general view of the composite binder; 200x magnification *(a)* and 400 x magnification *(b)*

Co-grinding of cement with ash and limestone can increase the activity of binding to 70.2 MPa. The increase in strength is due to the improvement of co-grinding of cement stone structure.

This is due to the fact that the floured active mineral components that contribute to an earlier binding of portlandite, intensify the process of hydration of clinker minerals [3,4].

At the same time, the larger particles act as nucleation sites and act as microfiller reducing shrinkage deformation and improving performance of the composite. A characteristic feature of the structure of cement stone with floured admixtures is substantially smaller number of microcracks.

The structure of the cement stone on the composite binder is denser than conventional Portland cement. It is a very dense packing of the grains in the total mass of new formations. Co-grinding of components leads not only to increase of the ultimate compressive strength, but also increases the rate of maturing of the samples with admixtures.

In order to determine the optimum particle size intergrinding of cement with hyperplastisizer, ash and limestone (in a ratio of composition number 7) to different specific surface: 500, 550, 600, 700, 800, 900 m^2 / kg was produced.

The results of determination of compressive strength of cement samples are shown in Table. 1.

Table 1. The results of determination of compressive strength (MPa) cement samples

Age of sample,	The specific surface area of the composite binder, m^2 / kg

d.	500	550	600	700	800	900
3	46.1	47.4	47.2	46.0	45.6	45.5
7	50.3	54.2	54.1	49.1	48.6	48.4
28	68.1	77.3	70.2	65.8	55.0	65.0

The optimal activity of the binder is 550-600 m^2 / kg. Increased activity in excess of these values has a negative effect on the structure. Using a binder with increased activity significantly speeds up the process of setting - setting of the mixture ends after 35-40 min., while the developing temperature at 95-97°C. Fast setting of airbrick prevents the formation of evenly distributed spherical particles in the macrostructure of cement stone. Electron microscopy revealed the presence of heterogeneity macrostructure and irregularly shaped cells [5,6].

The results to determine the sponginess of the samples of the cement stone are given in Table. 2. It follows that at the insignificant (less than 2%) difference in total sponginess samples differ by the distribution pattern of different diameter pores.

Table 2. Research porosity of the cement stone

Composition	Sponginess, %			
	technological (macroscopic level)	capillary (microscopic level / submicroscopic level)	gel (supramolecular level)	general
1	2.4	4.6/8.3	18.2	33.5
2	3.0	1.7/7.5	21.6	33.4
3	4.2	1.1/5.0	23.5	33.8
4	4.0	1.9/4.3	24.4	34.6
5	4.0	1.7/7.5	21.6	34.4
6	4.2	1.1/6.0	23.5	34.8
7	3.0	1.9/4.8	24.4	34.1

Analysis of the microstructure (Figure 2) showed that the cement stone with optimal dosage of components differs by denser matrix consisting of low basic calcium hydrosilicates, while the cement stone without any admixtures (Figure 2*a*) contains a highly basic calcium hydrosilicates and hexagonal plates of portlandite.

The highest technical effect is achieved due to the synergistic action of man-made pozzolanic admixtures (fly ash) and natural materials of sedimentary origin (limestone) when the content of cement - 55 wt.%, limestone - 5 wt.% and the ash - 40 wt.%. Such composites have a compressive strength of up to 57.3 MPa when the strength of the control of the composition (without any admixtures) is 50.5 MPa.

Earlier, it was revealed that the infusion of mineral supplements in the initial binding activates the hydration process. In composite binders hydration products redistribution of crystalline phases of both non-hydrated clinker minerals (S_3S, C_2S, C_4AF), quartz and calcite and hydration products (CH – portlandite, $3S_3A{\cdot}3CaSO_4{\cdot}32H_2O$ - ettringite) occurs [7,8].

Hydrate structure of the composite binder is represented by two types - the primary and secondary structures. The primary structure is represented by amorphous interporous space formed by the solution mechanism, wherein the composition of the products in the pore space depends on the chemistry of the large particles which surround a pore.

The results show a clear synergistic effect of components of the material composition of binder on the mineralogical composition of the hydration products and the rate of interaction of clinker minerals with water, on the morphology of hydration products and micro-porosity of the stone.

Replacement of the cement by different admixtures of natural and anthropogenic origin accelerates hydration and leads to formation of new products that enhance the density of the cement stone, and as a result, increase the strength and impermeability [9,10].

Realization of the potential capabilities of fiber-reinforced concrete is only possible when the optimum structure of the material is created. The formation of the structure is determined by the following main parameters: the type and quality of raw materials; technology of preparation of concrete mixtures; the quantitative relationship between the fiber-reinforced concrete components of the mixture.

As a result of tests on the device "Agama-2M" air permeability of concrete parameter a_c was found to be 0.0253 cm^3 / s. In Table. 7 of GOST 12730.5 corresponding mark on the permeability is W14 (Table. 3).

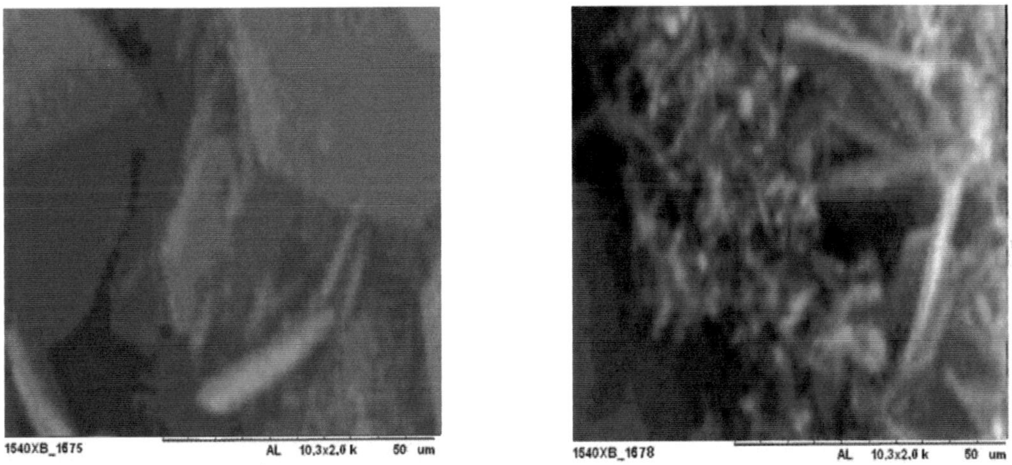

a) *b)*

Figure 2. The micrographs neoplasms: *a* - cement stone without admixtures; *b* - cement stone on the basis of the composite binder

Table 3. The permeability of fine concrete depending on the composition of the binder

№ Composition	Material consumption per 1 m^3							Sslump	Parameter air permeability of concrete a_c, cm^3/s	Mark on the water permeability *W*
	Binder, kg			Hyperplacticizer	Screenings crushing of granite, kg	Sand, kg	Water, l			
	Cement	Ash	Limestone							
1	415	-	-	-					0.0565	W10
2	210	178	27						0.0253	W14
3	200	186	29						0.0289	W14
4	190	194	31	1.2	1000	762	220	10-12	0.0402	W12
5	180	202	33						0.0465	W12
6	170	210	35						0.0423	W12

As test results show admixtures of fly ash and limestone at all dosages reduce water permeability of concrete. Thus, a clear link between the properties of concrete and structural features of the cement stone is revealed. Increase in the number of low basis hydrosilicates of calcium and increased gel and less capillary sponginess especially at the submicroscopic level determine the growth of the strength and reduction of the concrete permeability [11,12].

3. Results.

The results of the experiment on water vapor permeability for the created concrete are shown in Table 4. Obtained results show the better characteristics compared to the standard concrete (0.03 mg/(m-h Pa)).

Table 4. The vapor permeability of fine concrete depending on the composition of the binder

№ Composition	Material consumption per 1 m^3								Vapor permeability of concrete, mg/(m-h Pa)	
	Binder, kg									
	Cement	Ash	Limestone	Hyperplacticizer	Screenings crushing of granite, kg	Sand, kg	Water, l	Sslump	for dry climate	for wet climate
1	415	-	-	-					0.032	0.030
2	210	178	27						0.022	0.021
3	200	186	29						0.026	0.025
4	190	194	31	1.2	1000	762	220	10-12	0.027	0.026
5	180	202	33						0.030	0.029
6	170	210	35						0.032	0.030

Results of water absorption on the basis of fiber-reinforced concrete developed on designed composite binder are shown in Table. 5.

Table 5. The vapor permeability of fine concrete depending on the composition of the binder

№ Composition	Material consumption per 1 m^3								Water absorption		
	Binder, kg										
	Cement	Ash	Limestone	Hyperplasticizer	Screenings crushing of granite, kg	Sand, kg	Water, l	Slump	% by weight	кg/m^3	% by volume
1	415	-	-	-					6.1	148.23	14.8
2	210	178	27						2.5	60.75	6.1
3	200	186	29						2.6	63.18	6.3
4	190	194	31	1.2	1000	762	220	10-12	3.2	77.76	7.8
5	180	202	33						4.5	109.35	10.9
6	170	210	35						5.1	123.93	14.4

Rather low water absorption of the material is due to the structural features of its pore space described previously.

Besides lime powder prevents the penetration of moisture, reduces water saturation coverage that has a positive effect on service life. This also explains the rather low water vapor permeability value of the material.

As a result of an experiment on evaluating the diffusion constant, the average value of the thickness of the concrete neutralized layer *X is set, see* the following formula:

$$X = \frac{\sum_{1}^{n} X_i}{n}$$

where n = 10- number of measurements (Table. 6).

The standard deviation:

$$S = \sqrt{\frac{(X_i - X)^2}{n-1}}$$

The calculation is performed with the reliability of 95% and the probability of the risk $2\beta = 0.05$. Specification limits are determined by the formulas:

$$t_1 = X - L \cdot S \qquad t_2 = X + L \cdot S$$

Table 6. The average value of the thickness of the concrete neutralized layer *X, cm*

№ sample	X_i	X_i-X	$(X_i-X)^2$
1	7	-1.4	1.96
2	6.5	-1.9	3.61
3	11	2.6	6.76
4	10.5	2.1	4.41
5	7.5	0.9	0.81
6	8.5	0.1	0.01
7	10	1.6	2.56
8	9.5	1.1	1.21
9	7	-1.4	1.96
10	6.5	-1.9	3.61
All	84		26.9

$$X = \frac{\sum_{1}^{n} X_i}{n} = \frac{84}{10} = 8.4 \qquad S = \sqrt{\frac{(X_i - X)^2}{n-1}} = \sqrt{\frac{26.9}{10-1}} = 1.73$$

$$t_1 = X - L \cdot S = 8.4 - 3.39 \cdot 1.73 = 2.54 \quad \text{mm}$$

$$t_2 = X + L \cdot S = 8.4 + 3.39 \cdot 1.73 = 14.26 \quad \text{mm}$$

Diffusion permeability of concrete is calculated by the upper boundary of the tolerance, i. e, the depth of carbonation takes equal 14.3 mm.

According to the results of chemical analysis of the concrete reaction vessel of concrete m_0 in relative terms is calculated by the formula:

$$m_0 = 0.4C\rho f = 0.4 \cdot 210 \cdot 0.6 \cdot 0.6 = 30.24$$

where C - cement content, g in 1 cm^3 of concrete; ρ - the number of basic oxides in the cement based on CaO in weight relative values; f - a degree of neutralization of concrete, equal to the ratio of quantity of basic oxides, which undergo reaction with carbon dioxide to the total amount of cement (average $f = 0.6$).

The effective diffusion coefficient of carbon dioxide in concrete D', cm^2 / s, is calculated from the formula:

$$D' = \frac{m_0 X^2}{2C\tau} = \frac{30.24 \cdot 0.84^2}{2 \cdot 0.1 \cdot 14 \cdot 24 \cdot 3600} = 1.34 \cdot 10^{-4} \text{ cm}^2\text{/s}$$

The result revealed that the developed concrete provides an effective diffusion coefficient D', so we can conclude that the concrete layer protects buildings from toxic effects of styrofoam.

4. Conclusion

Developed composite binder consisting of cement (55%), acidic fly ash (40%) and limestone (5%) has been designed. It is obtained by co-milling to a specific surface of 550 kg / m^2 and it has an activity to 77.3 MPa and it can produce a more dense cement stone structure.

It was found that the combined effect of mechanical and chemical activation (the presence of limestone particles) increases the pozzolanic activity of acidic ashes. It has a catalytic effect on the reaction activity of the surface of ash and sand during mechanical conversion in vario planetary mill. Furthermore, the infusion of limestone increases the alkalinity of the concrete, which leads to the formation of more hydration products of cement per unit of time.

It was found that the addition of fly ash and limestone at all dosages reduces water permeability of concrete. Thus, it revealed a clear link between the properties of concrete and structural features of the cement stone. Increase in the number of low basis hydrosilicates of calcium and increased gel and less capillary sponginess especially at the submicroscopic level determine the growth of the strength and reduction of the concrete permeability.

Integrated study revealed that the concrete, developed on designed composite binder provides an effective diffusion coefficient D'. So we can conclude that the concrete layer protects buildings from toxic effects of expanded polystyrene. Rather low water absorption of the material (2.5% by weight) is due to structural features of its pore space of cement stone. Besides lime powder prevents from penetration of moisture, reduces water saturation coverage that has a positive effect on service life. This also explains the rather low water vapor permeability of the material values - 0.021 mg / (m-h-Pa).

References

[1] Ichiko Misumi, Jun-ichiro Kitta, Hiroyuki Fujimoto, Satoshi Gonda, Yasushi Azuma, Keisuke Maeda, Tomizo Kurosawa, Yoshiyasu Ito, Kazuhiko Omote, Yoshinori Nakayama and Hiroki Kawada 2012 *Meas. Sci. Technol.* **23** 015002

[2] M Tabuchi, H Tameoka, T Kawase and Y Takeda 2011 1757-899X **24** 012007

[3] H Nakao, Y Yamasaki, J Okamoto, T Sudayama, Y Takahashi, K Kobayashi, R Kumai and Y Murakami 2014 1742-6596 **502** 012015

[4] J del Cerro, J Manchado, F J Romero and M C Gallardo 2012 *Meas. Sci. Technol.* **23** 035003

[5] Camila Barreneche, Aran Solé, Laia Miró, Ingrid Martorell, A Inés Fernández and Luisa F Cabeza 2012 *Meas. Sci. Technol.* **23** 085606

[6] Md Jashim Uddin, N H Md Yusoff, O Anwar Bég and Ahamd Izani Ismail 2013 *Phys. Scr.* **87** 025401

[7] Hideyuki Nakano, Akira Kouchi, Shogo Tachibana and Akira Tsuchiyama 2003 0004-637X **592** 1252

[8] J M Cuevas, R Rubio, J M Laza, J L Vilas, M Rodriguez and L M León 2012 *Smart Mater. Struct.* **21** 035004

[9] Zhi Chen, Dai Mochizuki, Masato M Maitani and Yuji Wada 2013 *Nanotechnology* **24** 265602

[10] Yun Lou, Ling Wan, Yongzhong Ma, Huijuan Li, Qinghua Meng, Yuxia Kong, Weijie Zhu, Dapeng Wu and Limeng Cui 2013 *J. Radiol. Prot.* **33** 661

[11] M Eichmann, T Krause, D Flühs and B Spaan 2012 *Phys. Med. Biol.* **57** 421
[12] M Herranz, S Rozas, R Idoeta and N Alegría 2014 *J. Radiol. Prot.* **34** 133

Surface layer structure of AISI 1020 steel at different stages of dry sliding under electric current of high density

K A Aleutdinov[1,*], V Ye Rubtsov[1,2], V V Fadin[2] and M I Aleutdinova[2,3]

[1] National Research Tomsk Polytechnic University, Institute of Cybernetics, Lenin avenue, 30, Tomsk, 634050, Russia
[2] Institute of Strength Physics and Materials Science, pr.Akademicheskii, 2/4, Tomsk, 634021, Russia
[3] Seversk Technological Institute - branch of State Autonomous Educational Institution of Higher Professional Education «National Research Nuclear University «MEPhI», pr.Kommunisticheskii, 65, Seversk, 636036, Russia

*E-mail: aleut@ispms.ru

Abstract. Wear intensity of the sliding electric contact steel 1020/steel 1045 depending on sliding time is presented at the contact current density higher than 100 A/cm^2 without lubricant. It is shown that wear intensity of 1020 steel decreases at increasing of sliding time. Wear intensity is stabilized after some sliding time. This time (burn-in time) decreases at reduction of current density. Structural changes are realized in surface layer. Signs of liquid phase are observed on sliding surface. This liquid isn't a result of melting. It is established using Auger spectrometry that the contact layer contains up to 50 at.% of oxygen.

Keywords: structure changes of surface layer, contact spots, sliding current collection, wear intensity, chemical composition of sliding surface.

1. Introduction

It is known that high wear resistance of material during friction is reached due to structural stability of its contact layer [1]. The structure of surface layer (SL) can be stabilized by superficial hardening, for example, owing to chemical heat treatment, ionic implantation, etc. The structural state of sliding surface will be stable also in the case when external influence doesn't lead to its plastic deformation. It means that the contacting microvolumes will be deformed in the mode of elastic deformation or, more precisely, in the mode of high-cycle fatigue. Therefore creation of the strengthened SL and concrete microgeometry of a surface are expedient only for the friction modes, which don't break surface layer stability.

The heavy mode of friction causes plastic deformation of SL and its subsequent deterioration occurs in the conditions of low-cycle fatigue [2,3]. Surface material self-organizes in this case and its structural changes take place. Thus it is formed the layer of friction induced structures (FIS) as it is shown in [4]. This layer has parameters of structure (thickness, roughness, phase and chemical composition, etc.) changing its values according to changes of parameters of external influence on the tribosystem. Structure parameters achieve its equilibrium values during some time at every regime of normal (non

catastrophic) wear. The same time it is necessary for achievement of stable values of main outputting characteristics of tribosystem (wear intensity, friction coefficient, average contact temperature, etc).
The main mechanism of surface layer material deterioration is connected with its plastic deformation and its fragmentation [2,3,5]. Suppression of fragmentation and a transfer of SL plastic deformation on lower structural level can be carry out in material capable to well stresses relaxation in the vicinity of stress concentrators. The relaxation of stresses should occur due to local plastic shears. Such local shear instability at the meso-scale structural level has to provide the maximum shear stability at the macro-scale structural level, i.e. on the scale of FIS layer. Such structural state of a FIS layer will provide stable and minimal wear intensity.
Change of initial structure parameters is one of the main ways of management of these contact characteristics. Initial material structure is characterized by chemical (element) and phase composition, as well by stress state and plasticity, and by ability to form FIS layer, etc.
The heavy mode of friction can be set by means of electric current of high contact density. As a rule, current collection materials have the graphite containing composite structure where iron or copper [1] serves as a basis. Most of these composites show wear intensity higher, than cast AISI 1020 steel or copper under heavy conditions of dry friction with contact current density higher 100 A/cm^2 [6]. It means that composites form a low-plastic FIS layer which relax stresses more weakly, than a FIS layer of cast copper or steel 1020. Steels hardening by inducing of alloys elements or by solid phases form FIS layers with low plasticity. It leads to realization of high wear intensity in conditions of FIS layer plastic deformation under influence of electric current during dry sliding [6]. Hardening of material by deformation defects, i.e. peen creation, is a known way of initial structure hardening. This method is applied for SL hardening due to sliding of material during some time until burn-in state achieved and wear intensity will not depend on time at set input parameters of a tribosystem. It is of scientific interest to get some idea about efficiency of surface peen creation of material thanks to burn-in during dry sliding in electric contact with the high contact current density. Determination of FIS layer chemical composition and wear intensity in different points of sliding distance should be necessary stages of research in this direction. Carbon steel 1020 may serve as a model material.
The aim of present work is the study of interconnection of wear intensity and surface layer structure changes of carbon steel AISI 1020 under dry sliding and electric current of contact density higher 100 A/cm^2 in different sliding time moments.

2. Experimental details

The peen steel 1020 of hardness HB=2740 MPa served as a model specimen. Metallography studying is carried out applying optical microscope Neophot-21. The average chemical composition of surface layer is defined on Auger spectrometer "Shkhuna-2". Wear intensity is determined under dry sliding electric contact at alternating current (50 Hz), contact pressure p=0.13 MPa, sliding velocity v=5 m/s using tribometer SMT-1 by the scheme "pin-on-ring" (figure 1). Quenched AISI 1045 steel (50 HRC) served as a counterbody. Linear wear intensity is determined as $I_h=h/L$, where h is the specimen height change at sliding distance L=9 km. The contact current density is determined as $j=i/A_a$, where i is the current passing through the nominal contact area A_a (A_a=10 mm^2).

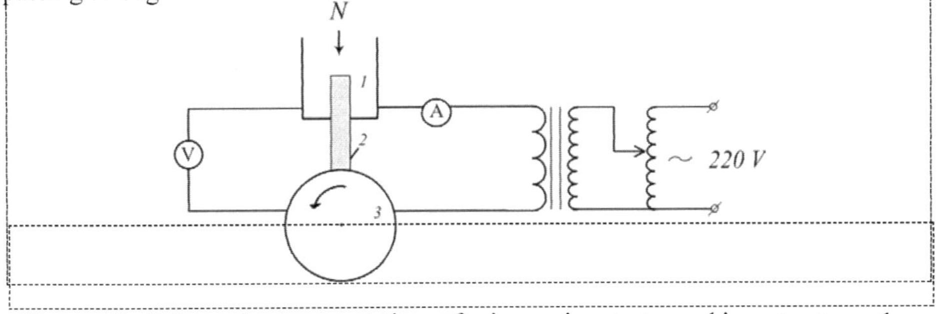

Figure 1. Schematic representation of pin-on-ring test machine structure: 1 - specimen holder, 2 – specimen, 3 - counterbody (steel 1045 HRC).

3. Results

Definition of sliding time sufficient for achievement of stable wear intensity at some set normal wear mode has to be the initial stage of peen studying without reaching of catastrophic wear regime. Normal dry wear of steels under electric current is realized usually till j=400 A/cm^2 [6]. The quasiequilibrium SL structural state of surface layer has to be achieved for shorter sliding time at low current density in comparison with a case of sliding at high current density. It becomes clearly from that the speed (curve slope) of specimen height change during sliding with j=270 A/cm^2 is less than that during sliding with j=400 A/cm^2 (figure 2, a). In other words the stable I_h at j=270 A/cm^2 has lower value than I_h at j=400 A/cm^2 (figure 2, b). It should be noted also that the SL structure can't be equilibrium at the surface peen beginning. It leads to the high height change speed of a specimen (figure 2, a) and to high wear intensity at low sliding time (figure 2, b).

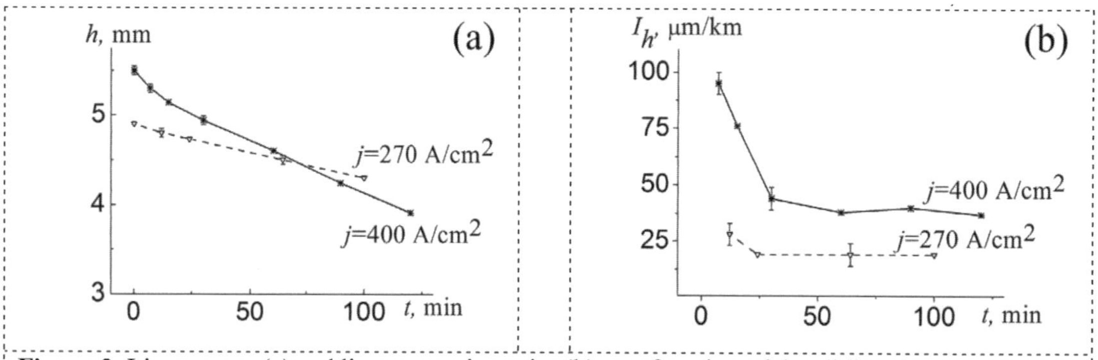

Figure 2. Linear wear (a) and linear wear intensity (b) as a function of the sliding time.

Surface layer structural changes are obvious as a formation of friction induced structure (FIS) layer (figure 3, a-b). One can see that thickness of this layer becomes higher 20 microns in the first 7 minutes of sliding and at sliding time increasing its thickness doesn't change considerably. The same layer appears also during sliding with lower current density.

Signs of liquid phase existence and traces of a surface layer plastic plowing by micro-asperity of counterbody are observed on sliding surface at j>100 A/cm^2 (figure 3, c). Traces of adhesive interaction aren't observed in an explicit form.

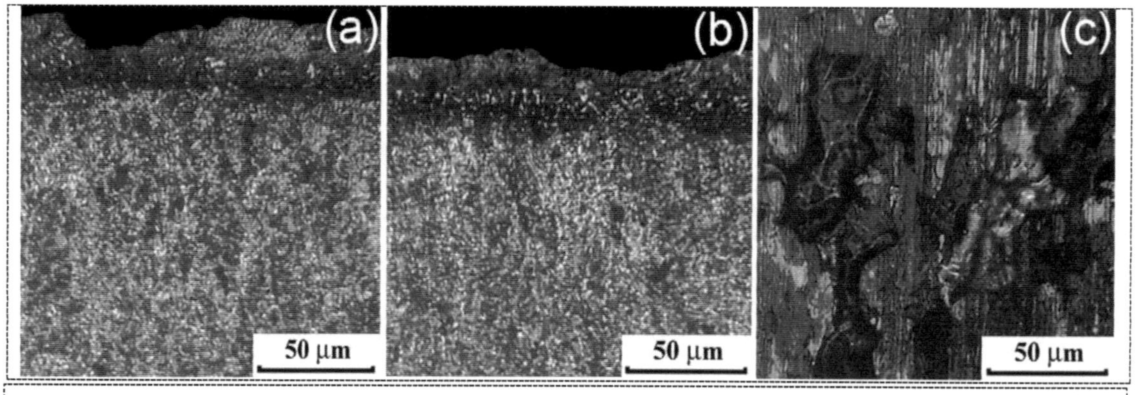

Figure 3. Cross-section of worn surface (a) after sliding at j=400 A/cm^2 for 7 minutes (a) and 120 minutes (b); the OM image of worn surface (c) after sliding at j=270 A/cm^2 for 24 minutes.

Oxygen concentration on sliding surface of steel 1020 reaches values 40-50 ат % (figure 4). It is visible that at initial sliding stage the oxygen concentration can be higher or lower iron concentration (figure 4, a-b). The equilibrium chemical composition of friction induced structures layer corresponds to the equal content of oxygen and iron (figure 4, c). Iron concentration at lower j=270 A/cm^2 exceeds oxygen concentration of (figure 4, d).

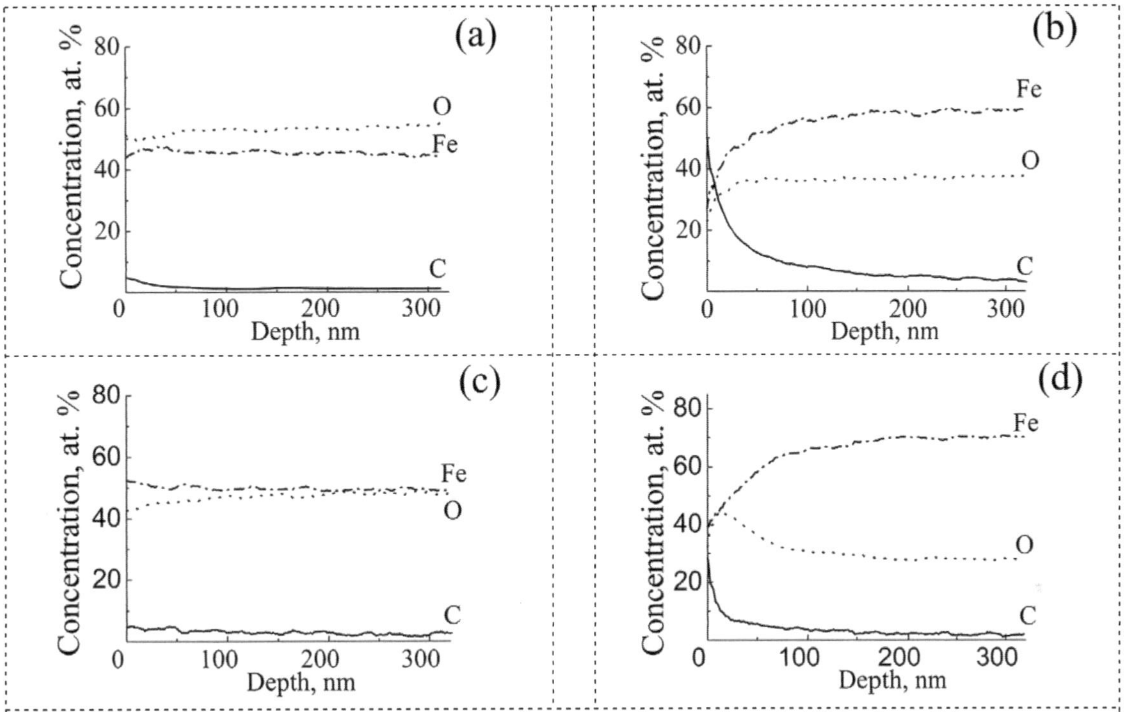

Figure 4. Distribution of chemical elements in contact layer of steel 1020 after sliding at contact current density j=400 A/cm^2 for 7 minutes (a), for 15 minutes (b), for 120 minutes (c) and after sliding at contact current density j=270 A/cm^2 for 24 minutes (d).

4. Discussion

Wear intensity reduction during burn-in process is caused by increasing of surface layer hardness. Deterioration of surface layer begins at exhaustion of FIS plasticity [3,5]. It means that there has to be an optimum ratio of FIS hardness and its plasticity reserve for achievement of the minimum wear intensity. This ratio can be reached for some sliding time. In the general case burn-in time depends on properties of initial structure, on the contact current density and on other input parameters of a tribosystem. Necessary hardness of FIS layer is provided, partly, with presence of the phase FeO, when about 40 at.% of oxygen are placed close to sliding surface (figure 4). If all these oxygen atoms form FeO oxide, so it means that more than a half of thin surface layer is occupied with FeO oxide. Low atomic oxygen concentration in some cases (figure 4, b and figure 4, d) in comparison with higher concentration of oxygen (figure 4, a and figure 4, *c*) in thin surface layer doesn't point to the low volume content of FeO oxide in FIS layer. In this case the absence of regularity in oxygen distribution on sliding surfaces is a consequence of formation of the contact surface sectors having

sharply various structural states. Such distinction of chemical element composition in different places of sliding surface is a usual occurrence independent of initial structure of friction pair materials.
Main ways of external energy dissipation in contact realize as creation of structural defects in surface layer (i.e. FIS layer formation and creation of wear particles) and as thermal streams forming. If high concentration of structural defects is shown in the form of cracks [7] or as severe adhesion of worn surface [8], these are the signs of catastrophic wear. In this case wear particles have the big sizes or big sectors of a plastic flow are observed on sliding surface. But signs of viscous liquid emergence (figure 3, c) shouldn't be considered as a result of catastrophic wear realization. In this case it is necessary to take into account that FIS layer contains a large amount of vacancies. These vacancies provide anomalously high diffusion, first of all, in micro-volumes of contact spots vicinity. As a result it is formed the thin contact layer of high-excited atoms [9] where diffusion can reach the level of the diffusion corresponding to melt of material. But this layer is in a crystalline state. In this case the stress relaxation on these surface sectors can be carried out in accordance with mechanism of viscous liquid deformation [5] when deformation defects are not formed. I.e. signs of viscous liquid existence aren't caused by high temperature in contact. This layer of the strong excited atoms may contain to 50% of oxygen (figure 4), but this oxygen can't be completely connected in FeO oxide since contact electric conductivity in this case has to be low because of high specific resistance of FeO oxide.
Absence of obvious interrelation between wear intensity reduction (figure 2), chemical element composition (figure 4) and optical micro-structure images (figure 3) means that surface layer hardening (i.e. a FIS layer hardening) for some time happens thanks to other mechanism, namely, owing to realization of local plastic micro-shears and formation of deformation defects. Accumulation of defect amount to some limit concentration cause the stress relaxation by formation of cracks and wear particles. It means that the leading mechanism of surface layer deterioration and wear is based on low-cyclic fatigue.

5. Conclusion

Dry sliding of steel 1020 against steel 1045 under the influence of electric current with a contact density higher 200 A/cm^2 leads to changes of surface layer structure. It causes the formation of a friction induced structures layer higher 20 microns thick. This layer contains to 50 at.% of oxygen in the vicinity of sliding surface that points to formation of FeO oxide. Obvious dependence of the oxygen content and FIS layer thickness on the contact current density and on sliding time isn't observed.
Wear intensity decreases during burn-in time to some limit that is conditioned on appearance of deformation defects and contact surface hardening. This wear intensity limit and burn-in time decrease at reduction of contact current density. The stress relaxation in surface layer occurs due to local plastic micro-shears. At the same time other types of a stress relaxation are realized, namely local liquid phase formation and emergence of the high-excited atomic states.

Acknowledgments

The work was carried out according to Project No. III.23.2.4 of SB RAS Fundamental Research Program III.23.2 and supported by Russian Foundation for Basic Researches project No.13-08-00076

References

[1] Fedorchenko I M, Pugina L 1980 *Composition Sintered Antifriction Materials* (Kiev: Nauk.Dumka)
[2] Kragelsky I V, Dobychin M N and Kombalov V S 1982 *Friction and Wear Calculation Methods* (New York: Pergamon Press)
[3] Bowden F P, Tabor D 1973 *Friction: An Introduction to Tribology* (New York: Anchor)
[4] Yao B, Han Z, Li Y S, Tao N R and Lu K 2011 *Wear* **271** 1609–16
[5] Panin V E, Pochivalov Yu I 2010 *Phys. Mesomech.* **13** 27-34 [in Russian]

[6] Aleutdinova M I, Fadin V V, Kolubaev A V and Aleutdinova V A 2014 *Friction and Wear Research (FWR)* **2** 22-8
[7] Jankauskas V, Antonov M, Varnauskas V, Skirkus R, Goljandin D 2015 *Wear* **328-329** 378-90
[8] Zhenga R G, Zhanb Z J, Wanga W K 2010 *Wear* **268** 72–6
[9] Panin V E, Grinyaev Yu V 1990 *Structural levels of plastic deformation and destruction* (Novosibirsk: Science. Siberian office) [in Russian]

Modification of Surface Layers by Surfacing Intermetallic Coatings with Variable Properties

D N Makeev[1], O V Zakharov[1], A N Vinogradov[1], A V Kochetkov[2]
[1] Yuri Gagarin State Technical University, Saratov, Russia,
[2] Perm National Research Polytechnic University, Perm, Russia

E-mail: zov17@mail.ru

Abstract. The paper considers the possibility of forming coating layers for parts within wide limits of microhardness. The technology uses surfacing of intermetallic coatings provided by a unique experimental setup. Theoretical and experimental dependence of the coating layer microhardness on the filler concentration using the changes in the speed of the filler wire feed and current intensity were determined.

1. Introduction

One of the fundamental problems in modern engineering is improvement of reliability of machines and mechanisms which is connected with upgrading the durability of their components. For this purpose it is necessary to improve the quality of the mating surfaces, primarily those operating under conditions of rolling or sliding friction. With continuous reduction of raw material resources and the need for increasing energy efficiency of the production process, manufacturing of components with advanced types of covering becomes of major importance [1-3]. The coating allows for changes in the technology, and utilization of work pieces made of inexpensive materials characterized by poorer physical and mechanical properties, and subsequently, allows for modification of the working surfaces using high-strength alloys. The surfacing provides the possibility for coatings with the thickness ranging from 0.1 to 10 mm. The given process facilitates reconditioning expensive parts, such as workpieces for vehicles, which are subjected to the highest wear while in operation.

At this stage, development of new approaches in the design of construction materials and creation of efficient technologies for their fabrication are needed. Modification of the coating layers using the surfacing of intermetallic coatings might be one of the required methods [4-7]. The given technology was first developed at Yuri Gagarin State Technical University of Saratov and went through the experimental pre-check needed to produce new and restore worn-out parts of vehicles.

2. Theoretical Estimates

An overview of existing surfacing technologies with further machining showed that the basic studies focus on resolving the following problems: improving the quality of the surfacing metal, enhancing alloying process, increasing the surfacing rate, reducing the costs of surfacing and further machining processes. The quality of the surfacing layer, to a great extent, can be defined by the following characteristics: structure refinement, reduction of the heat-affected zone, reduction of specific heat input, and, consequently, reduction of internal stresses and deformation in the post-surfacing period.

While surfacing the coating is provided by melting the filler and the near-surface layer of a work piece, their fusion and further crystallization over the covered work piece. The welded joint crystallizes as separate layers with the thickness rate at hundredths of a millimetre. The layered structure of the crystallized metal is due to periodic shutdowns caused by delays of temperature reduction in the welding bath before the crystallization front simulated by the latent heat release. The process of heat elimination of heat into the base metal is followed by crystallization of the melt layer in the coating metal. The intermittent nature of primary crystallization of the melt metal bath impacts the formation of stratified inhomogeneity where the lower part of the layer is characterized for concentrated ratio of alloying elements, while the upper part of the layer if compared to the middle part is characterized for depletion of these elements. When under cooling, the melt discharges crystalline grains with high concentration of refractory components, whereas the concentration of fusible components is rising in the non- solidified part of the melt. Thus, the surfacing metal becomes chemically inhomogeneous.

Two competing processes develop during solidification of the metal: formation of chemical inhomogeneity of the material and diffusion leading to the levelling of the compound in different areas of crystalline grains. However, full homogenization of the metal does not occur, which results in the development of zonal segregation and intradendritic heterogeneity. Segregation of the various chemical elements has its own peculiar features influencing each other. For example, an increase in the concentration of carbon in the surfacing metal strengthens segregation of manganese, molybdenum, silicon and chromium. Thus, alloying the surfacing metal using several elements results in their sophisticated interaction on the pattern of metal inhomogeneity. Chemical inhomogeneity of the surfacing metal is also connected with the surfacing technology itself and characteristics of the materials used in the coating process. The main reason for inhomogeneity at single-layer surfacing provided by separate beads is higher intensity of metal alloying in the overlapping areas. With increasing the duty cycle of the flux cored wire or powder-coated tape, the scattering of the material decreases and conditions for the welding beads formation improve, therefore inhomogeneity of the surfacing metal decreases.

Physical inhomogeneity of the surfacing metal is connected with certain defects in the crystal lattice. The majority of defects occur during the dendritic crystallization typical of the surfacing metal. Dendrite branches are characterized for a variety of spatial orientation, while crystalline grains have block structure strengthened due to additives and dislocation between the branches of dendrites. At melt cooling during the crystallization process, we can observe the volume contraction of the coating material. The shrinkage effect creates significant stresses within the crystallized metal. If the shrinkage effect exceeds the existing resistance of material, then hot cracks occur in the coating. Metal cooling is accompanied with the increase of cracks under the stress caused by uneven heating. The changes in the chemical and physical inhomogeneity of the surfacing metal, cause the changes in its physical-mechanical properties.

In the course of preliminary studies, we found the possibility for directional formation of properties of intermetallic coatings during the surfacing procedure. The possibility can be realized by means of introduction of a separate filler wire into a specified point of the welding bath. As a result, heat removal conditions are created leading to reduction of the heat-affected zone and internal stresses. Thus the elements of the filler wire alloy the metal of the joint.

Using the above data, we suggested a hypothesis which relates the possibility of quality management of the surfacing layer using the changes of machining conditions. To conduct the experiments, a unique setup was developed. Preliminary studies proved the fact of microhardness changes in the coating layer ranging from 28-32 to 45-55 HRC.

3. Experimental Procedure

To obtain the intermetallic coating, we used the surfacing method with the flux layer and additional grounded aluminium filler wire (patent RU 2403138). The surfacing was performed with the 30ХГСА base wire and $AlMg_3$ filler wire. For wire feeding (base and filler) we utilized the surfacing head (patent RU 2494843). Figure 1 shows the surfacing scheme with two wires – the base and the filler, where 1 – the base wire, 2 – the filler wire, 3 – the work piece, and 4 – the flux.

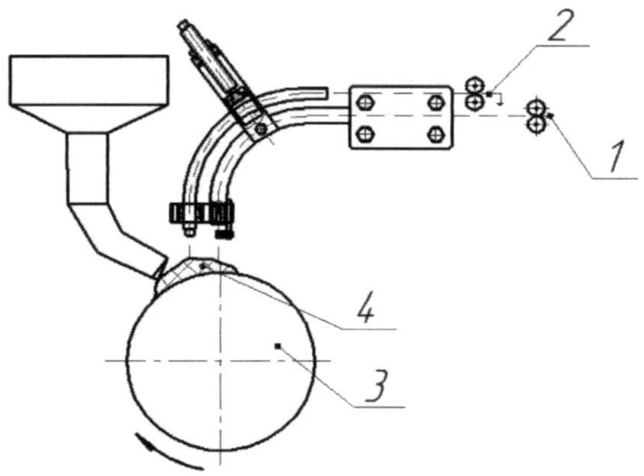

Figure 1. The surfacing scheme with two wires.

The aim of the experiment study was to define the influence of the alloying constituent ratio (aluminium) on the microhardness of welded coating layers, find out empirical dependence of microhardness on the surfacing mode, and research the distribution of the hardness profile over the depth of the surfacing layer.

To determine the influence mechanism of the analyzed factors on the results of intermetallic coatings, we used the interpolation models developed on the basis of power functions. Microhardness of the coating layer Hμ was used as the factor of dependency on the percentage of the filler wire *n*, % and current strength feeding the base wire *I*, *A*.

After transformations according to the formulas shown above, we obtained the following empirical dependency:

$$H\mu = 17500 \cdot n^{0.2} \cdot I^{-0.38}. \qquad (1)$$

According to the formula (1), the microhardness increases depending on the increase of ratio of the filler wire n, whereas the increase of the current strength I leads to an insignificant decrease in the microhardness. The chart (Figure 2) shows the influence of each factor on the microhardness of the welded surface.

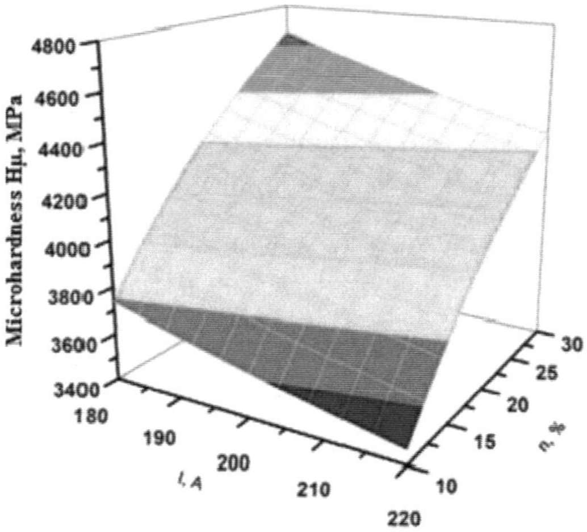

Figure 2. Dependence of microhardness of the welded surface on ratio of the filler wire *n* and current strength *I*.

When increasing the filler wire feed speed, the percentage of filler constituents in the welding bath increases. Thus, the number of intermetallic compounds increases, and consequently, results in the increase of microhardness of the coating layers.

When increasing the current strength, the temperature of the welding bath increases, which facilitates evaporation of the constituent alloying materials in the base wire and constituent materials in the filler wire, since the melting temperature is significantly lower than the melting temperature of steel. Thus, the increase of the current strength causes decrease of microhardness of intermetallic coatings.

4. Results and Analysis

Let us consider the microhardness behaviour of the welded intermetallic samples along the depth of the surfacing layer.

The chart (Figure 3) shows that microhardness of sample 2 is higher than that of sample 1, which is interconnected with the percentage of the filler material in the samples. The more the percentage of the filler material, the higher the microhardness of the surface. This is proved by the results of the secondary ion-ion emission, which demonstrates that sample 2 has 2 or 2, 5 times more aluminium compared to sample 1.

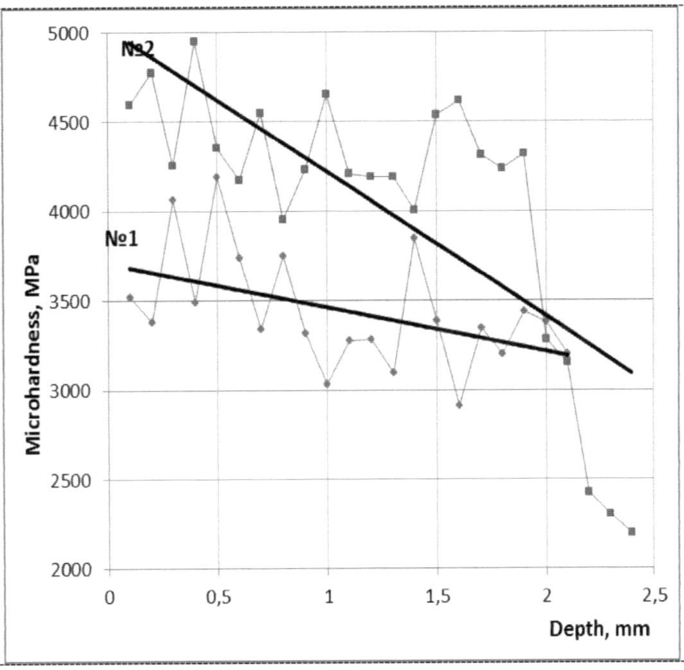

Figure 3. Microhardness dependence of the welded layer, samples 1 and 2.

Concentration ratio of other alloying materials in the samples is the same, since these materials make the constituent part of the base wire. The feeding speed and current strength of the wire remains unchanged during over the whole stage of the experiment.

The chart (Figure 3) shows that microhardness of sample 4 is higher than that of sample 3, since the percentage of the filler material in sample 4 is higher than that in sample 3. The maximum microhardness ratio of sample 2 is insignificantly higher than in sample 4.

This is connected with the current strength in the base wire, since the increase of the current strength results in the temperature increase in the welding bath, and consequently, leads to evaporation of alloying materials. Therefore, concentration of the aluminium in sample 4 is 1.3% less than in sample 2.

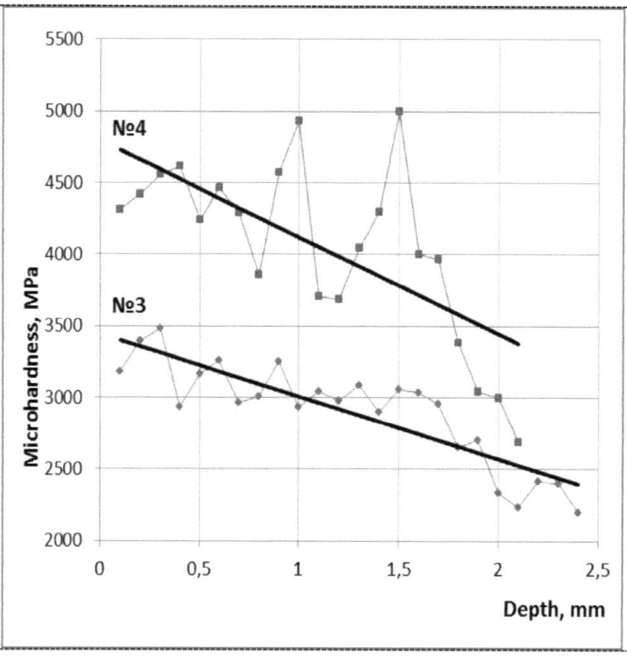

Figure 4. Microhardness dependence of the welded layer, samples 3 and 4.

5. Conclusion

The research proves the hypothesis referring the possibility of directional changes in the properties of the surfacing layers by means of surfacing intermetallic coatings. For the samples and real vehicle parts, we obtained the changes of surface microhardness ranging from 28-32 to 45-55 HRC depending on the surfacing conditions.

Using a complete factorial experiment, we obtained empirical dependency of microhardness in the welded coating layers on the filler concentration and current strength. The given data prove that the developed method of surfacing intermetallic coatings used for coating new parts and refacing worn-out parts can be recommended for application.

References

[1] Suryanarayana C, Ivanov E and Boldyrev V V 2001 The science and technology of mechanical alloying *Materials Science and Engineering: A* **304-306** 151-158

[2] Canakci A. Varol T, Erdemir F, Ozkaya S and Mindivan H 2014 Microstructure and properties of Fe-Al intermetallic coatings on the low carbon steel synthesized by mechanical alloying *The International Journal of Advanced Manufacturing Technology* **73** 849-858

[3] Zhaolin Zhana, Yedong Hea, Deren Wanga and Wei Gao 2006 Low-temperature processing of Fe-Al intermetallic coatings assisted by ball milling *Intermetallics* **14** 75-81

[4] Zadorozhnyy V, Kaloshkin S, Kaevitser E and Romankov S 2011 Coating of metals with intermetallics by mechanical alloying *Journal of Alloys and Compounds* **509** 507-509

[5] Krasnowski M and Kulik T 2010 Nanocrystalline Al-Fe Intermetallics – Light Weight Alloys with High Hardnes *Intermetallics* **18** 47-50

[6] Cezary Senderowski 2014 Nanocomposite Fe-Al Intermetallic Coating Obtained by Gas Detonation Spraying of Milled Self-Decomposing Powder *Journal of Thermal Spray Technology* **23** 1124-1134

[7] Yang D M and Tian B H 2013 Microstructure and Mechanical Properties of FeAl Coating Deposited by Low Pressure Plasma *Applied Mechanics and Materials* **333-335** 1916-1920

IOP Conf. Series: Materials Science and Engineering **116** (2016) 012024 doi:10.1088/1757-899X/116/1/012024

Effect of hydrogen on plastic strain localization and fracture of steels

M V Nadjozhkin[1], A G Lunev[1], Yu V Li[1], S A Barannikova[1,2,3]

[1] Institute of Strength Physics and Materials Science, SB RAS, Tomsk, 634055, Russia
[2] Tomsk State University, Tomsk, 634050, Russia
[3] Tomsk State University of Architecture and Building, Tomsk, 634003, Russia

E-mail: bsa@ispms.tsc.ru

Abstract. The effect of interstitial hydrogen atoms on the mechanical properties and plastic strain localization patterns in tensile tested specimens of low-carbon steels have been studied using a double exposure speckle photography technique. It is found that the mechanical properties of low-carbon steels are affected adversely by hydrogen embrittlement. The deformation diagrams were examined for the deformed samples of low-carbon steels. These are found to show all the plastic flow stages: the linear, parabolic and pre-failure stages would occur for the respective values of the exponent *n* from the Ludwik-Holomon equation.

1. Introduction

Previously, we presented experimental data in [1] according to which plastic strain development in solids exhibited a localized character over the entire process. This phenomenon is especially clearly manifested on a macroscopic scale, where the patterns of strain localization are related to the deformation hardening operative on the corresponding stages of straining. Embrittlement due to hydrogen (H) involves a vast loss of mechanical properties with the following characteristics such as, for example, decrease of ductility and fracture tension with the increase of H concentration. As is pointed out in [2, 3], this phenomenon poses a serious practical problem, the solution of which determines the durability and safety of operation of a steel structure.

It was established in [4, 5] that the significant progress in understanding of Hydrogen-Enhanced Localized Plasticity (HELP) mechanism has been recently achieved in studies based on the so-called hydrogen-induced localized softening effect. Main idea of this approach takes into consideration material softening in micro-scale caused by enhanced dislocation mobility in the presence of hydrogen. This phenomenon has been evidenced in a variety of metals with different types of dislocations [6]. The main aim of this investigation was to elucidate the effect of dissolved hydrogen on the macroscopic plastic flow localization patterns in tensile strained steels.

2. Experimental details

The investigations were performed for polycrystals of low-carbon steel (Fe-0.07%C). The samples had a working part with dimensions of 50·5·1 mm and were tensile strained at 300 K on an Instron Model 1185 testing machine at a mobile clamp velocity of 8.3·10-6 m/s. The loading curves of the tested materials are illustrated in Fig. 1. The method for studying the hydrogen diffusivity and the interaction of hydrogen with lattice defects was proposed in [7]. Based on these observation, the steel samples were electrolytically saturated with hydrogen in a thermostatted three-electrode electrochemical cell with graphite anode, operating at a controlled constant cathode potential of U = –600 mV (relative to silver chloride reference electrode) in a 1 N sulfuric acid solution containing 20 mg/l thiourea. The hydrogenation was effected at 323 K for 24 h after preliminary purging the solution with nitrogen. The current–voltage curves were recorded using an IPC-Compact potentiostat. Prior to mechanical testing, the samples were kept in liquid nitrogen.

The stress-strain diagram was obtained simultaneously with measuring the fields of the displacement vectors $\boldsymbol{r}(x, y)$ with the aid of double-exposure speckle photography (DESP) technique. A special device was also designed in [1]; it had field of vision ~100 mm; spatial resolution comparable to optical microscopy ~1…2 μm and real-time mode of operation. This enabled reconstruction of displacement vector fields $\boldsymbol{r}(x, y)$ for the sample surface. On the base of this data, the plastic distortion tensor is evaluated for the deforming sample in the coordinates x, y and z, i.e. longitudinal (ε_{xx}), transverse (ε_{yy}), shear ($\varepsilon_{xy} = \varepsilon_{yx}$) and rotation ($\omega_z$) components. This technique can visualize localized plastic flow nuclei, using the spatial distributions of plastic distortion tensor components; the kinetics of nuclei motion can be determined from the temporal evolution of nuclei.

3. Experimental results

We have studied the evolution of patterns of the plastic flow macrolocalization in low-carbon steel polycrystals in the initial state and upon the saturation of the electrolytic with hydrogen.

The stress–strain (σ–ε) curve of plastic flow measured in tension for low-carbon steel polycrystals in the initial (hydrogen-free) state, exhibit a tooth and a flow trough (I) that extends up to εtot = 0.04 and is followed by the stages of (II) linear deformation hardening with a constant coefficient of θII≈ 920 MPa; (III) parabolic (Taylor's) hardening with a power index of n = ½ and (IV) prefracture with n < 1/2, which extends up to εtot = 0.42 (Figure 1, curve 1).

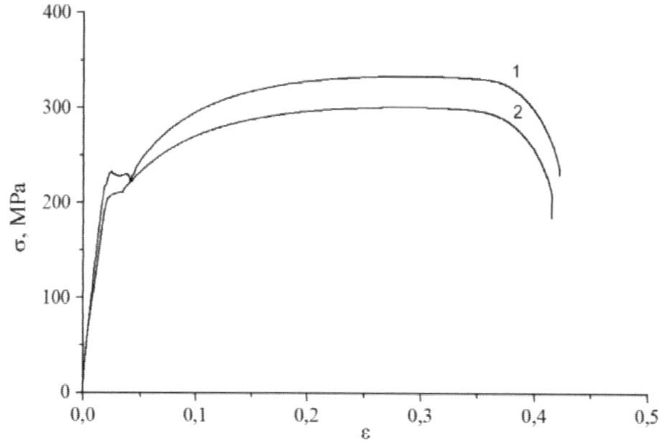

Figure 1. The flow curves of low-carbon steel polycrystals (a): 1 – free of interstitial impurity atoms and 2 – upon the electrolytic saturation with hydrogen at U = –600 mV in 1 N aqueous H_2SO_4 for 24 h at T = 323 K

The σ(ε) curves of low-carbon steel polycrystals saturated with hydrogen exhibit a flow trough (I) with a constant coefficient of $\theta_I \approx$ 450 MPa that extends up to ε_{tot} = 0.05 and is followed by the stages of (II) linear deformation hardening with a constant coefficient of $\theta_{II} \approx$ 730 MPa; (III) parabolic (Taylor's) hardening with a power index of n = 1/2; and (IV) prefracture with n < 1/2, which extends up to ε_{tot} = 0.41 (Figure 1, curve 2). The entire plastic flow curve shifts downward relative to that for the low-carbon steel polycrystals in the initial (hydrogen-free) state. Thus, the hydrogenation of low-carbon steel polycrystals led to a decrease in the yield and strength stresses.

The measurements of local strain distributions using the DESP method showed that the strain is macroscopically localized at all stages of plastic flow in low-carbon steel polycrystals. At the yield plateau Lüders band propagation is observed which separated the strained and unstrained parts of the sample. An analysis of these patterns revealed that, in the initial (hydrogen-free) state of polycrystals, the tensile plastic strain at the stage of yield plateau (I) in the deforming material there travels four solitary fronts of plastic deformation at different constant rates $V_{aw} \sim 10^{-5}$ m/s (Figure 2a).

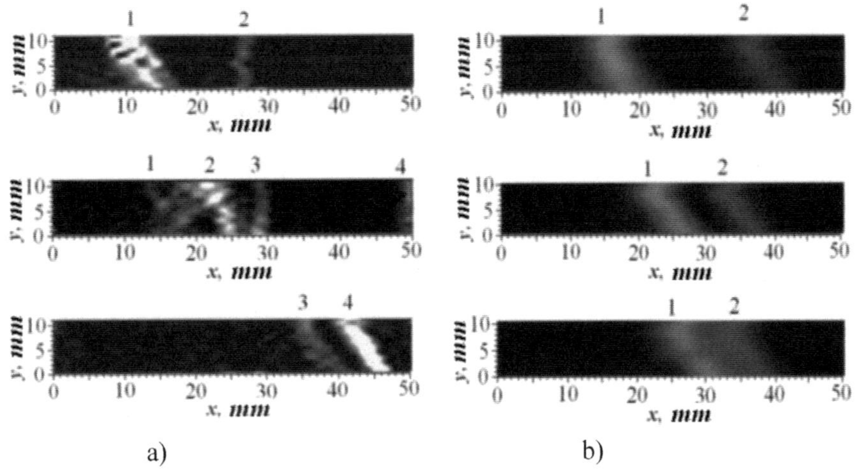

Figure 2. Pattern of strain distribution at the yield plateau (I) in low-carbon steel polycrystals in the initial state (free of interstitial impurity atoms) (a) and upon the electrolytic saturation with hydrogen at $U = -600$ mV in 1 N aqueous H_2SO_4 for 24 h at $T = 323$ K (b) at a total strain of $\varepsilon_{tot} = 0.005\ldots0.03$

The tensile plastic strain at the stage of linear deformation hardening (II) is concentrated within local zones spaced by a distance of $\lambda = 8\pm1$ mm (Figure 3a). These local strain zones (autowaves) moved at a velocity of $5.3\cdot10^{-5}$ m/s, which was determined from the slope of the X(t) curves that plot the positions X of maxima of the ε_{xx} strain component on the axis of straining as functions of time t. At the stage of parabolic deformation hardening (III), the system of equidistant strain localization zones becomes stationary, then these immobile foci of plastic strain localization start to move consistently with a tendency to merge together at the middle of a sample, where the fracture takes place (Figure 4 a).

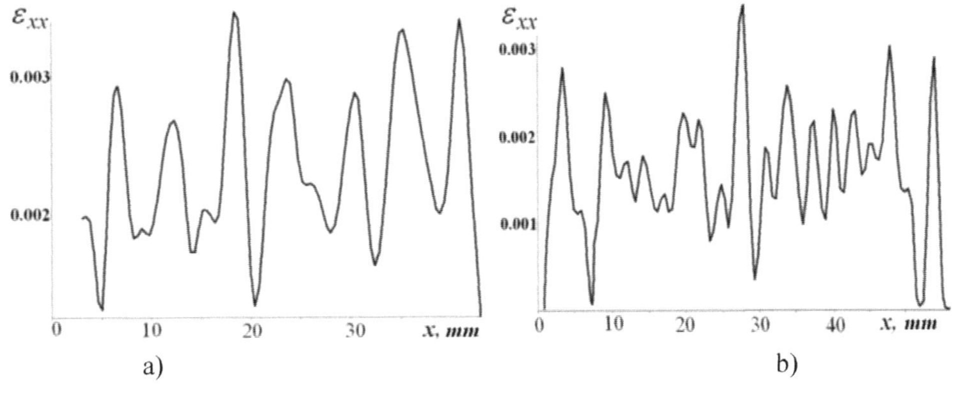

Figure 3. Pattern of strain distribution plotted as the local elongation component $\varepsilon_{xx}(x)$ along axis of tension at stage of linear deformation hardening (II) in low-carbon steel polycrystals in the initial state (free of interstitial impurity atoms) at a total strain of $\varepsilon_{tot} = 0.067$ (a) and upon the electrolytic saturation with hydrogen at $U = -600$ mV in 1 N aqueous H_2SO_4 for 24 h at $T = 323$ K at a total strain of $\varepsilon_{tot} = 0.075$ (b)

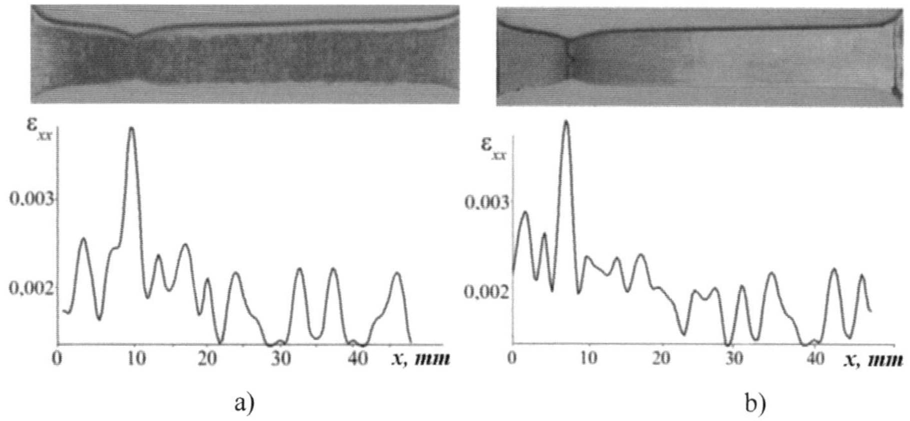

Figure 4. Pattern of strain distribution plotted as the local elongation component $\varepsilon_{xx}(x)$ along axis of tension at the prefracture stage (IV) in low-carbon steel polycrystals in the initial state (free of interstitial impurity atoms) at a total strain of $\varepsilon_{tot} = 0.36$ (a) and upon the electrolytic saturation with hydrogen at $U = -600$ mV in 1 N aqueous H_2SO_4 for 24 h at $T = 323$ K at a total strain of $\varepsilon_{tot} = 0.35$ (b)

An analysis of the distributions of local strain ε_{xx} in tensile tested low-carbon steel polycrystals saturated hydrogen showed that, in this case at the flow trough (I) with a constant coefficient of $\theta_I \approx$ 450 MPa in the material there emerge two deformation fronts moving in opposite directions (Figure 2b). The fronts' motion rates are $1.4 \cdot 10^{-4}$ and $5.5 \cdot 10^{-5}$ m/s. The point, at which the fronts meet, marks off the end of stage I. At the stage of linear deformation hardening (II), the pattern of plastic strain localization represented a set of wide zones, each of which contains two to three related local strain foci. The zones exhibited a characteristic spacing $\lambda = 6.0 \pm 1$ mm (Figure 3b). These local straining zones (autowaves) also moved at a velocity of 8.3·10-5 m/s. At the stage of parabolic deformation hardening (III), the system of strain localization zones became stationary. At the prefracture stage (IV), the immobile foci of plastic strain localization started moving consistently with a tendency to merge into a high-amplitude focus of localized straining (Figure 4 b), where a neck-like narrowing of the sample cross section was formed. According to [8] in the course of deformation the space-time periodic pattern of nuclei is finally replaced by a single localized plasticity nucleus for single crystals of Fe–18Cr–12Ni–2Mo austenite steel upon the electrolytic saturation with hydrogen.

4. Conclusion

Thus, we have elucidated the effect of hydrogen on the macroscopic patterns of plastic strain localization in tensile tested polycrystals of low-carbon steel (*Fe*-0.07%*C*). It is established that the hydrogenation enhances the localization of straining leads to significant changes in the characteristics distances between local straining zones. According to [5] it can be a result of interdislocation interactions and the generation of defects. The mechanism of hydrogen-stimulated plastic strain localization is still under discussion. Therefore, the wave patterns of localized plasticity appear to be useful for a detailed analysis of plasticity exhibited by real metals and alloys. The examination of localized deformation patterns suggests that the location and time of future fracture can be determined long before a macroscopic neck forms in the test sample.

Acknowledgement

The work was performed in the frame of the Tomsk State University Academic D.I. Mendeleev Fund Program and the Program of Fundamental Research of State Academies of Sciences for the period 2013-2020 yrs.

References

[1] Zuev L B, Danilov V I, Barannikova S A, Gorbatenko V V 2009 *Phys. Wave Phen.* **17** 1
[2] Hirth J P 1980 *Met. Mat. Trans. A* **11** 861
[3] Ramunni V P, De Paiva Coelho T, de Miranda P E V 2006 *Mat. Sci. Eng. A* **435–436** 504
[4] Birnbaum H K, Sofronis P 1994 *Mat. Sci. Eng. A* **176** 191
[5] Sofronis P, Liang Y, Aravas N 2001 *Eur. J. Mech. A: Sol.* **20** 857
[6] Robertson I M 2001 *Eng. Frac. Mech.* **68** 671
[7] Yagodzinskyy Y, Todoshchenko O, Papula S, Hänninen H 2011 *Steel Res. Int.* **82** 20
[8] Barannikova S A, Nadezhkin M V, Mel'nichuk V A, Zuev L B 2011 *Tech. Phys. Lett.* **37** 793

Influence of Anodization Time and Voltage on the Parameters of TiO_2 Nanotubes

R V Chernozem[1], M A Surmeneva[1] and R A Surmenev[1]

[1] Department of Experimental Physics, National Research Tomsk Polytechnic University, Tomsk, Lenina Avenue 30, Russia

Abstract. A vertically aligned titania nanotube layer was obtained by electrochemical anodic oxidation in the electrolyte contained 0.4 wt% solution of NH_4F in 54 ml of ethylene glycol and 5 ml of deionized water, after titanium was chemically cleaned/etched with a mixture of HCl, H_2O and HNO_3 solution for removing the natural oxide films. The morphology and composition of the titania nanotube layer were examined by scanning electron microscopy (SEM) and energy-dispersive X-ray spectroscopy (EDS). The anodization of TiO_2 nanotubes was done using 60 V for 240 min and 30 min, and 30 V for 30 min. The diameter of the titania nanotubes was about 52–156 nm, the wall thickness about 32-53 nm and the height about 0.9–6.3 μm. The pore size of TiO_2 nanotubes influences the dissolution rate of CaP thin films and Young's modulus, which is significantly lower than that of the Ti substrate. Our future challenge will be investigation of the microstructure and mechanical behavior of titania nanotubes with CaP film.

1. Introduction

Titanium and its alloys are increasingly used in orthopedic and dental implants due to their excellent mechanical properties, high corrosion resistance, and good biocompatibility. However, titanium is a bioinert material, therefore, investigation of hybrid bioactive coatings formation on the Ti surface is required [3]. Titanium nanotubes can be used as one of the components for these coatings.

The advantage of using TiO_2 nanotubes is that they can be grown directly on the Ti surface by a cost-effective technique as anodic oxidation. In addition, the nanoindentation results indicate that the Young's modulus of the TiO_2 nanotubes is significantly lower than that of the Ti substrate. Also, it is well known that important disadvantage connected with hydroxyapatite (HA) coating is poor adhesion strength at the HA/Ti interface. It is suggested that TiO_2 nanotubes with a 3-D micro/nonporous structure may enhance apatite formation when compared to dense TiO_2 [2].

There are a few methods for HA coatings fabrication on the surface of titanium oxide nanotubes. Various methods are available to prepare such coatings. Radio frequency (RF) magnetron sputtering has several advantages. It can produce uniform and reproducible thin films with the properties that can be tuned by changing the operational parameters. Whereas sputter deposited calcium phosphate (CaP) coatings have much to offer in this regard, in the as-deposited state they are inherently amorphous and readily undergo rapid dissolution [1]. The authors of the study [1] reported that the nanotubes parameters can affect the dissolution rate of CaP films deposited on the nanotubes arrays and attendant cell behaviour.

To our knowledge, a systematic investigation of the microstructure and mechanical behaviour of self-assembled, vertically aligned nanotubes coated with HA film has not yet been reported. The preparation of the vertically aligned titanium nanotubes is needed for a design of the hybrid HA/TiO_2 coating on Ti substrate.

The purpose of this study is to obtain titania nanotubes with different dimensions and investigate the effect of anodization parameters, such as anodization time and voltage, on the titanium nanotubes dimensions and growth rate in a NH_4F electrolyte solution.

2. Materials and methods

2.1. Fabrication of titanium nanotube layer

TiO_2 nanotubes were fabricated using the technique of anodization. The anodization process involved is as follows: two electrode systems comprising Ti foil (2 x 5 cm) as the anode and cathode. The samples were chemically cleaned/etched with a mixture of HCl, H_2O and HNO_3 solution for removing the natural oxide films, in the proportion of 1:1:2 respectively, and after that the samples were rinsed in the deionized water and dried in air.

The electrolyte contained 0.4 wt% solution of NH_4F in 54 ml of ethylene glycol and 5 ml of deionized water. The anodization was performed using constant current voltage power. The distance between the anode and cathode was 2.5 cm. After anodization all the samples were washed in de-ionised water and dried in the environment prior to further investigation.

2.2. Characterization

The morphological study of non-detached titania nanotubes was done using a field emission scanning electron microscope (SEM, Quanta 200D). The length, inner diameter and wall thickness of the tubes were measured by segmenting SEM micrographs and using image analysis software (xT microscope Control). Micrographs for quantitative analysis were taken from at least three different locations on each sample. Characterization of the coating/substrate cross sections was carried out by fracturing the coating by bending, and examination of the detached TiO_2 coating. SEM was used to examine both the bottom of the coating as well as the Ti substrate. The elemental analysis was performed using energy-dispersive X-ray spectroscopy (EDXS, Quanta 200D). The images are presented in Figures 1-3.

3. Results and discussion

Figure 1, 2 and 3 reveal a top view and cross-sectional SEM images of TiO_2 nanotubes anodized at 60 V for 240 min, 30 min and at 30 V for 30 min, respectively, in NH_4F -containing electrolyte. A uniform structure of thin-walled TiO_2 nanotubes was obtained. The anodization resulted in vertically oriented titania nanotubes arrays.

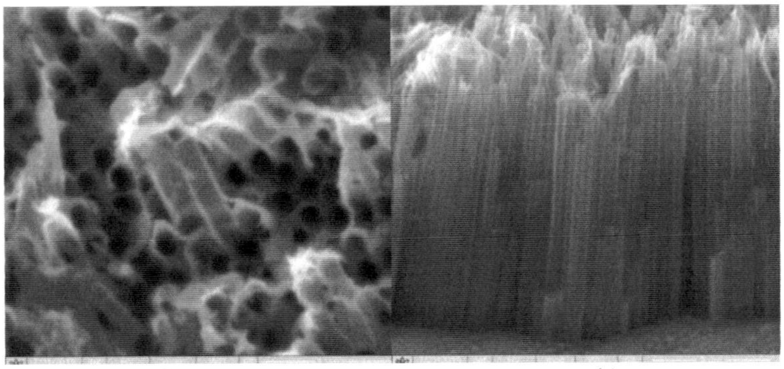

a) b)

Figure 1. (a) Top view and (b) cross-sectional SEM images of TiO_2 nanotube arrays formed by anodic oxidation of Ti in NH_4F - containing electrolyte, at the potential of 60 V for 240 min

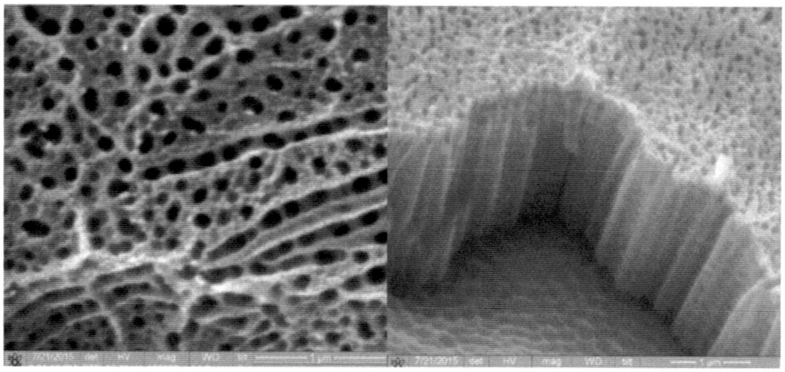

a) b)

Figure 2. a) Top view and (b) cross-sectional SEM images of TiO_2 nanotube arrays formed by anodic oxidation of Ti in NH_4F -containing electrolyte, at the potential of 60 V for 30 min

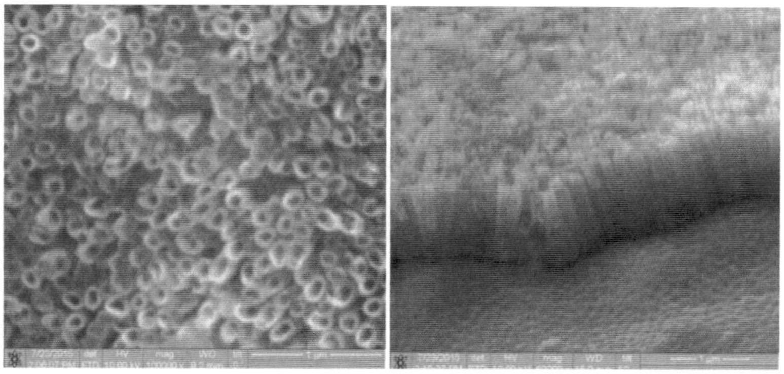

a) b)

Figure 3. a) Top view and (b) cross-sectional SEM images of TiO_2 nanotube arrays formed by anodic oxidation of Ti in NH_4F -containing electrolyte, at the potential of 30 V for 30 min

Table 1 shows the quantitative measurements of the important dimensions of the nanotubes. An increase in anodization time and voltage had a significant effect on nanotubes diameter, wall thickness and length. An increase in tube dimensions with increasing in the anodization time, from 30 min to 240 min or increasing anodization voltage, from 30 min to 60 V, was observed. The alteration of the corresponding length (0.9–6.3 μm) and wall thickness (32–52 nm) of the nanotubes was achieved by varying the anodization time (30–240 min) and voltage (30–60 V).

Table 1. Dimensions of TiO_2 nanotubes fabricated via anodic oxidation of Ti in NH_4F-containing electrolytes.

Anodization voltage (V)	Anodization time (min)	Nanotubes dimensions (nm)		
		Tube length	Inner diameter	Wall thickness
60	30	1410±60	98.23±15.92	52.43±25.62
60	240	6270±360	155.95±27.96	32.78±7.89
30	30	909±50	52.96±10.12	50.70±10.50

SEM images of the cross-sections allowed to reveal the coating thickness. It appeared that there was a significant increase in the coating thickness with time between 30 and 240 min. However, the results showed that the average growing rate for the anodization time of 30 and 240 min is 47.0 and 26.1 nm per min, respectively. But the average growing rate for the anodization voltage of 30 and 60 V was 30.1 and 47.0 nm per min, respectively. The growth process of the nanotubes was discussed and several possible mechanisms for the formation of nanotubes were proposed in [2], [5] and [6].

The authors of the paper [2] showed that pore size influenced the mechanical properties and elastic modulus. Nanoindentation results yielded progressively higher values of elastic modulus for thinner films [2].

Mutreja et al. showed that prolonged availability of the as-deposited CaP can be provided by the use of a substrate surface comprising titania nanotubes. Moreover, nanotubes pore size and length affected the rate of CaP dissolution and attendant cell behaviour. The authors of the study [1] reported that the highest levels of ALP activity of osteoblast measured at 21 days occurred on CaP thin films deposited onto titania nanotubes with an average pore size of 30 ± 7.5 nm and tube length of 1.3 ± 0.2 μm.

EDS pattern of TiO_2 nanotubes prepared at the electron energy of 15 keV is shown in Figure 4.

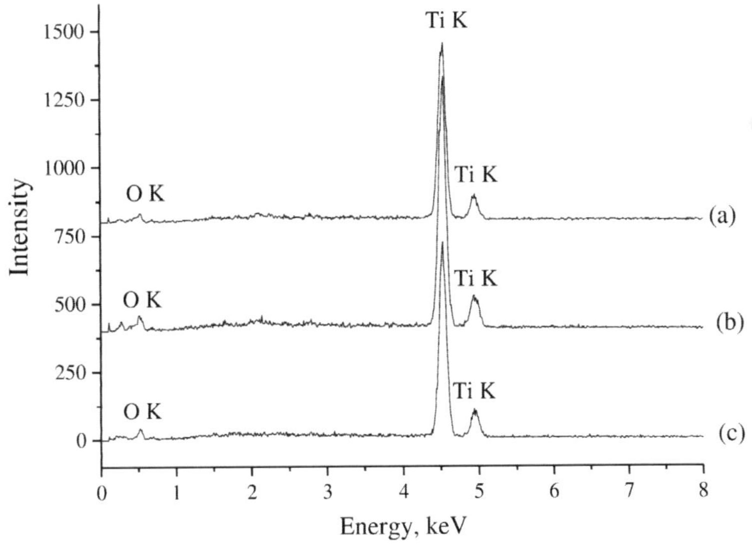

Figure 4. EDS pattern of titanium nanotubes were obtained at: a) 60 V for 240 min, b) 60 V for 30 min, c) 30 V for 30 min.

4. Conclusion

The electrochemical anodization method allows to produce amorphous TiO_2 nanotubes. Annealing results in their transformation into crystalline nanotubes. Based on a comprehensive study of the surface features prepared using a range of processing voltage and time settings, nanotubes with three significantly different pore sizes and lengths were obtained by performing electrochemical anodization at 60 V for 240 min, 30 min and 30 V for 30 min, respectively.

The quantitative microstructure characterization and elemental composition of TiO_2 nanotubes on Ti substrate are reported. The following conclusions can be drawn:

1 Nanotubes were prepared using anodic oxidation, in the NH_4F-containing electrolyte.
2 The dimensions of titania nanotubes increased with increasing in the anodization time. However, the average growth rate decreased with increase in the anodization time.

3 Increase in the anodization voltage leads to the increase in the nanotubes dimensions and their growth rate.
4 The variation of anodization conditions did not affect the chemical composition of TiO_2 nanotubes.

The investigation of the microstructure and mechanical behavior of vertically aligned titania nanotubes coated with HA films will be investigated in the future.

Acknowledgements
This research was supported by the Russian President's Stipend SP-6664.2013.4, grant MK-485.2014, the State order NAUKA #11.1359.2014/K and government contract # 14.587.21.0013.

References

[1] Mutreja I, Kumar D, Boyd A R and Meenan B J 2013 Titania nanotube porosity controls dissolution rate of sputter deposited calcium phosphate (CaP) thin film coatings *RSC Adv.* **3** 11263-11273
[2] Crawford G A, Chawla N, Das K, Bose S and Bandyopadhyay A 2007 Microstructure and deformation behavior of biocompatible TiO2 nanotubes on titanium substrate *Acta Biomater.* **3** 359–367
[3] Feng B, Chu X, Chen J, Wang J, Lu X and Wneg J 2010 Hydroxyapatite coating on titanium surface with titania nanotube layer and its bond strength to substrate *J. of Porous Mater.* **17** 453-458
[4] Seung-Han Oh, Finones R, Daraio C, Chen L and Jin S 2005 Growth of nano-scale hydroxyapatite using chemically treated titanium oxide nanotubes *Biomaterials.*
[5] Skokuhfar T, Sinha-Ray S, Sukotjo C, Yarin A L 2013 Inercalation of anti-inflammatory drug molecules within TiO_2 nanotubes *RSC Advances* **3** 1730-1738
[6] Macak J, Tsuchiya H, Schmuki P 2005 High-Aspect-Ratio TiO2 Nanotubes by Anodization of Titanium *Angewandte Chemie International Edition.* **44** 2100-2102

IOP Conf. Series: Materials Science and Engineering **116** (2016) 012026 doi:10.1088/1757-899X/116/1/012026

The study of the thermal stability of Ni_3Al nanoneedles using computer simulation

G M Poletaev[1], M D Starostenkov[1], D V Novoselova[1] and V Y Tsellermaer[2]
[1]Altai State Technical University, Lenin st. 46, Barnaul, 656038, Russia
[2]Siberian State Industrial University, Kirov st. 42, Novokuznetsk, 654007, Russia

E-mail: gmpoletaev@mail.ru

Abstract. The study of the thermal stability of Ni_3Al nanoneedles in the dependence on the sharpness and crystallography orientation of the needle axis was held using molecular dynamics method. It is shown that the most stability nanoneedles in the conditions of thermal influence are the nanoneedles with the axis along [111] direction. Violation of the nanoneedles stability associated with their blunting, which is caused by surface diffusion. Stable radius of curvature of the needles depends on the angle of the tip and orientation of the needle axis. It is 0.6-1 nm at low temperatures.

1. Introduction

At present, the direction of researches related to the development of new materials and nano-devices have been allocated to one of the most important problems of modern materials science. The sizes of functional devices that are comparable in size to the ensemble of atoms or molecules, their properties are defined not only by physical and chemical characteristics, but also by their geometrical dimensions, structure and form. The influence of interfaces (free surface, interphase and grain boundaries) is growing with a decrease in the size of the structural elements of material.

One of the examples of such nanoobjects is needle-like nanocrystals of refractory metals which are used, for example, as probes for high-resolution scanning tunneling microscopes, autoemission cathodes for vacuum electronics. Because of a small size, as other nanoobjects, the nanoneedles have a very high proportion of free surface, i.e. the percentage of atoms contained in the surface layers with more potential energy than atoms in the volume of the material. Furthermore, the needles have a high curvature of the surface. Due to these two factors, the form of the nanoneedle is relatively unstable under the conditions of thermal influence - the activation energy of surface diffusion is low [1, 2], so that the process of blunting of the nanoneedle intensively occurs at the increase of temperature.

This work is devoted to the study of the thermal stability of the nanoneedles of Ni_3Al in the dependence on the crystallographic orientation of the axis of needles and its severity. Choosing Ni_3Al intermetallic with $L1_2$ ordered structure is connected with its unique properties: high melting point, low density, heat resistance, high corrosion resistance [3]. The main issue decided in this paper was to find the most stable form and structure of the Ni_3Al nanoneedle.

[1]gmpoletaev@mail.ru

IOP Conf. Series: Materials Science and Engineering **116** (2016) 012026 doi:10.1088/1757-899X/116/1/012026

2. Description of the model

The simulation was held using the method of molecular dynamics. The needle was created as a cone (Figure 1), which was cut from the Ni_3Al ideal crystal. Three orientation of the axis of the nanoneedle were taken into consideration: [100], [110] and [111] (it is known that the greatest strength of crystals is implemented along the directions with low indexes). The main variable parameters were two-dimensional angle of the needle tip Θ and height of the needle H (Figure 1a). Figure 1 shows examples of the starting structure of the nanoneedles with the tip angle of 45° and the axes along the [100] (Figure 1a) and [111] (Figure 1b) directions.

Several atomic planes in the base of the cone were rigidly fixed, i.e. were motionless during molecular dynamics experiments (in Figure 1 fixed atoms are painted in dark gray color). It was necessary to hold the base form of the needle. There are no conditions to be imposed on other atoms, i.e. side surfaces of the needles were free.

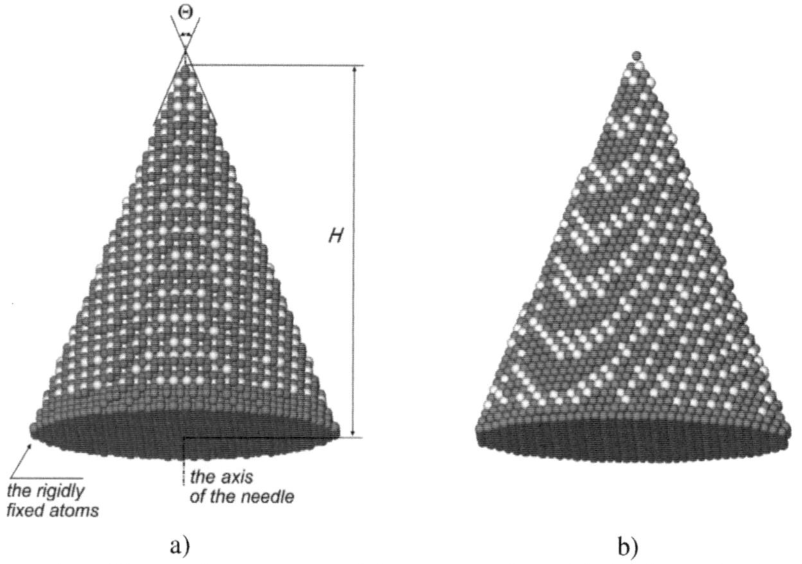

Figure 1. Examples of the starting structure of the nanoneedles with the tip angle of 45° and the axes along the [100] (a) and [111] (b) directions. Ni – dark atoms, Al – light ones. Dark gray atoms in the base of the cones are the atoms which during the computer experiment remained motionless.

The pair Morse potential function was used for a description of the interatomic interactions in the work. Parameters of the potentials were taken from [4]. Application experience of these potentials shows that with their help it is possible to describe various properties of metals and alloys [5-8]. Potential parameters describing the interaction of atoms of one type (Ni-Ni and Al-Al) in [4] were determined from the properties of the pure metals by the values of the sublimation energy, lattice parameter and bulk modulus. While determining the parameters of the atoms interaction of different types (Ni-Al) as an energy characteristic the values of the energies of formation of shear antiphase boundaries in Ni_3Al intermetallide were used. These Morse potentials satisfy the condition of the equilibrium of the alloy lattice, describe the anisotropy of the energies of antiphase boundaries in a proper way and the basic property of intermetallic is to keep the order until the melting temperature [4].

The temperature in the model was set by the initial velocity of the atoms according to the Maxwell-Boltzmann distribution [6]. In this case the total kinetic energy was corresponded to predetermined temperature, and the total momentum of the calculation block was zero. The time step in the molecular dynamic experiments was 5 fs ($5 \cdot 10^{-15}$ s).

3. Results and discussion

Stability of the nanoneedles, as mentioned above, was investigated for three orientations of the axes: [100], [110] and [111]. Two-dimensional angle of the tip was varied from 30° to 90°. Thermal stability was determined by the dependence graph of the nanoneedle temperature *T* on the time of computer experiment *t*. Structural transformation, consisting in the needle blunting, was accompanied by an temperature increase : the more intensively the blunting occurred, the more intensively the temperature raised.

Figures 2-4 shows changes in the shape of the needles with the tip angle of 30°, 45°, 60° in the process of the computer experiment during 50 ps and the corresponding dependences of the temperature on the time *T(t)*. The figures present results only for the needles with the axes along [100] and [111] directions. Needles with the [110] axis were significantly less stable than the [100] and [111] needles for all considered angles of the tip. Blunting of [110] needles took place without any additional initiation (without thermal influence), even for large values of the tip angle Θ.

At the tip angle of 30° the needles with the axes along the [100] and [111] directions also became blunt spontaneously without additional initiation. In Figure 2 it can be seen that during the whole 50 ps computer experiment temperature increase occurred: intensively at first, then more smoothly. Moreover, the relatively larger increase in temperature and change in the needle shape were observed in the case of the needle with the axis along [100]. Needles with the axis along [111] were the most stable, as with other values of the tip angle.

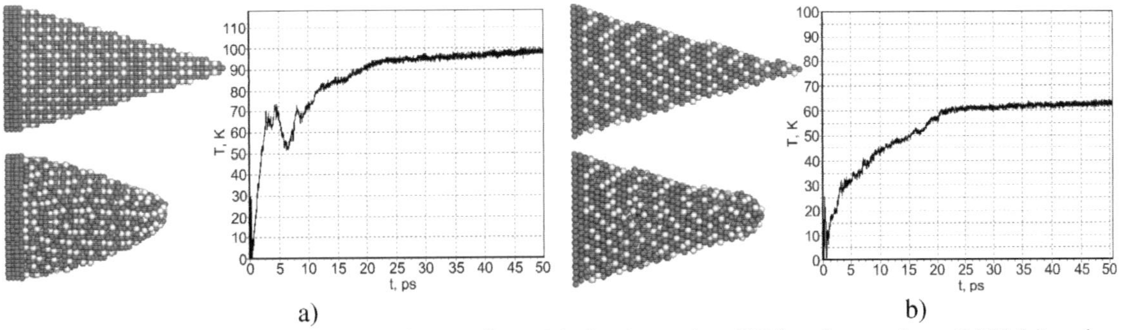

Figure 2. Changes in the shape of the needles with the tip angle of 30° and axes along [100] (a) and [111] (b) directions in the process of the computer experiment during 50 ps and the corresponding dependences of the temperature on the time *T(t)*.

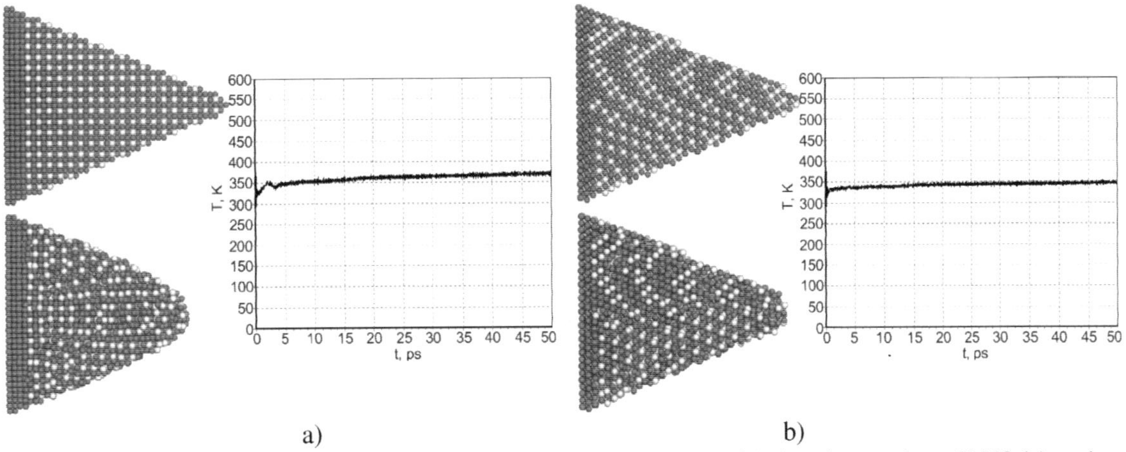

Figure 3. Changes in the shape of the needles with the tip angle of 45° and axes along [100] (a) and [111] (b) directions in the process of the computer experiment during 50 ps and the corresponding dependences of the temperature on the time *T(t)*.

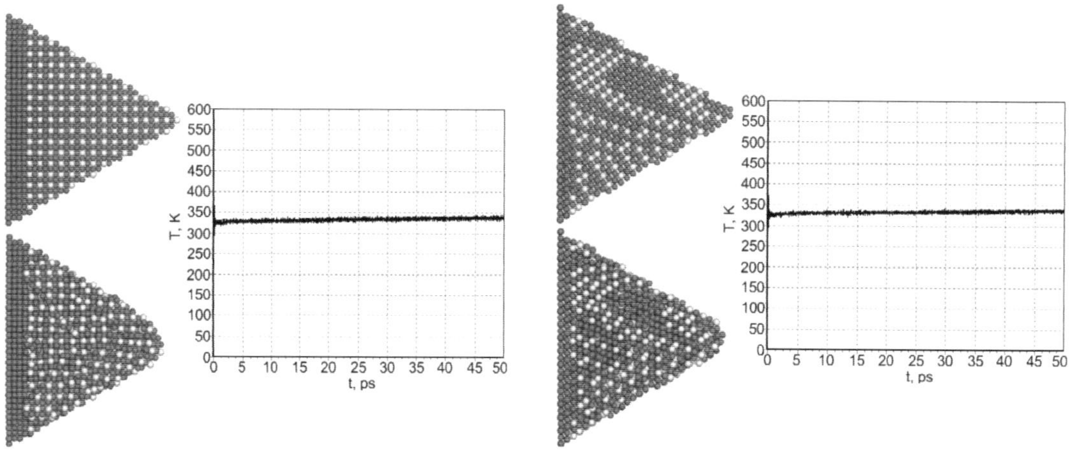

a) b)

Figure 4. Changes in the shape of the needles with the tip angle of 60° and axes along [100] (a) and [111] (b) directions in the process of the computer experiment during 50 ps and the corresponding dependences of the temperature on the time *T(t)*.

With the growth of the tip angle Θ, as expected, the stability of the needle increased. Figure 3 and Figure 4 depict changes in the shape of the needles for the angles 45° and 60° at the starting temperature of 300 K, i.e. at room temperature. In the case of the [100] 45° needle (Figure 3) the needle shape is still unstable – the temperature increase temperature and needle blunting were observed during the whole computer experiment, but in other cases the temperature has not been changed. Some blunting, especially at the beginning of the simulation, it took place, but then it (as well as surface diffusion) was relatively slow, due to what can be said about the relative stability of these forms of the needles. First needles blunting with the angle Θ=60° to the radius of curvature of about 0.7-1 nm remained unchanged in subsequent molecular dynamic experiments within 50 ps at higher temperatures: for needles with the [100] axis up to 500 K for [111] axis up to 800 K.

About a perfect stability of the needles can speak clearly only when considering the already slightly blunted needles to the radius of curvature of about 0.6-1 nm at low temperature conditions, such as, for example, the needles depicted in Figure 4. Stability and sharpness of the needle with considered sizes at normal temperature cannot be guaranteed. The activation energy of surface diffusion is comparatively very low [1, 2], it proceeds intensively even at normal temperatures. In this connection, the radius of the needle tip for industrial cantilevers for scanning microscope is in the range of 5-90 nm, laboratory - a maximum from 1 nm [9, 10].

The smallest radius of curvature of the needles (i.e. minimum blunting) was observed at the simulation of the needles with the tip angle close to the "geometrically ideal", i.e. when the tip was created on the basis of the pyramid that is the following atomic plane was created by one monatomic step less than the previous. For the needles with [100] axis the tip angle is 90°, for [111] – 78.37°. Such needles are the most stable and resistant to the thermal influence.

4. Conclusion

Using the molecular dynamics simulations it is shown that the most stability nanoneedles in the conditions of thermal influence are the nanoneedles with the axis along [111] direction. Needles with the [100] axis are relatively less stable, and needles with [110] axis are the least stable of the considered ones. In all cases with the growth of the angle of the tip the stability increased. It was found that the tip angles corresponding to the highest thermal stability are close to the "geometrically ideal" angles, when the following atomic plane was created by one monatomic step less than the previous one. For the needles with [100] axis is 90°, for [111] – 78.37°. In this case the surfaces of the needles have fewer movable atoms which are capable of participate in the surface diffusion. Violation of the nanoneedles stability associated with their blunting, which is caused by surface diffusion, which in its turn, is more intensive with increasing in density of monatomic steps and breaks on them at the free surfaces of the nanoneedles. Due to the active process of surface diffusion it is practically very difficult to achieve the atomic sharpness of the nanoneedles. According to the results of this work, stable radius of curvature of the needles at low temperatures, depending on the angle of the tip and orientation of the needle axis, is 0.6-1 nm. In practice [9, 10] the nanoneedles laboratory prepared with a radius of curvature from a few nanometers.

Acknowledgment
The reported study was partially supported by Ministry of Education and Science of the Russian Federation, project No.166 of program "Formation of state tasks to higher education institutions in terms of scientific research", and by RFBR, research projects No. 13-02-00301_a, No. 14-02-98000-r_sibir_a and No. 14-08-90416-Ukr_a.

References
[1] Oura K, Lifshits V G, Saranin A A, Zotov A V, Katayama M 2006 *Introduction to the surface physics* (Moscow: Nauka) (in Russian)
[2] Pratton M 2000 *Introduction to the surface physics* (Izhevsk: SRC "Regular and chaotic dynamics") (in Russian)
[3] Druzhkov A P, Perminov D A, Stepanova N N 2010 *Physics of the Solid State* **52** 2005
[4] Tsaregorodtsev A I, Gorlov N V, Demyanov B F, Starostenkov M D 1984 *Fizika Metallov i Metallovedenie* **58** 336 (in Russian)
[5] Starostenkov M D, Sinyaev D V, Rakitin R Y, Poletaev G M 2008 *Solid State Phenomena* **139** 89
[6] Poletaev G M, Starostenkov M D 2010 *Physics of the Solid State* **52** 1146
[7] Rakitin R Yu, Poletaev G M, Aksenov M S, Starostenkov M D 2005 *Technical Physics Letters* **31** 650
[8] Poletaev G M, Dmitrienko D V, Diabdenkov V V, Mikrukov V R, Starostenkov M D 2013 *Physics of the Solid State* **55** 1920
[9] Mazeran P E, Odoni L, Loubet J L 2005 *Surface Science* **585** 25
[10] Labuda A, Grutter P H 2011 *Review of scientific instruments* **82** 013704

Metal-ceramic composite development based on its modelling results

E S Dvilis, O L Khasanov, A O Khasanov, M S Petyukevich
Department of nanomaterials and nanotechnologies, National Research Tomsk Polytechnic University, 30 Lenin Ave., Tomsk, 634050, Russia.

E-mail: khasanov@tpu.ru

Abstract. The modeling (and its experimental verification) of packing and deformation of the composites consisted of aluminum-magnesium alloy AMg6, B_4C powder and W nano-powder has been performed. The powder compositions were determined using discrete element modeling of the composite particles packing based on the particle size distribution functions of real powders. The models of maximum mixture packing densities have been rendered.

1. Introduction

The development of nuclear energy applications, space exploration, and the use of radiation in medicine depends on the solution of various problems of the protection of structural materials, electronic components and products, and biological tissues from their exposure to ionizing radiation. Therefore, the design and development of new materials to improve the effectiveness of anti-radiation protection is an important modern area of research. It includes the reduction of the weight and dimensions of radiation-shielding materials.

Light metal alloys have good mechanical properties, but when subjected to radiation, they are susceptible to swelling and structural changes. These defects can be prevented by creating metal-matrix composites containing particulate fillers with the required radiation-shielding properties. The use of radiation-absorbing nanoparticles of heavy metals and ceramic materials as fillers dispersed within metal-matrix composites improves these composites' absorption properties with respect to the neutron, gamma and X-rays [1, 2].

It is shown in [3] that the use of Al/Mg matrix filled with nano-scale particles of radiation-absorbing ceramic and metallic materials (BN, B_4C, Pb, and W) increases the rate of neutron absorption factor by 50% and the gamma radiation scattering coefficient by 30-40%.

To manufacture this kind of a composite various approaches have been used [4, 5]. In particular, the method of hot extrusion has been applied [3]. The composite included 60-65wt% of aluminum-magnesium alloy (grades AMg6 or V95), 15-20wt% of W nano-powder, and B_4C (or BN) – the rest.

The present work objectives include the modeling (and its experimental verification) of the process of the consolidation of the composites consisted of aluminum-magnesium alloy AMg6 (65 wt%), B_4C powder (15 wt%) and W nano-powder (20 wt%), as well as the optimization of the composite content to achieve higher density.

2. Powder materials and experimental procedures

The characterization of initial powders and their mixtures has been performed by SEM and EDS analysis (JSM-7500F JEOL; LEO EVO 50 Zeiss), XRD analysis (XRD-7000S Shimadzu), particle size distribution analysis using laser diffraction technique (SALD-7101 Shimadzu), and specific surface analysis by BET method (Sorbi META).

The powders were mixed by ball milling in a roller mill using zirconia milling balls. The densification behavior of the dry mixed composite powder has been analyzed using a computer controlled hydraulic precision press (IP-500M-auto ZIPO). The consolidation of the mixed composite powder was conducted using pressureless sintering in a vacuum furnace VHT 8/22GR Nabertherm. The sizes of green samples were 15 mm in diameter and 3-6 mm in height. The mechanical properties of the consolidated specimens were determined by nanoindentation (DUH 211S Shimadzu) of the polished surfaces of sintered samples.

AMg6 powder: the commercial magnesium aluminum alloy powder AMg6 (AA5556 alloy analogue) consists of particles with a shape close to spherical, with a wide spread in particle sizes from 1 μm to 70 μm ($d_{50} \approx$ 15 μm). The conducted XRD and EDS analyzes of the AMg6 powder indicate less than 2% impurities. The broadening of the XRD reflexes confirms the presence of submicron fractions in the powder. The degree of agglomeration of the powders has been assessed in terms of the ratio of the mean particle size obtained by laser diffraction method and the average particle size obtained by the BET method. AMg6 powder agglomeration degree is equal to 2 (Table 1), *i.e.*, this powder was weakly agglomerated.

B_4C powder: the commercial boron carbide powder produced by "OKB-BOR" (Russia) has a nearly equiaxed particle shape with a wide spread in particle sizes from 0.1 μm to 10 μm ($d_{50} \approx$ 1.22 μm). The amount of fraction of nanoparticles up to 10% has been determined. The conducted XRD and EDS analyses did not reveal any impurities in the powder. B_4C powder agglomeration degree is equal to 4.43 (Table 1).

W powder: tungsten powder was prepared by electric explosion of tungsten wire [12]. The powder particles have a spherical shape and wide bimodal size distribution (from 0.05 to 0.4 μm and from 0.4 to 9 μm). The powder contains 20% nano-sized (80-200 nm) and 20% submicron (200-900 nm) fractions, and therefore the degree of agglomeration of the W powder is 9.3 (see Table 1). No impurities in tungsten powder were detected.

The summary of the dimensional characteristics of the starting powders is provided in Table 1. The analysis of the obtained data indicates that initial raw powders have fairly wide particle size distribution, but their mixture can be optimized with respect to the concentration of the components to form a continuous alloy matrix and to increase the packing density of the particulate filler. Given the morphological characteristics of the powders, the maximum density of packing can be expected when filling the space between tightly packed particles of the matrix alloy by a mixture of boron carbide particles with tungsten particles' agglomerates. Parameter d_m in Table 1 has been determined for spherical particles using specific surface value S_{BET} (BET data) and theoretical density ρ, for respective materials [6]:

$$d_m = \frac{6}{\rho S_{BET}} \quad (1)$$

Table 1. Average particle/agglomerate sizes of the components in the powder mixture

Component	Average size (micron) of structural elements according to different analysis				Agglo-meration degree
	XRD (crystallite size)	SEM (agglomerate size)	Laser Diffraction (agglomerate size)	BET (particle size d_m)	
AMg6	0.2	15.4	13.6	6.8	2

B_4C	0.04	1.8	1.22	0.28	4.43
W	0.09	1.1	1.07	0.12	9.3

3. Optimization of powder mixtures

Nearly spherical and equiaxed particle shape of the used powders enables the modeling of their packing and optimization of their mixing ratio in terms of the values of the average number of inter-particle contacts (the coordination number N_c) and of the packing density. For this purpose, the particle size distributions obtained by laser diffraction have been approximated by matching functions. These functions were used to define the relevant classes of particles in the framework of the discrete element model environment of the S3D PorouStructure^{TM} software utilizing the Ichikawa algorithm (with central packing). The total number of particles in a representative model set (Figure 1) varied from 20000 to 72000.

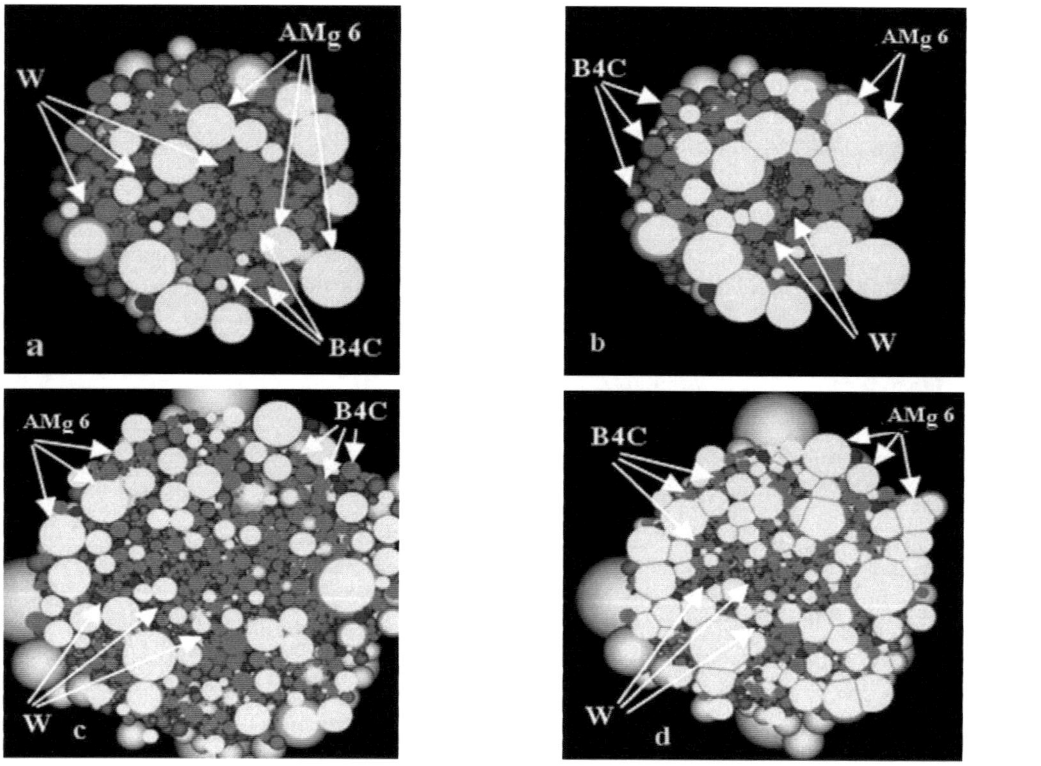

Figure 1. Models of packing for the composites with minimum permissible (a, b) and the optimal density of the matrix alloy and particulate fillers, (c, d) in a free-laying state (a, c) and after plastic deformation of the particles (b, d): AMg6 particles – blue; B4C - gray; W – dark blue.

The comparative analysis of the results of the modeling of the packing of the composite structure AMg6 (65 wt.%) + B_4C (15 wt.%) + W (20 wt.%) was carried out using S3D Evolution^{TM} code. For this simulation, the compaction parameters were associated with the deformation of the particles of the matrix alloy and tungsten, subjected to pressures from 40 MPa to 800 MPa (including at temperatures up to 600 °C). The particles of boron carbide were considered to be non-deformable (Figure 1b, d).

The conducted packing's statistical analysis based on the criteria of the maximum density and of the ensured continuous contact of the aluminum matrix particles showed the following. The minimum composition forming a contiguous aluminum alloy matrix rendering the optimal dense filling of the dispersed particles of tungsten and boron carbide was found to be 65% AMg6 + 15% B_4C + 20% W (Figure 1a, b). For such a mixture in the initial underformed packing state, the alloy particles' average partial coordination number (the number of contacts between alloy particles only, excluding tungsten and boron carbide particles) is in the range of the minimum allowable values ($1 < N_c < 2$), and the space between them can be filled by the optimal content of tungsten and boron carbide particles with a predetermined size distribution. However, the mixing ratio rendering the maximum packing density of the particles has a different composition: 74% AMg6 + 6% B_4C + 20% W (Figure 1c, d). For this composition, the partial coordination number of the matrix alloy particles ranges from 3 (in a loose powder state) to 4 (after plastic deformation of the particles), and the partial density of the alloy particle packing varies from 43% (Figure 1c) to 67% (Figure. 1d), respectively.

The quantitative results of the modeling for the two above-mentioned mixtures are presented in Table 2. According to the table, the increase in the partial coordination number from 1.78 to 4.29 in a continuous matrix AMg6 can be achieved by increasing the content of the powder alloy by 9 wt.% (from 65% to 74%). In this case, the strength of the composite can be increased by at least 2.5 times, assuming that the matrix material is the only component in the consolidated mixture capable of bearing mechanical load, and the strength of the porous material is directly proportional to its density. An improvement of the strength of such a composite is also possible when using a fine powder of boron carbide which would effectively fill the space between the densely packed particles of the matrix alloy.

Table 2. Parameters of deformed packed structures of composite's particles

Characteristics of composite with optimal particles density packing and minimum possible content of the matrix alloy (after deformation 25%)					
Component (theoretical density, g/cm^3)	Content,% wt	Content,% vol	Partial packing density	Average coordination number for the component particles	Partial coordination number for particles in the powder component
AMg6 (2.6)	65	77.9	0.276	95.04	1.78
B_4C (2.52)	15	18.8	0.498	6.23	4.42
W (19.25)	20	3.3	0.057	5.17	0.83
Mixture (3.14)	100		0.782	6.49	6.49
AMg6 (2.6)	74	89.2	0.67	124.7	4.29
B_4C (2.52)	6	7.5	0.254	6.34	3.28
W (19.25)	20	3.3	0.052	5.23	1.54
Mixture (3.14)	100		0.89	7.23	7.53

4. Analysis of composite material consolidation

The powder compressibility characterization has been performed by plotting the compaction curves *ρ(P)* obtained for the conditions of uniaxial quasi-static pressing with loading-reloading cycles and using an approximation by a logarithmic equation in a dimensionless form [7, 8]:

$$\rho = b \cdot \ln\left(\frac{P}{P_{\mathrm{cr}}}\right) + 1 \tag{2}$$

where ρ is the relative density of the green body; P is the applied compaction pressure; P_{cr} is the critical pressure (pressure enabling compaction up to full density of the powder material); b is a material constant defining densification pressure sensitivity. Due to the limitations on the values of the relative density ($0 < \rho \le 1$), equation (2) is valid for a limited range of the values of the applied pressure: $\frac{P_{cr}}{\exp(1/b)} < P \le P_{cr}$.

The results presented in Figure 2 (with approximation certainty not less than 99.9%) show that, for the conventional quasi-static pressing in a rigid die, the critical pressure value P_{cr} for the tested composite is 5.219 GPa, which exceeds the strength limit of hardened steel – the structural material of the pressing die. While the AMg6 alloy powder without a particulate filler may be densified at low temperatures down to a non-porous state under the critical pressure of 1.262 GPa using a special tungsten carbide die of the collector type [8], the achievement of the critical pressure of 5.219 GPa for pressing bulk components of complex shapes made of the studied composite material appears to be impossible.

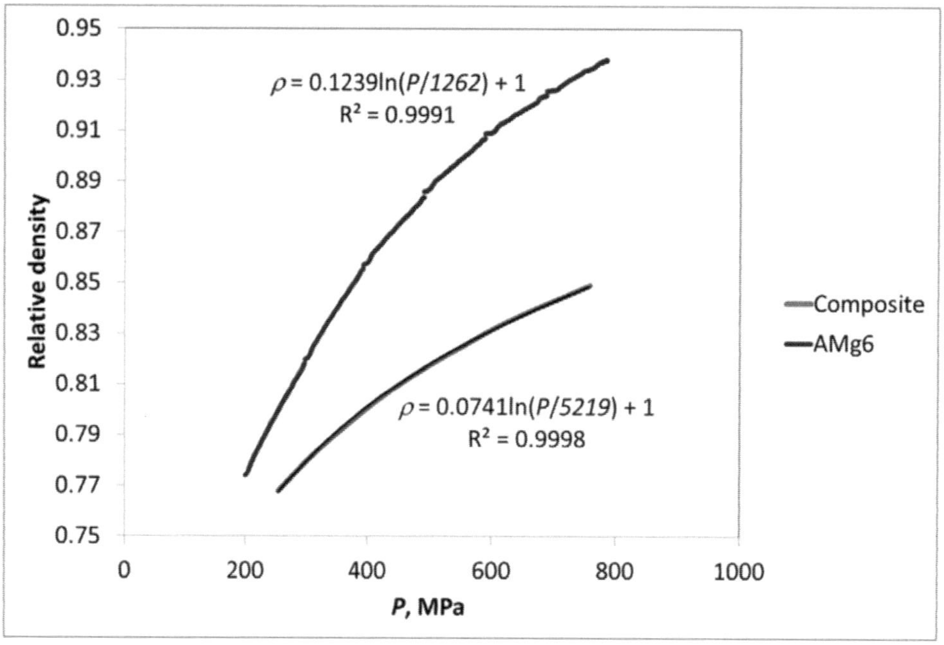

Figure 2. Pressure-density dependence of AMg6 matrix alloy and composite (65%AMg6+15%B4C+20%W) powders compacted by conventional quasi-static pressing in a die

At pressure of 750 MPa the composite powder (65% AMg6 + 15% B_4C + 20% W) samples with a density of 85.3% were obtained by the conventional cold quasi-static pressing. During free sintering of these samples in a vacuum at temperatures of 450, 500, 550 and 600 °C the linear dimensions of the processed specimens remained unchanged up to the melting point of the aluminum-magnesium matrix. This fact indicates a non-significant contribution of the pressure-less solid state diffusion-based mass transport into the densification of the studied powder composites.

For the powder with an optimized composition (74% AMg6 + 6% B_4C + 20% W) at the same pressure of 750 MPa the density of 91.6% has been achieved, which confirms the results of the simulation-based optimization of powder mixtures (Figure 1,c, d; Table 2). Density after conventional compacting at 800 MPa was equal to 92.6% and 95.1 % with using powerful ultrasonic action (PUA) [8]. After sintering for 2 hours at 590 °C, the density was equal to 93.4% (without PUA) and 96.8 % (with PUA).

It should be noted that for the tested composite the conventional quasi-static pressing may be applied also in the case of the yield strength reduction of the AMg6 alloy even under ambient temperature conditions. One can use a well-known acoustic-plastic effect by utilizing powerful ultrasonication during materials compression, which can significantly increase the concentration and mobility of dislocations of the crystal lattice of the matrix material, thereby reducing its yield strength. In particular, according to [9], the effect of ultrasound on the deformation process of aluminum at an oscillating stress amplitude of 0.2 kg/mm^2 (~1.9 MPa) and under static loading of 1 kg/mm^2 ~(9.8 MPa) increases the steady creep rate almost 10 fold, and the yield strength decreases by half.

Additional effects can be achieved by using carbide materials for the collector type dies, which enable the achievement of the pressure levels higher than the tensile strength of the material of such dies [8].

The equation (2) will be utilized not only for the description of the plastic deformation of the AMg6 matrix alloy, but also as one of the possible models for the analysis of the consolidation of the studied composite material, containing boron carbide particles, which are essentially non-deformable in the specified pressure range.

5. Conclusions

1. Discrete element modeling of the composite particles' packing based on the particle size distribution functions of real powders describes well the mixing of powder components and enables the determination of the powder compositions rendering maximum mixture packing densities.
2. The optimal content of the studied composite has been determined (74%AMg6+6%B_4C+20%W). This content provides the maximum dense particle packing for all the components and strong alloy matrix with the coordination number greater than 4 and packing density of 0.67. The experimentally obtained density of this composite material subjected to the conventional quasi-static pressing (under pressure of 800 MPa) was achieved to be 92.6% and 95.1% after PUA. After sintering for 2 hours at 590 °C, the density was equal to 93.4% (without PUA) and 96.8% (with PUA).

Acknowledgements
The work has been supported by FTP project #RFMEFI57514X0003.

References

[1] Artemyev V A 1997 *JTP Lett.* **1** 5-9
[2] Bogachev I, Olevsky E, Grigoryev E, Khasanov O 2014 *JOM 66* 1020
[3] Gorshenkov M V, Kaloshkin S D, Tcherdyntsev V V, Danilov V.D, Gulbin V N 2011 *Defect and Diffusion Forum* **309-310** 249-254
[4] Muktepavela F, Manika I, Mironov V, 1997 *Mater. Design* **18** 257-259
[5] Mironov V, Filippov O, Boiko I Eston. 2010 J. Eng. **16** 142-149.
[6] Gregg S J, Sing K S W 1982 *Adsorption, Surface Area and Porosity* (Academic Press)
[7] Khasanov O L, Dvilis E S, 2007 *J. Europ. Ceram. Soc.* **27** 749-752
[8] Khasanov O L, Dvilis E S 2008 *Adv. App. Ceram.* **107** 135-141
[9] Kyulemin A V 1978 *Ultrasound and Diffusion in Metals*, (Moscow: Metalurgiszdat)

Modeling and comparative analysis of packing of the polyfractional mixture of B_4C powder

E S Dvilis, O L Khasanov, A O Khasanov, M S Petyukevich, D K Iazykov

Department of nanomaterials and nanotechnologies, National Research Tomsk Polytechnic University, 30 Lenin Ave., Tomsk, Russia

Email: khasanov@tpu.ru

Abstract. Modeling of the real polyfractional powder mixtures has been carried out using the data obtained by SEM, XRD and grain-size analysis. Parameters of the dense packing of powder mixture particles were defined. The influence of every fraction on the packing density of powder mixtures has been deduced.

1. Introduction

The main problem of powdered microstructure modeling, using dense packing of equivalent particles, is the shortage of initial data about the grain-size composition, i.e. knowledge incompleteness of the spatial positioning of microparticles and its geometric shape. Basically the real shape of the powder particles is idealized in a model by elemental bodies (usually by spheres) [1]. As a result of particle packing modeling, with satisfactory accuracy, imitating the microstructure, the task adds up to the search of the value of model input parameters responsible for the characteristics of grain-size composition [2, 3].

2. Experiments

High-clean boron carbide powder mixtures with 5-15 wt.% of nano additive content (B_4C-PIHT1, B_4C-PIHT2, B_4C-PIHT3, B_4C-PIHT4) were obtained by combination of mechanical and chemical methods [4]. The initial powder were milled, then purified of the impurities of enriched milled nanopowder particles, rinsed and dried. After that, the mixture was deagglomerated and finally high-clean powder with specified nanosized fraction was obtained.

Averaged values of powder mixture morphologic parameters, obtained by laser diffraction (LD), BET, SEM and XRD methods, are presented in the Table 1. According to SEM data, the powders consisted of equiaxed particles with a broad particle size distribution and practically did not form agglomerates.

According to the XRD pattern analysis, PIHT powder mixtures contained not more than 3 wt.% of H_3BO_4. The content of iron in the different samples varied from 0.16 wt.% to 0.51 wt.% with the mean value of 0.32 wt.%. There were not less than 79 wt.% of boron content in the powders.

The analyses of XRD diffractograms with the separation of peak broadenings on the components of different boron carbide fractions (Figure 1) showed the average size of coherent-scattering region of only submicron powder fraction. These results satisfactorily correlate with the results of nanosize fraction contents obtained by laser diffraction method (Table 1).

The summary results of the powder materials study are presented in the Table 1.

Table 1. Summary results of the powder materials study

	PIHT-1(1)	PIHT-2(2)	PIHT-3(3)	PIHT-4(2)
Specific surface area, S_{ss}, m^2/g	3.71	5.94	5.11	5.20
Particle size (BET) *d*, μm	0.64	0.40	0.47	0.46
Particle size (SEM) *d*, μm	2.6	2.9	4.2	2.7
Particle size (LD) D, μm	1.87	2.74	3.81	2.56
Agglomeration degree, *nD*	2.93	6.86	8.10	5.56
B_4C content, wt.%	97.3	99.8	97.8	98.9
H_3BO_3 content, wt.%	2.7	0.2	2.2	1.1

3. Results and discussion

Grain-size composition and SEM images of the particle of synthesized powders are presented in Figure 2.

Based on the experimental data of grain-size analysis, the modeling of optimal particle packing has been carried out in order to optimize the process of PIHT powder consolidation. The modeling objective of the densest polyfractional PIHT powder particles, when the fine particle fractions effectively fill the spaces between larger particles, was being solved.

According to SEM analysis, since PIHT powder particles have equiaxed shape, it is possible to formalize them in a packing model as a class set of spherical particles of appropriate grain-size composition. This fact had been taken into account in order to perform the modeling.

$$Qv = \frac{A}{w \cdot r \cdot \sqrt{2\pi}} \cdot exp\left(\frac{-\left[\ln \frac{r}{<r>} \right]^2}{2w^2} \right) \quad (1)$$

where Qv – percent fraction content of particles with radius r, A – amplitude value of approximating class, w – class standard deviation, $<r>$ - mean values of particle class radius.

Quantitative variables of modeled objects, obtained by discrete element method using S3D PorouStructure software (Ichikawa's algorithm, central packing), were used as an evaluation criterion of powder possibility in forming dense powder packing. Multimodal behavior of synthesized powder particle grain-size distribution has been taken into consideration for model construction of packings (Figure 2). In order to describe experimental data of granulometry, the sum of normal logarithmical distributions with the profile, described by equation (1), has been used.

Data treatment of the grain-size distribution results, obtained by LD, showed, that investigated powders consist of 4-5 particle classes with logarithmically normal and quite broad size range (from tens of nanometer to tenths of microns, Figure 2). Let us denote these fractions for all powders by class serial number from 1 to 5, with the increase of mean grain size (Table 2). In addition, parameters of the constituent classes of this powder distribution function (mean particle radius r (μm), standard deviation w) were defined. The transition from diameter to radius of the particles is governed by the format of input parameters, which are used in discrete element packing model.

Table 2. Resulting parameters of synthesized powder packings

Parameter	Powder	Class 1	Class 2	Class 3	Class 4	Class 5	mixture
Mean coordination number N_c	PIHT-1(1)	6.0	244.1	4659.7	88554.5	541091.1	6.2
	PIHT -2(2)	5.0	25.6	86.3	2174.7	459.0	6.1
	PIHT -3(3)	4.7	20.5	234.8	-	1337.0	6.2
	PIHT -4(2)	5.3	34.7	1231.0	-	95.3	6.2
Relative packing density	PIHT -1(1)	0.138	0.081	0.108	0.529	0.045	0.901
	PIHT -2(2)	0.054	0.098	0.141	0.488	-	0.779
	PIHT -3(3)	0.052	0.253	0.381	-	0.066	0.752
	PIHT -4(2)	0.123	0.137	0.520	-	-	0.780

Comment. Class color corresponds to the color of the presented below plots and images of packing models.

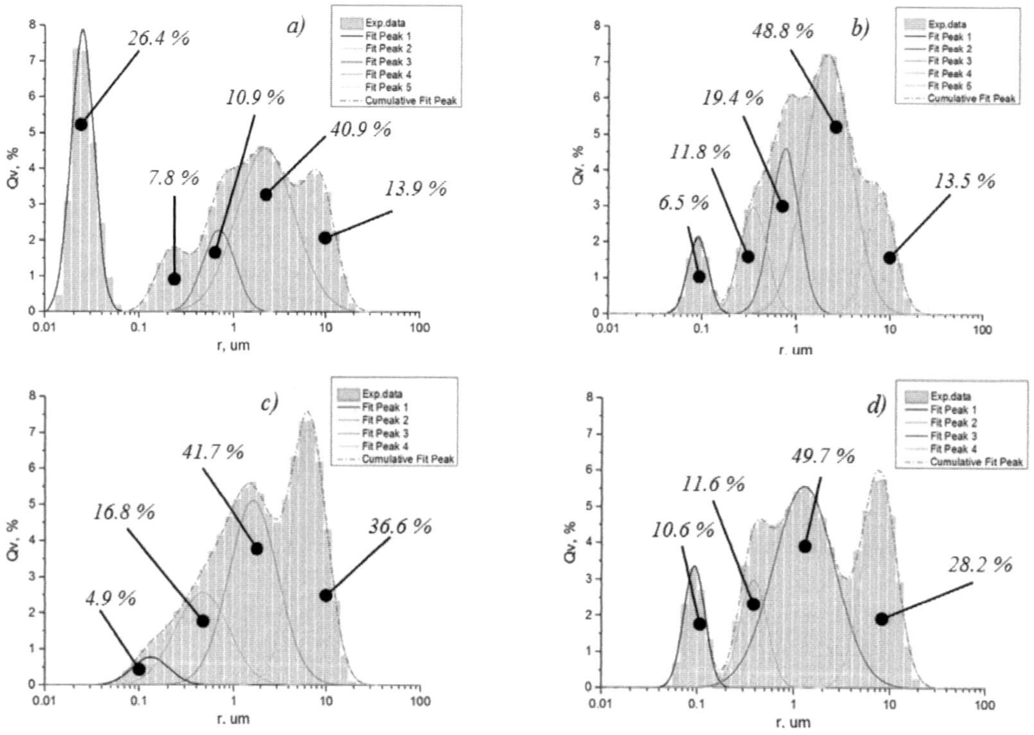

Figure 2. Results of grain-size analysis data expansion of the powder particles on the class fractions of logarithmically normal distribution: *a)* – mixture PIHT-1(1); *b)* – mixture PIHT-2(2); *c)* – mixture PIHT-3(3); *d)* – mixture PIHT 4(2).

PIHT-1(1) powder have distinctly more submicron and nanofraction content and they are shifted more to the smaller particle size (Figure 2 a). The behavior and distribution ranges of the PIHT-2(2) larger (micron) fractions are sufficiently close to the PIHT-1(1). In addition, the PIHT-2(2) micron fractions are considerably overlapped, forming close to normal distribution in a range from 0,2 to 20 μm. There are two factors which can favor the process of powder packing and consolidation. First, the relatively high individual content of submicron fraction (from 6 to 20%) in PIHT-1(1) and PIHT-2(2), PIHT-4(2) (Figure 2 a,b,d). Second, the presence of large particle distribution close to uniform in a range of two orders of magnitude. At the same time, fine fraction acts not only as a filler of the free space between large particles, but also forms a buffer active layer on the boundaries between large particles. This buffer layer eases large grain intermotion (imitation of plastic deformation of powder body) under the prepressing pressure and provides activation processes of high-temperature sintering due to the well-known effect of nanosized state.

Packing models, build by discrete element method using obtained parameters of distribution functions, consisted of representative set of particles in amount from 600000 to 200000. The images of typical packing sections, where each particle class has its own color, are presented in Figure 3.

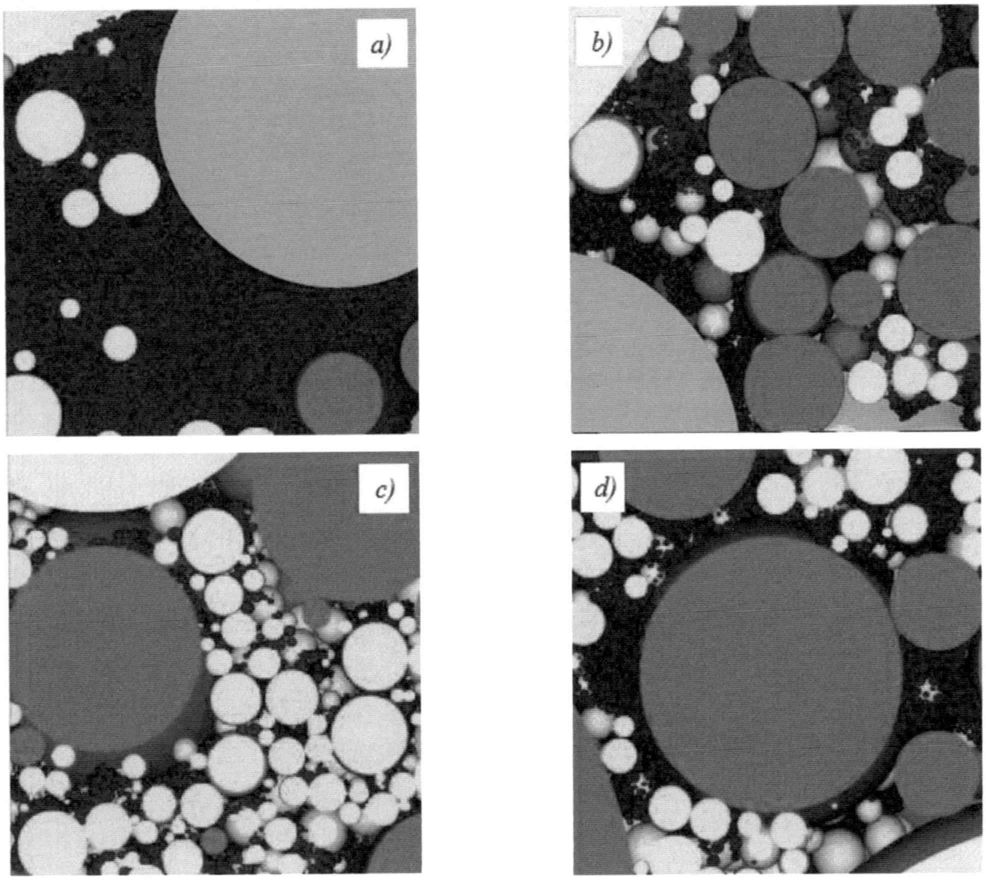

Figure 3. Section fragments of particle packings of different PIHT powder classes, obtained by discrete element modeling using data of granulometric analysis: *a)* – mixture PIHT-1(1); *b)* – mixture PIHT-2(2); *c)* – mixture PIHT-3(3); *d)* – mixture PIHT-4(2).

Comparative analysis of the obtained results (quantitative parameters of packings, Table 2) allows to make several conclusion regarding the optimization of grain-size composition of boron carbide powders, produced from low-quality and waste products by developing methods. These conclusions are of a great importance for further production of high-quality consolidated articles. One clear example of the influence of grain-size composition and parameters of powder particle classes on the packing characteristics is the accumulation curve of contribution of these classes to the compaction, presented in Figure 4. In particular, PIHT-1(1) sample, where nano fraction effectively fills the spaces between large particles of wide and continuous fraction composition, has the most dense packing (other conditions of the model being equal).

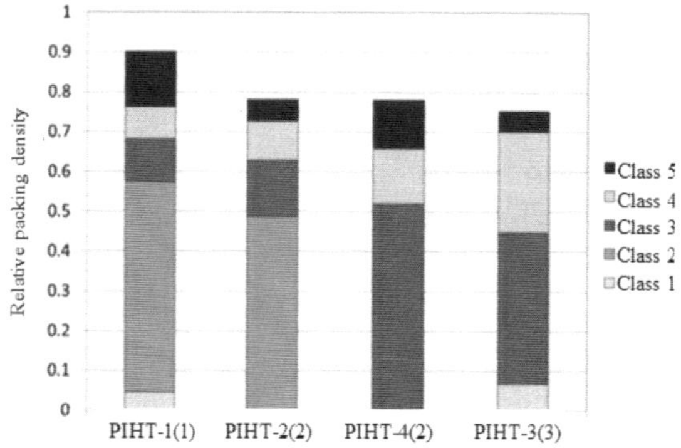

Figure 4. Accumulative density plot of PIHT powder particle packing

The obtained results allow to make several conclusions regarding optimization study.

1. Particle classes with highest content make the main contribution to the packing density. At the same time, class 5 with the maximum particle size influences the less effectively on the density increase.

2. The contribution of the finest particle fraction (class 1) to densification of all powder considerably depends on the difference of mean size and the size of the next class. The higher this difference, the most effectively submicron and nanoparticles of class 1 fills the spaces between particles of large fractions.

3. The interruption of continuous distribution (mainly the absence of class 4 in the PIHT-3(3) and PIHT-4(2) samples) negatively affects the packing density, which can not necessarily be compensated by the presence of individual finely dispersed fraction. Moreover, some trend of the directly proportional dependence of every class contribution to the densification on its percentage content in the packing is observed (Figure 5).

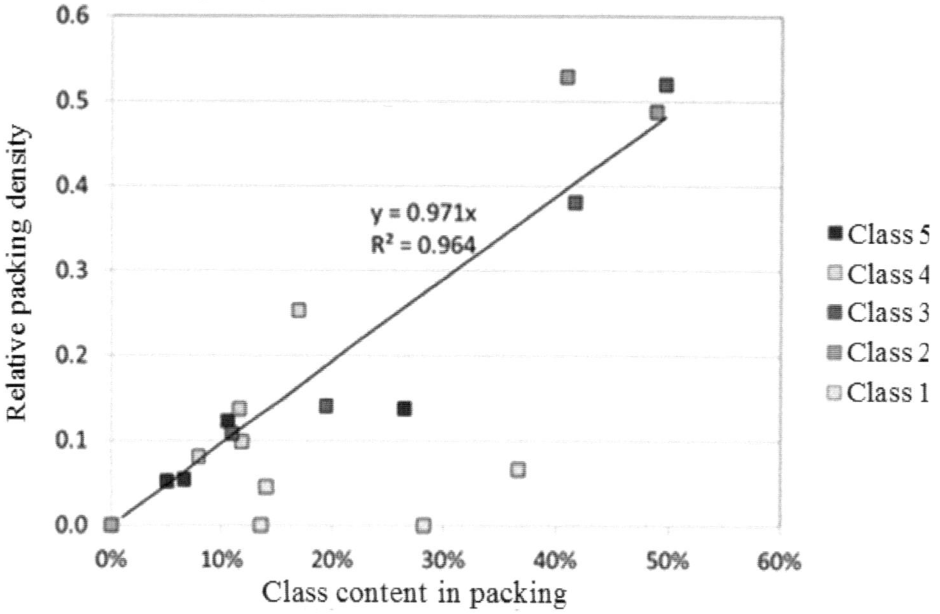

Figure 5. Correlation between relative class density and its content in model packing.

Considerable deviation from such dependence is observed only for the large particle class (class 5), which can be described by its relatively small content in the model packing (up to 100 units versus half a million particles in the other classes). However, the mentioned feature can be used in order to preliminary evaluation of particle contribution of different fractions to the density of consolidated powders.

Acknowledgements
The work has been supported by FTP project #RFMEFI57514X0003.

References

[1] Califice A, Michel F, Dislaire G, Pirard E 2013 Influence of particle shape on size distribution measurements by 3D and 2D image analyses and laser diffraction *Powder Technology* **237** 67-75
[2] Ilić M, Budaka I, Vučinić Vasić M, Nagode A, Kozmidis-Luburić U, Hodoliča J, Puškarc T 2015 Size and shape particle analysis by applying image analysis and laser diffraction – Inhalable dust in a dental laboratory *Measurement* **66** 109-117
[3] Tinke A P, Carnicer A, Govoreanu R, Scheltjensc G, Lauwerysena L, Mertensa N, Vanhouttea K, Brewstera M E 2008 Particle shape and orientation in laser diffraction and static image analysis size distribution analysis of micrometer sized rectangular particles *Powder Technology* **186** 154-167
[4] Kazantsev O A, Sivokhin A P, Shirshin K V, Gur'yanov O P, Samodurova S I 2010 Synthesis of N-alkylacrylamides from commercial fractions of higher olefins by the Ritter reaction *Russian Journal of Applied Chemistry* **83** 1062-1068

IOP Conf. Series: Materials Science and Engineering **116** (2016) 012029 doi:10.1088/1757-899X/116/1/012029

Sintering of oxide and carbide ceramics by electron beam at forevacuum pressure

E Dvilis[1], O Tolkachev[1], V Burdovitsin[2], A Khasanov[1], M Petyukevich[1]

[1]Department of nanomaterials and nanotechnologies, National Research Tomsk Polytechnic University, 30 Lenin Ave., Tomsk, 634050, Russia
[2]Tomsk University of Control Systems and Radioelectronics, 40 Lenin Ave., Tomsk, 634050, Russia

Email: dvilis@tpu.ru

Abstract. Novel approaches for electron beam sintering of zirconia and silicon carbide ceramics have been investigated: application of forevacuum pressure plasma-cathode to compensate the charge induced by the electron beam on the green compact surface, and previous dry powder compaction under powerful ultrasound assistance. Dense YSZ ceramics with submicron and micron grains have been consolidated by electron beam sintering after powder compaction using ultrasound.

1. Introduction

Electron Beam Sintering (EBS) is a kind of Field Assisted Sintering. However to provide a temperature available for sintering of ceramics it is necessary to use electron beam in high vacuum, and with megaelectronvolt (MeV) energy [1]. Hence, there are significant limitations for wide application of EBS: high power consumption; shielding effect for dielectric materials due to charge accumulation on the sample surface, which leads to deceleration and deflection of the electron beam; formation of micro- and macroscopic defects in sintered ceramics [2].

The specified problem may be resolved by application of the electron beam with energy 5-15 keV in a forevacuum, i.e. in the pressure range of 5-20 Pa. For this purpose the electron beam source is the forevacuum pressure plasma-cathode [3, 4], which provides the generation of dense plasma owing to hollow-cathode discharge. The plasma compensates the charge induced by the electron beam on the dielectric target surface. In this case the floating potential near the target is much less than accelerating voltage and impact of the electron beam on the dielectric ceramics is similar to effect on metals [5].

The aim of the work was to carry out testing of EBS method in forevacuum with electron beam having kiloelectronvolt energy in order to sinter yttrium-stabilized zirconia (YSZ) and silicon carbide (SiC) ceramics.

2. Powder compaction and sintering. Experimental techniques

The raw powders were commercial YSZ nanopowder TZ-3YS (TOSOH) with average particle size of 200 nm and commercial SiC powder (Neomat) with average particle size of 1 μm.
To prepare green compacts for EBS the dry powder pressing without any binders and lubricants using both conventional uniaxal compaction and uniaxal compaction under powerful ultrasound assistance (PUA) [6] have been used. Compaction pressure was 480 MPa. In case of PUA compaction the

resonant ultrasound frequency was 21 kHz, power of ultrasonic generator was 4 kW. Green pellets after compaction had the diameter of 14 mm, thickness about 4 mm and green density up to 59%.

Setup of EBS installation is shown in Figure 1. On opposite walls of a vacuum chamber (1) two identical electron guns (2) supplied with focusing and deflecting coils (3) were established. Their heat protection was provided by shields (4). Two identical electron beams (5) from two opposite directions provided uniform heating of green pellet (6), which was located on a graphite crucible (7) in special holder and settled down in the plane of a perpendicular axis of the electron beams. For visual observation through the window (8) two mirrors (9) were mounted in the vacuum chamber. Measurement of a sample temperature was carried out by contactless method using RAYTEK 1MH pyrometer (10).

EBS modes have been varied depending on ceramics type and for purpose of their optimizing. Accelerating voltage of the electron beam U varied from 10.5 keV to 12 keV; beam current I was in the range 200 mA – 280 mA. Sintering duration (exposition of green compacts under electron beam at sintering temperature) varied from t_e = 5 sec to t_e = 80 min. EBS energy for sintering E (MJ) has been calculated by the equation $E = UIt_e$.

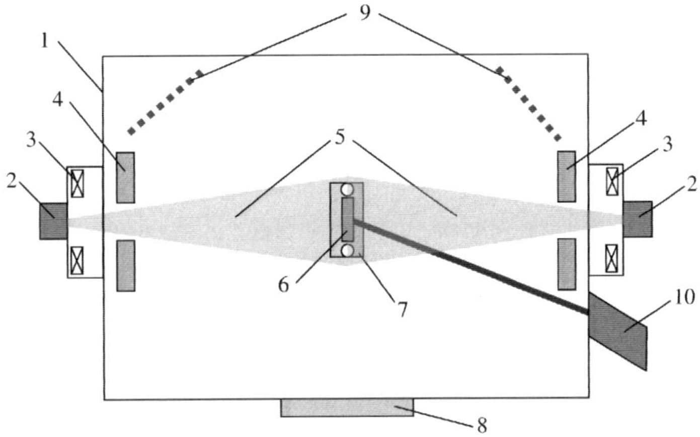

Figure 1. Setup of installation for electron beam sintering: 1 – vacuum chamber, 2 – electron guns with the hollow cathode, 3 –focusing and deflecting coils, 4 – heat shields, 5 –electron beams, 6 – green pellet, 7 – graphite crucible, 8 – observation window, 9 – mirrors, 10 –pyrometer.

Also YSZ green pellets have been consolidated by the conventional sintering in air (Nabertherm LHT 02/18 furnace).

The microstructure and mean grain size d_g versus relative density ρ of sintered specimens have been examined by SEM analysis (JSM-7500FA JEOL; "ImageJ" code). Phase composition has been studied by XRD analysis (XRD-7000S Shimadzu) and Vickers microhardness by the nanoindentation technique (DUH-211S Shimadzu).

3. Results and discussion

3.1. YSZ ceramics

YSZ samples were sintered by EBS at temperature 1400-1450 °C and EBS energy was variable from 5.5 MJ to 12.3 MJ.

For all EBS modes YSZ melting, recrystallization were avoided and XRD analysis shown that sintered samples had pure tetragonal phase of ZrO_2 (Figure 2).

It was found an extreme behavior of the relative density ρ and grain size d_g of YSZ ceramics vs EBS energy (Figure 3): at E = 8.4 MJ (t_e = 40 min) mean grain size had maximum value $d_g \approx$ 5 µm, as well as maximum density ρ = 95% (Figure 4a), but at E = 11 MJ (t_e = 60 min) d_g was significantly decreased down to 2.5 μm, however density remains high, about ρ = 95.2%. Following increasing EBS energy leads to lower ρ and d_g values. This behavior indicates an opportunity of optimization of the EBS modes to achieve the higher YSZ density at lower mean grain size.

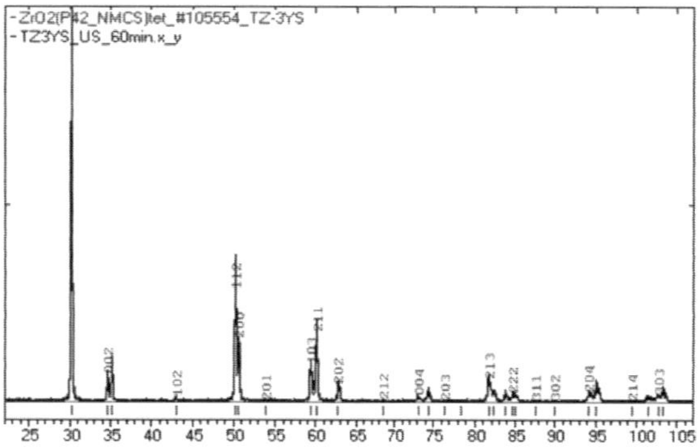

Figure 2. XRD pattern of YSZ ceramics sintered by EBS technique.

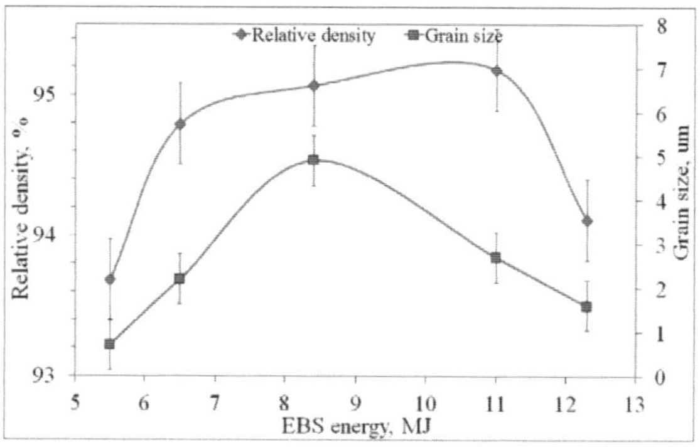

Figure 3. Relative density and grain size of YSZ ceramics vs EBS energy.

Conventional sintering (1400 °C, 60 min) showed ρ = 97%, but large grain size d_g = 22 μm.
EBS after conventional uniaxial pressing and after compaction under PUA showed that in case of PUA density can be achieved up to 97.3% and compaction modes influence on the grain size. EBS at t_e = 5 sec after conventional pressing leads to d_g = 4.8 μm, but at t_e = 60 min the mean grain size was 1.5 μm. The same EBS modes, but after PUA compacting, provided d_g = 2.4 μm and submicron grain scale of 0.7 μm correspondingly.

Figure 4. Microstructure of YSZ ceramics (a) after EBS at E = 8.4 MJ; (b) after EBS at E = 5.5 MJ.

Microhardness of YSZ ceramics consolidated by EBS after conventional pressing was 13.7 GPa, but after PUA compacting at the same EBS modes it was increased up to 15.9 GPa.
So it was shown a possibility to decrease YSZ mean grain size at high density and increase the microhardness by application of dry powder compaction under powerful ultrasound assistance.

Important peculiarity of YSZ ceramics microstructure after EBS is presented in Figure 5: the porous surface layer with thickness of about 11 μm and mean pore size comparable with d_g.

Figure 5. Porous surface of the YSZ ceramics after EBS.

Formation of this porous surface can be explained by the modeling of YSZ consolidation by EBS using the assumption of anisotropic sintering due to electron charge accumulation on the surface grains. Such charging leads to slowing down of sintering in normal direction to the surface. Figure 6a shows the surface model of initial green compact with monolayers (S3D-Evolution code, Smartimtech) and Figure 6b, 6c - simulation of anisotropic sintering and porous space of sintered ceramics, having satisfactory correlation with the SEM image of YSZ ceramics after EBS (Figure 6d).

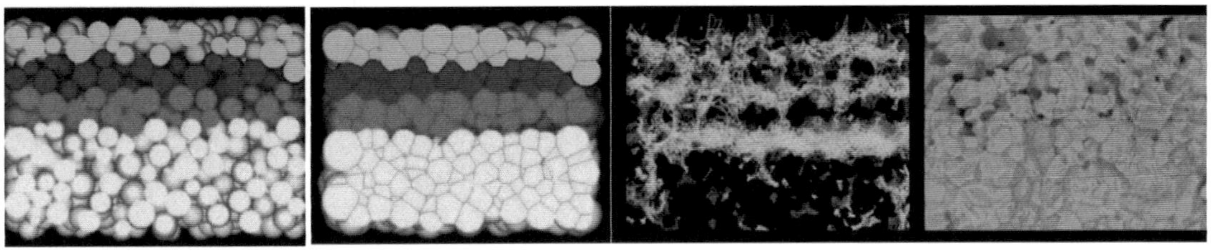

a) *b)* *c)* *d)*

Figure 6. Simulation of anisotropic sintering *(a, b)*, porous space *(c)* and SEM image of YSZ ceramics after EBS *(d)*.

3.2. SiC ceramics

EBS of SiC green pellets in a range from 1700 °C to 1950 °C demonstrated the mass loss (ablation) and density decreasing even at slight shrinkage (Figure 7). The possible reason of such behavior is additional Joule heating in SiC owing to relatively high specific resistance at EBS temperature 1700 °C (5 Ω·cm) in comparison with specific resistance of YSZ at 1400 °C (1 Ω·cm).

Different microstructure of sintered SiC samples along their spall and pellet surface has been found by the SEM analysis (Figure 8a-d) which can be explained by nonuniform distribution of EBS temperature from the pellet center to edge due to nonuniform Joule heating.

Thus, more thorough research is essential for optimization of EBS modes in order to sinter dense SiC ceramics.

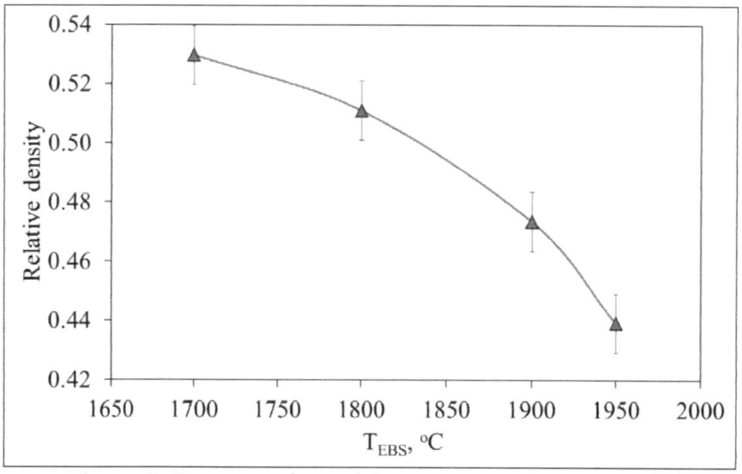

Figure 7. Porous surface of the YSZ ceramics after EBS.

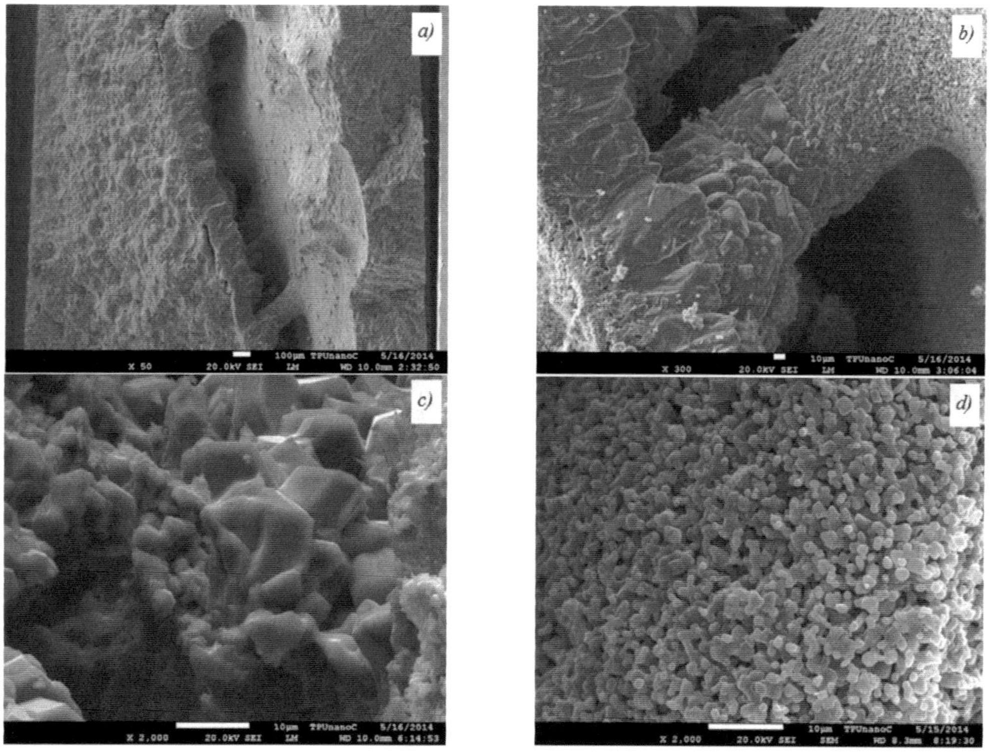

Figure 8. Microstructure of ceramics:*(a)* cavities in the SiC sample spall after EBS due to ablation; *(b)* cavities in the central area of pellet surface; *(c)* SiC recrystallized junctions in necks in the central area of pellet surface; *(d)* small SiC grains in the edge area of pellet surface.

4. Conclusions

1. Electron beam sintering in forevacuum can be effective method to sinter dense YSZ ceramics with small grains. Submicron grain scale can be achieved using EBS at exposition duration 60 min after dry YSZ nanopowder compaction under powerful ultrasound assistance.

2. Porous surface layer on the YSZ ceramics can be caused by anisotropic sintering.
3. EBS of SiC ceramics is accompanied by the ablation due to high SiC specific resistance.

Acknowledgements
The work has been supported by the RFBR grant No.14-08-00775, State assignment "Science" and FTP project #RFMEF157514X0003.

References

[1] Annenkov Yu M 1996 *Russian Physics Journal* **39** 1146-59
[2] Sun C N, Gupta Mool C and Karen M B 2010 *J. Am. Ceram. Soc.* **93** 2484–6.
[3] Burdovitsin V A and Oks E M 2008 *Laser and particle beams,* **26** 619–35.
[4] Burdovitsin V, Dvilis E., Zenin A, Klimov A, Oks E, Sokolov V and Kachaev A 2014 *Advanced Materials Research* **872** 150-6
[5] Burdovitsin V, Klimov A and Oks E 2009 *Technical Physics Let.* **35** 511-3
[6] Khasanov O L and Dvilis E S 2008 *Adv. Appl. Cer.* **107** 135-41
[7] Munro R G 1997 *J. Phys. Chem. Ref. Data.* **26** 1195-1203

The Study on Thermal Expansion of Ceramic Composites with Addition of ZrW_2O_8

E S Dedova[1,2,3], V S Shadrin[2,3], M Y Petrushina[4] and S N Kulkov[1,2,3]
[1] Tomsk Polytechnic University, avenue Lenin, 30, 634050, Tomsk, Russia
[2] Institute of Strength Physics and Materials Science SB RAS, pr. Akademicheskii, 2/4, 634055, Tomsk, Russia
[3] Tomsk State University, avenue Lenin, 36, 634050, Tomsk, Russia
[4] Novosibirsk State University, st. Pirigova, 2, 630090, Novosibirsk, Russia

E-mail: lsdedova@yandex.ru

Abstract. The studies on structure, phase composition and thermal properties of (Al_2O_3 – 20 wt% ZrO_2) – ZrW_2O_8 ceramic composites obtained using nanosized, initial powders were conducted. Homogeneously distributed white particles on the polished surface of composites were observed. Phase composition of the composites was represented with corundum, monoclinic ZrO_2 and two modifications of ZrW_2O_8 (tetragonal and cubic). Linear thermal expansion coefficient values of the composites were determined. The difference in experimental and calculated coefficient of thermal expansion values for composites obtained may be attributed to phase transformations, features of the structure, internal stresses due to thermal expansion mismatch, which contribute significantly to thermal expansion of the ceramic composites.

1. Introduction

Development of new composites with specific thermal properties, for example low thermal expansion, for today is a significant and perspective research task. Recently one can observe a growing number of studies on the creation of ceramic composites with controlled coefficient of linear thermal expansion to provide dimensional stability of the object during the operation from very low to high temperatures. The solution of the problem lies in a combination of materials with positive and negative coefficients of thermal expansion (CTE). Zirconium tungstate is considered as most effective material for such application due to its isotropic negative thermal expansion coefficient ($\alpha=-9\cdot10^{-6}$ C^{-1}) in a wide temperature range from -273 to 770 °C [1]. Ceramic materials with unique mechanical, thermal properties, with a near zero CTE can be applied in various fields such as rocket engineering, high-precision optics, electronics, oil and gas industry and others. However, the interaction between zirconium tungstate and ceramics during thermal treatment is not fully investigated. Phase transitions or formation of internal stresses can lead to a change of physical and mechanical properties of the material. A detailed study on these processes is required in order to obtain material that retains mechanical properties on sufficient level and the required value of the coefficient of thermal expansion.

The main aim of the work is to study the structure and thermal properties of the (Al_2O_3 - 20 wt.% ZrO_2) ceramic composite with addition of zirconium tungstate ZrW_2O_8.

2. Materials and Experimental Procedure

The powder mixtures of Al_2O_3 – 20 wt.% ZrO_2 and ZrW_2O_8 were used as initial components. Zirconium tungstate was obtained using hydrothermal method based on low-temperature decomposition of the precursor $ZrW_2O_7(OH_{1,5}Cl_{0,5})2H_2O$ [2]. The mixing of Al_2O_3 and ZrO_2 powders was carried out in a ball mill for 25 hours. The volume fraction of zirconium tungstate powder in the powder mixture was 10 wt. %. Samples were prepared by uniaxially cold pressing the mixtures in a cylindrical shaped steel press mold with diameter 7 mm under 10 MPa. (Al_2O_3 - 20 wt.% ZrO_2) – ZrW_2O_8 ceramic composites were obtained by free sintering the green compacts in air at 1550 °C.
Structure and phase composition were studied using XRD analysis on "DRON" diffractometer with CuKα radiation. Particle shapes and sizes were characterized using optical and scanning electron microscopy (SEM). Thermal expansion was investigated using quartz mechanical dilatometer in temperature range 100 - 600 °C.

3. Results and Discussion

3.1. ZrW_2O_8 powder

According to SEM analysis the ZrW_2O_8 powder synthesized was represented with elongated particles with block structure, Figure 1. The mean size of individual elongated particles in the transverse direction was 200 nm, the block size –50 nm [3].
Thermal behavior of zirconium tungstate was described in detail in [4]. ZrW_2O_8 has cubic crystal structure in temperature range from 25 to 540 °C. The further increase of temperature causes rearrangement of atoms in crystal structure of ZrW_2O_8 intended to form sublattices of zirconium and tungsten oxides. The rearrangement is a predecessor of the decomposition process of zirconium tungstate. The decomposition process of zirconium tungstate into constituent oxides ends at 850 °C.
The lattice parameter (a) of cubic ZrW_2O_8 decreased with increase of temperature from 25 to 1000 °C indicating the negative thermal expansion of the material. The linear thermal expansion coefficient of ZrW_2O_8 in temperature range from 20 to 200 °C was equal to $\alpha = -9.4\cdot10^{-6}$ C^{-1} and $\alpha = -3.8\cdot10^{-6}$ C^{-1} in temperature range from 200 to 1000 °C, Figure 2. Increase of the CTE value at 200 ± 25 ° C corresponds to a phase transition from a low temperature (α-) to high-temperature (β-) modification of zirconium tungstate.

Figure 1. SEM image of ZrW_2O_8 powder.

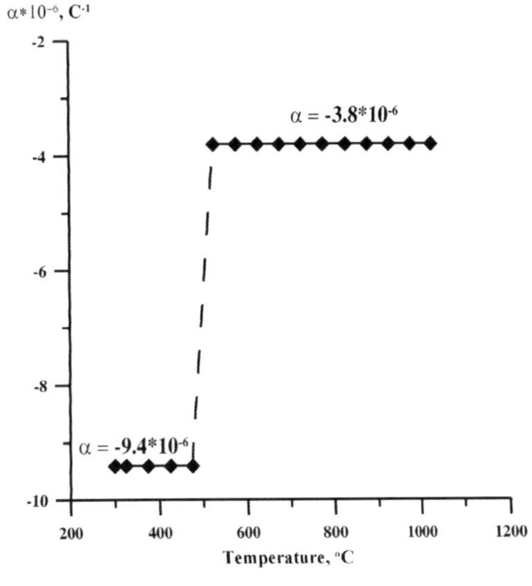

Figure 2. Lattice parameter-temperature dependence.

3.2. (Al_2O_3 - 20 wt.% ZrO_2) – ZrW_2O_8 ceramic composites

Nanosized plasmochemically synthesized alumina and zirconia powders were used to obtain the (Al_2O_3 - 20 wt.% ZrO_2) – ZrW_2O_8 ceramic composites.

The morphology of ZrO_2 powder was represented with nanocrystals, spherical particles with mean size 0.1 μm and agglomerates without regular shape with mean size 5 μm , Figure 3. The specific surface area, which is the total surface area of a material per unit of mass, was 20 m^2/g. Bulk density, which is density of a loose material, was 0.7 g/cm^3

Phase composition of the powder was represented with cubic, tetragonal and monoclinic ZrO_2 with prevalence of cubic modification (75%). The mean coherent diffracting domain (CDD) size was 20 nm for cubic ZrO_2, 30 nm for monoclinic ZrO_2 and 15 nm for tetragonal ZrO_2.

According to SEM analysis Al_2O_3 powder was represented with agglomerates formed by particles with irregular shapes and spherical particles with mean size of 1.3 μm, Figure 4.

Figure 3. SEM image of ZrO_2 powder.

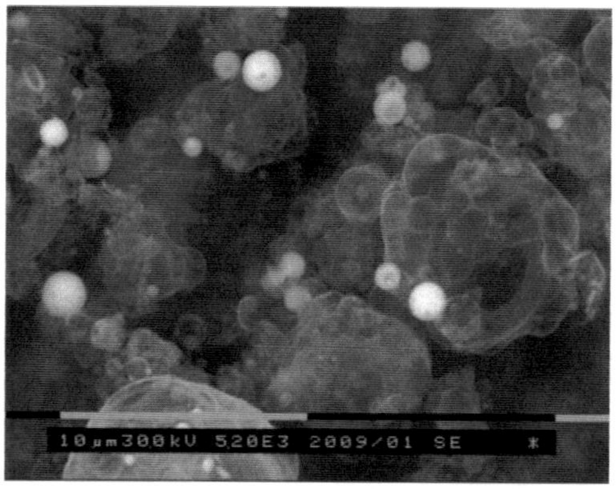

Figure 4. SEM image of Al_2O_3 powder.

According to XRD analysis in the composites obtained alumina is represented with α-modification (corundum), zirconia is represented with monoclinic modification, and two modifications of ZrW_2O_8 – cubic and tetragonal (*t*) were identified, Figure 5. On the X–ray diffraction pattern also appeared peaks corresponding to tetragonal ZrO_2 and WO_3. For cubic ZrW_2O_8 a = 9.8019 Å, the lattice parameter increased by 7% compared with the literature (a = 9.1540 Å) [1]. The presence of tetragonal ZrW_2O_8 was an unexpected result. Tetragonal modification of zirconium tungstate at the moment is reported only in [5] and there is no information about its thermal expansion behavior. Apparently during the cooling process cubic zirconium tungstate decomposes to *t*-ZrW_2O_8, ZrO_2 and WO_3. Thus the decomposition of ZrW_2O_8 to the oxides may proceed through intermediate *t*-ZrW_2O_8 phase.

It was established experimentally that the thermal expansion coefficient of the sintered Al_2O_3 –20 wt.% ZrO_2 ceramic without addition of ZrW_2O_8 was $9.7{\cdot}10^{-6}$ C^{-1}. CTE value calculated using the mixture rule: $a = \Sigma aiVi$, here ai are thermal expansion coefficients for each component and Vi are the volume concentrations. Using the experimental data for Al_2O_3 ($\alpha = 9.7{\cdot}10^{-6}$ C^{-1}) and ZrO_2 ($\alpha = 11{\cdot}10^{-6}$ C^{-1}) CTE of the ceramic composite was calculated - $\alpha = 10.1{\cdot}10^{-6}$ C^{-1}. The experimental data are in a good agreement with calculated values of CTE for Al_2O_3 – 20 wt.% ZrO_2 ceramic composite.

The relative elongation – temperature relationship for (Al_2O_3 - 20 wt.% ZrO_2) – 10 wt.% ZrW_2O_8 ceramic composite is presented in Figure 6. This relationship can be divided into 2 areas with different slope to the horizontal axis. In this case thermal expansion coefficient is the slope of approximation line to the x axis. The CTE of (Al_2O_3 - 20 wt.% ZrO_2) – 10 wt.% ZrW_2O_8 ceramic composite was

$9 \cdot 10^{-6}$ C^{-1}. The temperature of deviation from linear law on the double logarithmic plot was 210 ± 5 °C.

The CTE value calculated using the mixture rule was $7.8 \cdot 10^{-6}$ C^{-1}. The reasons for the slope change and mismatch between experimental and calculated data can be phase transitions, internal stresses, formed as a result of thermal expansion mismatch of the components. All this can make a significant contribution to the thermal expansion of the material. To understand completely the thermal behavior of (Al_2O_3 – 20 wt.% ZrO_2) - ZrW_2O_8 additional studies are required.

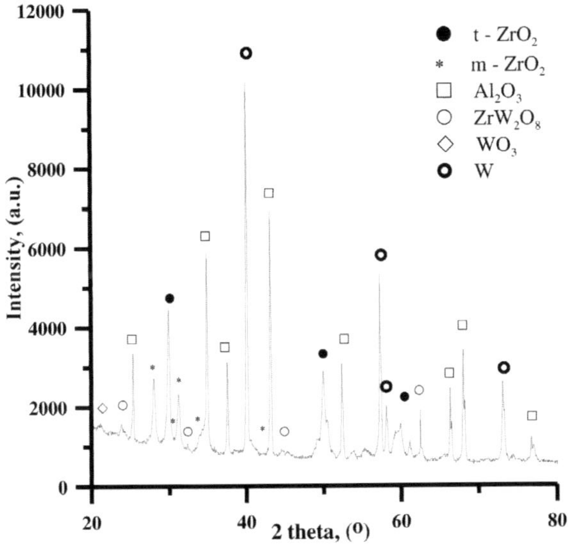

Figure 5. XRD pattern of (Al_2O_3 - 20 wt.% ZrO_2) – 10 wt % ZrW_2O_8.

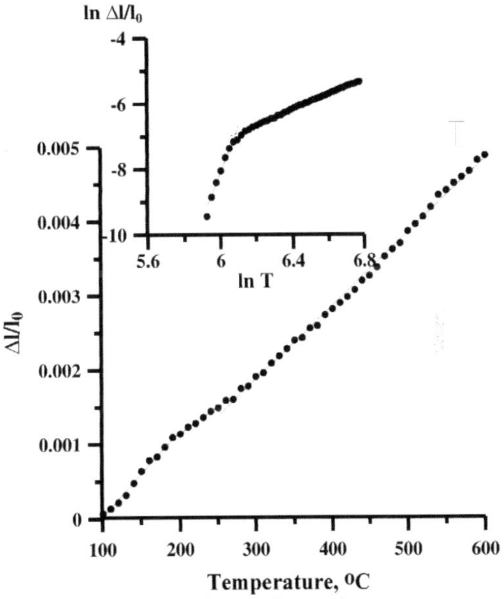

Figure 6. Relative elongation–temperature dependence of (Al_2O_3 - 20 wt.% ZrO_2) –10 wt % ZrW_2O_8.

4. Conclusion

The morphology of the zirconium tungstate powder obtained by hydrothermal method is represented with elongated particles with mean transverse size 0.2 μm.

Zirconium tungstate remains kinetically stable up to 540 °C. After that further increase of temperature up to 850 °C causes decomposition into constituent oxides through rearrangement of atoms in crystal structure of ZrW2O8 intended to form sublattices of zirconium and tungsten oxides.

Phase transition from low-temperature α- to high-temperature β-ZrW_2O_8 modification at 200 °C is accompanied by change of CTE of the material from – $9.4 \cdot 10^{-6}$ C^{-1} for α- ZrW_2O_8 to $-3.8 \cdot 10^{-6}$ C^{-1} for β-ZrW_2O_8.

Phase composition of the ceramic composite was represented with corundum, monoclinic ZrO_2 and two modifications of ZrW_2O_8 – cubic and tetragonal. According to dilatometric studies the CTE of (Al_2O_3 - 20 wt.% ZrO_2) – ZrW_2O_8 ceramic composite was $\alpha = 9 \cdot 10^{-6}$ C^{-1}. The difference between experimentally obtained data and calculations can be determined by several factors including phase transitions, formation of internal stresses due to thermal expansion mismatch.

Acknowledgement

The research was conducted with partial financial support by an agreement with the Ministry of Education № *14.575.21.0040 (RFMEFI57514X0040)*.

References

[1] Evans J S O 1999 *J Chem Soc.*
[2] Gubanov A I, Dedova E S, Plyusnin P E 2014 *Thermochimica Acta.* ***597*** *19*
[3] Dedova E S, Shadrin V S, Gubanov A I, Kulkov S N 2014 *Inorganic Materials: Applied Research. 55 471*
[4] Dedova E S, Kulkov S N, F Pedrasa 2014 *AIP Conf Pro* ***1623*** *99*
[5] Ekstrom T, Tilley J 1976 *J Solid State Chem.* ***19*** *227*

IOP Conf. Series: Materials Science and Engineering **116** (2016) 012031 doi:10.1088/1757-899X/116/1/012031

Activation of consolidation processes of alumina ceramics

S V Matrenin, B S Zenin and R V Tayukin
Tomsk Polytechnic University, 30 Lenin Avenue, Tomsk, 634050, Russia

E-mail: msv@tpu.ru

Abstract. The methods for activating sintering ceramics based on Al_2O_3 by mechanical activation in the planetary mill, by adding in the mixture of nanopowders (NP) Al, Al_2O_3, and submicron powder TiO_2, and by applying the technology of spark plasma sintering (SPS) are developed. It has been shown that adding the nanopowder up to 20 wt. % Al_2O_3 in a coarse powder α-Al_2O_3 activates the sintering process resulting in increased density and hardness of the sintered alumina ceramics. Substantial effect of increasing density of alumina ceramics due to adding the submicron powder TiO_2 in the compound of initial powder mixtures has been established.

1. Introduction

Compacting ceramic powders can be carried out using cold pressing with static one- or two-sided application pressure; hot pressing; cold or hot isostatic pressing in hydro- or gazostats; hot extrusion; slip casting; magnetic pulse, impact molding and explosive compacting; ultrasonic compaction. Nevertheless, the complexity and poor performance technology of hot and hot isostatic pressing [1-2] allowing to obtain materials with high strength properties [3] prevent the wide practical use of solid alumina ceramics. A relatively simple technique of uniaxial pressing with following sintering generally cannot produce ceramics with the high level of mechanical properties [4]. Therefore the problem of activating processes of consolidation alumina ceramics is of practical importance.

The aim of this work is to develop methods to activate the sintering process of the ceramics based on Al_2O_3 powders by mechanical treatment in a planetary ball mill and adding to the mixture of nanopowders (NP) Al, Al_2O_3 and submicron powder TiO_2 as well as application of spark plasma sintering (SPS) technology.

2. Experiment

Industrial nanocrystalline oxide powders (NP) Al_2O_3, Al_2O_3-ZrO_2-Y_2O_3 of the ultra-dispersed type obtained by a plasma-chemical synthesis (PCS) were used. Chemical composition of the powders is given in table 1.

Table 1. Chemical composition of initial plasma-chemical oxide powders (mol %).

Number	Al_2O_3	ZrO_2	Y_2O_3
1	100	-	-
2	80	19	1

Besides plasmachemical nanopowders, the coarse-grained powder Al_2O_3 – pure alumina, technical grade alumina and white electrocorundum (see table 2) were used in the work. White electrocorundum is widely used in the alumina ceramics technology. Its sintering temperature is not less than 1800°C [5]. To reduce the temperature the submicron powder TiO_2 with a particle size of 0.5...2 μm in the amount of 1.5 wt.% was added to the white alumina powder.

Table 2. Chemical composition of white electrocorundum.

Type	Mass fraction (%), no more			
	Al_2O_3	SiO_2	Na_2O	Fe_2O_3
25A	99.1	0.1	0.26	0.06

Electroexplosive NP Al was used as an activating additive. Average surface diameter of particles of the NP (d_{av}=60000/ρS_{sp}., where ρ – density, g/cm^3, S_{sp} – specific surface of the powder, cm^2/g) determined with a help of gas adsorption (method BET) was not exceed 140 nm.

Sieve analysis of the coarse powders pure alumina and technical alumina was carried out using the analyzer A 20. Powder samples of 50 g weight were used for analysis. The frequency of a vibrator was 70 Hz, the sieving time – 10 minutes. After sifting fractions were weighed and content of each was calculated. The bulk and tap density, flowability of investigated powders were determined.

Oxide powders were annealed in air atmosphere in a high temperature resistance furnace at 1450°C for one hour for phase transition from γ-Al_2O_3 to α-Al_2O_3.

To improve processing characteristics and enhancing the activity of annealed powders they were mechanically activated in the planetary ball mill Activator 2SL for 20 minutes at a rotation speed of the grinding vessels of 30 Hz. Zirconia balls were as grinding bodies. As a result of the mechanical action there is a significant plastic deformation at the vicinity of interparticle contacts. In order to define an activating effect of the mechanical treatment on the consolidating process the white electrocorundum was processed at different modes (rotation frequency of the grinding vessels f – 20 and 30 Hz, the time of treatment τ – 10, 20, 30 and 40 minutes for each frequency).

After the treatment the powder mixtures were sieved through a sieve 0045 for 10 minutes using vibratory drive C.1 to obtain the fraction < 45 μm and plasticized with an aqueous solution of carboxymethyl cellulose (CMC) at the rate of 5 wt. % CMC and 95% a powder. After granulation and drying the plasticized powders were molded using uniaxial pressing in a steel mold. The obtained compacts were cylinders of diameter 10 ± 0.01 mm and a height of 5 ± 0.01 mm.

Sintering of the compacts was carried out in a resistance furnace at a high temperature regime: a heating rate – 10 dg/min, temperature – 1600°C and isothermal holding time – one hour cooling with a furnace.

Treated non-plasticized plasma chemical NP Al_2O_3 was consolidated using the SPS method in the device SPS-515S (Sumitomo). Sintering mode is shown in table 3.

Table 3. SPS mode.

Composition	Pressure (MPa)	Holding time (min)	Sintering temperature (°C)
NP Al_2O_3	40	5	1400

Spark plasma sintering (SPS) is considered to be a promising effective method of consolidation of powder materials [6, 7]. Compared with traditional methods of consolidation of powder materials such as furnace sintering and hot pressing (HP) SPS allows to obtain high density sintered materials at a lower temperature for a short time of isothermal holding.

After SPS samples were cylinders with a diameter 15.0 ± 0.1 mm and a height of 2.0 ± 0.2 mm. The density of sintered samples ρ was determined by using hydrostatic weighing in 96% ethanol with an accuracy of ± 0.001g. The relative density of the samples Θ was also calculated.

Microhardness was determined with a help of the device PMT-3, hardness HRA – with a stationary Rockwell hardness tester, the load was 1000 mN (100 g).

The study of technological properties according to the above techniques showed the almost complete absence of flowability and very low bulk density of oxide NP. Coarse-dispersed powders Al_2O_3 have a satisfactory level of characteristics presented in the table 4.

Table 4. Technological characteristics of initial powders.

No.	Powder composition	Bulk density, (g/cm^3)	Tap density, (g/cm^3)	Flowability, (g/s)
1	NP Al_2O_3	0.07	0.06	-
2	NP 80% Al_2O_3-19% ZrO_2-1% Y_2O_3	0.14	0.14	-
3	Coarse pure Al_2O_3	1.47	1.65	0.4
4	Technical alumina	1.50	1.66	0.4

Thus, the study of plasma-chemical properties of NP Al_2O_3 suggests practical impossibility of their use in the initial state for the corundum ceramics technology.

The results of sieve analysis of the initial coarse-dispersed oxide powders are presented in table 5. It can be seen that the powders have a similar fractionated distribution of particle.

Table 5. Granulometric composition of the coarse-dispersed powders.

Particle fraction	Content of fraction X, %	
	Coarse pure Al_2O_3	Technical alumina
+025	0	0
-025+020	1.0	3.3
-020+016	6.1	4.2
-016+0125	9.6	10.1
-0125+008	15.3	13.8
-008+0063	24.0	23.8
-0063+0045	23.9	24.6
-0045	20.1	19.8

To improve technological characteristics plasma-chemical and coarse-dispersed powders based on Al_2O_3 were activated in a planetary mill for 20 minutes at the rotation frequency of grinding vessels – 30 Hz. This mode is optimum for oxide powders to treat [8]. The data are given in table 6.

Table 6. Technological characteristics of treated powders.

No.	Powder composition	Bulk density, (g/cm^3)	Tap density, (g/cm^3)	Flowability, (g/s)
1	NP Al_2O_3	0.82	1.23	0.2
2	NP 80% Al_2O_3-19% ZrO_2-1% Y_2O_3	0.95	1.26	0.2
3	Coarse pure Al_2O_3	1.22	1.47	0.4
4	Technical alumina	1.36	1.49	0.4

During the treatment spherical particles of plasma chemical powders were failed and united in hard agglomerates. Processing of coarse-dispersed powders Al_2O_3 significantly increased the content of fine fractions (-0063) – up to 60%.

The most effective mode of mechanical activation of the white electrocorundum powder in a planetary ball mill Activator 2SL was defined: the rotation frequency of grinding vessels f – 30 Hz, time of treatment – 40 min. During the treatment by this mode there was practically no coarse-dispersed fractions (+010), the content of medium sized fraction was reduced (-010 + 008) and the output of fine fraction (-0063) was significantly increased – more than 75% (table 7). Bulk density of the treated powder was 1.23 g/cm^3.

For further experiments the Al_2O_3 powder of pure alumina grade had been eliminated since the properties of corundum ceramics sintered from this powder have been investigated in detail in [9].

Table 7. Granulometric composition of the white electrocorundum (the rotation frequency of grinding vessels *f*–30 Hz).

Particle fraction	Time of treatment, min			
	10	20	30	40
	Content of fraction X, %			
+020	0.5	0.1	0	0
-020+014	4.4	2.5	2.1	0
-014+010	2.2	2.1	2.0	1.4
-010+008	18.5	14.4	9.7	10.1
-008+0063	8.8	13.3	12.6	13.3
-0063+0045	53.7	51.4	48.8	47.7
-0045	11.8	16.2	24.8	27.5

Treated powder compositions 1, 2 and 4 (table 6) and the powder of white electrocorundum were dispersed on a fraction –0063 and mixed with activating additives in a planetary mill at the rotation frequency of grinding vessels *f* – 30 Hz for 40 minutes. Then the mixture was plasticized according to the procedure described above. Compositions of obtained powder mixtures are given in table 8.

Table 8. Compositions of powder mixtures.

No.	Composition, wt. %					
	NP Al_2O_3	NP 80% Al_2O_3-19% ZrO_2-1% Y_2O_3	Technical alumina	White electrocorundum	NP Al	TiO_2
1	100	0	0	0	0	0
2	95	0	0	0	5	0
3	98.5	0	0	0	0	1.5
4	0	100	0	0	0	0
5	0	95	0	0	5	0
6	0	0	100	0	0	0
7	5	0	95	0	0	0
8	10	0	90	0	0	0
9	20	0	80	0	0	0
10	0	0	95	0	5	0
11	0	0	98.5	0	0	1.5
12	0	0	0	100	0	0
13	10	0	0	90	0	0
14	0	0	0	95	5	0
15	0	0	0	98.5	0	1.5

Compacts were obtained from these mixtures by sintering under the given above modes. Dependencies of density and microhardness of the ceramics sintered from the technical alumina powder on the content of NP Al_2O_3 additives are shown in Figure 1. Adding NP Al_2O_3 in the technical alumina increased density and microhardness of sintered ceramics. This activating effect can be explained by the increase of the area of interparticle contacts which is caused by the adding of NP Al_2O_3. The mechanism of sintering activation due to increased structural and surface activity of NP

Al_2O_3 which can be defined by the crystalline structure defects, size and shape of particles. The most significant increase in density was observed in the ceramics containing 5...20 wt. % of additive NP Al_2O_3.

In this paper, experiments were conducted to study the influence of the NP Al additive on a structure and physico-mechanical properties of ceramics sintered from NP Al_2O_3. The problem of additional contribution to the activation of NP Al_2O_3 sintering by adding of NP Al is of a practical interest. Decrease in density of sintered ceramics was revealed when adding nanosized aluminum to the NP Al_2O_3. This, in turn, led to decrease in hardness of the samples (Figure 2).

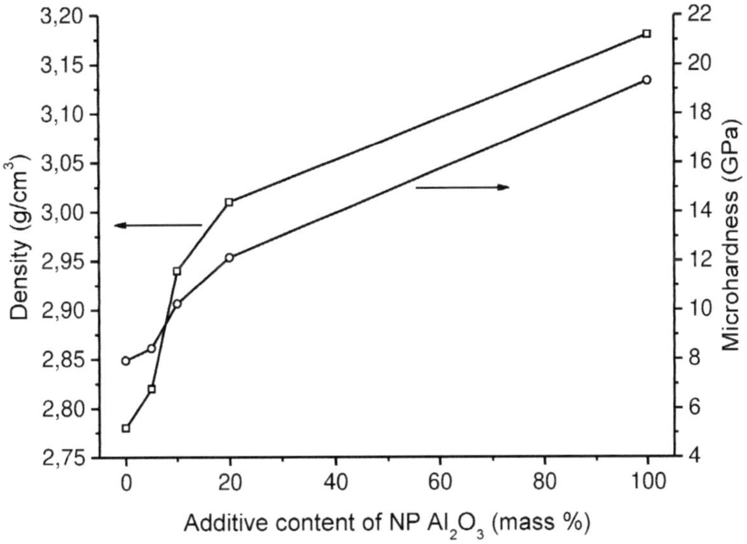

Figure 1. Dependence of the density and microhardness of alumina sintered from the coarse-grained powder Al_2O_3 of a technical alumina type on the additive content of NP Al_2O_3.

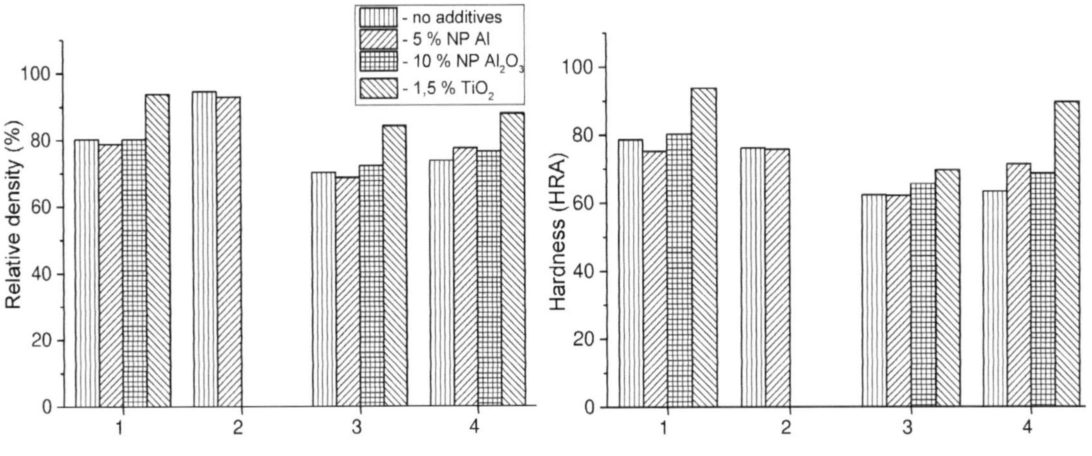

Figure 2. Dependence of relative density and hardness of ceramics on the content and chemical **composition of additives: 1, 2, 3, 4 – ceramics sintered from NP Al_2O_3, NP 80% Al_2O_3-19% ZrO_2-1% Y_2O_3,** technical alumina and white corundum, respectively (for composition (1) the first and third columns are equivalent, for composition (2) NP Al_2O_3 and TiO_2 powder have not added).

Such dependencies can be explained by the increased porosity of sintered ceramics due to oxidation of the additive of Al NP to α-Al_2O_3 during sintering. The oxidation was accompanied by greatly reduced specific volume of a final product of the added additive caused significant difference density of Al (2.7 g/cm^3) and α-Al_2O_3 (3.96 g/cm^3) that has been observed in some experiments.

The density of corundum ceramics obtained by consolidation of the treated unplasticized, plasmachemical NP Al_2O_3 by the SPS method made 3.63 g/cm^3 (92% of theoretical density α-Al_2O_3), hardness – 93 HRA.

The greatest activating effect had the additive of TiO_2 in the Al_2O_3 powder. The density of the sintered ceramics containing 1.5 wt. % TiO_2 reached 3.48 g/cm^3. The partial substitution for Al^{+3} by Ti^{+4} ions caused oxygen vacancies. It should be noted that the number of atoms in the unit cell of the crystalline lattice of the solid solution TiO_2 in α-Al_2O_3 was reduced. It also confirms that the TiO_2 forms a solid solution of subtraction in corundum. The crystalline lattice of a solid solution of subtraction has increased diffusive ability, and thus activates the sintering of corundum.

3. Summary

It is shown that adding the nanopowder Al_2O_3 to the content of 20 wt. % to the coarse-dispersed powders α-Al_2O_3 of grades technical and white alumina activated a sintering process of corundum ceramics that led to increase in its relative density and hardness from 70.2 to 72.2% and from 62.3 to 65.6 HRA for technical alumina, and from 73.7 to 76.5% and from 63.3 to 68.6 HRA for white alumina.

The activating influence of adding of Al nanopowder in the coarse-dispersed powder of α-Al_2O_3 on the sintering process was not revealed. When adding of nano-dispersed Al in the NP Al_2O_3 decrease in the density and the hardness of sintered ceramics due to the additive oxidation up to α-Al_2O_3 was observed. This led to reduced specific volume of the final product of the added additive caused by significant difference of Al and α-Al_2O_3 density, and decrease in density of sintered ceramics. In doing so, the main advantage of NP additive as sintering activator was not implemented, it is the possibility to form a large number of interparticle contacts at very low its content in a sintered compact.

The efficiency of SPS method to obtain dense corundum ceramics from the investigated powders is proved. The greatest activating effect had the adding of TiO_2 powder (1.5 wt. %) in corundum powders. During sintering it is formed a solid solution subtraction of TiO_2 in α-Al_2O_3, the lattice of which has a high diffusivity and activates sintering process.

References

[1] Kiparisov S S and Padalko O V 1988 Equipment of Power Metallurgy Enterprises (Moscow: Metallurgy)
[2] Hunold K, Kempten 1985 Hot Isostatic Pressing of High Temperature Ceramics *Interceram.* **39** 2 40-43
[3] Lukin E S, Makarov N A and others 2003 Strong and Super-Strong Ceramics Based on Aluminum Oxide and Partly Stabilized Zirconium Dioxide *Glass and Ceramics* **9** 32–34
[4] Smirnov A I 1990 Structural Ceramics *Itogi nauki i tekhniki* (All-Russian Institute of Scientific and Technical Information Series Powder Metallurgy) pp 64–106
[5] Balkevich V.L. 1984 *Technical ceramics* (Moskow: Stroyizdat) p 256
[6] Orru R, Licheri R, Locci A, Cincotti A, and Cao G 2009 Consolidation/synthesis of materials by electric current activated/assisted sintering *Materials Science and Engineering* **63** 127-287
[7] Tokita M 2004 Mechanism of Spark Plasma Sintering *J. Material Science* **45** 78-82
[8] Matrenin S V, Belokrylova A O and Ovechkin B B 2012 Research on the Influence of Oxide Powders Mechanical Activation on Properties and Structure of Sintered Ceramics *Izvesiya vuzov. Physics* **55** 5/2 205-208
[9] Matrenin S V, Ilyin A P, Tolbanova L O, Zolotareva E V 2010 Activation of the Oxide Ceramic Sintering by Nanodispersed Powder Additives *Bulletin of the Tomsk Polytechnic University* **317** 3 24-28

Method for phase boundary structure control of laminated materials; destruction process investigations of nanostructured coatings with predetermined phase boundary texture

T A Konstantinova*, **A I Mamaev, A K Chubenko and V A Mamaeva**
Tomsk State University, 36 Lenina ave., Tomsk, 634050, Russia

E-mail: konstantinova.ta9@gmail.com

Abstract. New surface texturing method by means of microplasma coating deposition with the following etching of the coating was shown and described. The method of step by step microplasma texturing was proposed to control the phase boundary of laminated materials. Micrographs of nanostructured inorganic non-metallic coating surface were obtained and analyzed before and after mechanical deformation. The nature of cracks formation and growth was investigated.

1. Introduction

The problem of new functional laminated materials creation and development and their application methods is relevant topic nowadays. The relevance of laminated materials design is directly related to durability issues of these materials under external impacts including temperature and mechanical loads.

Laminated materials are widely used in industries such as electronics (multilayered boards), protecting coatings for aviation and aerospace.

The method of coating formation (nanostructured inorganic non-metallic NIN coatings) by means of high-energy flows localization at the phase boundary with emergence of microplasma processes in solutions allows creating composites of different structures and functions [1] including multilayered materials. Currently known papers in this area are dedicated to single-layered coatings formation [2, 3].

The formation of predetermined phase boundary texture provides increased durability of the coating under external impacts. Microplasma method allows forming of laminated materials with predetermined phase boundary texture. Thus phase boundary texturing results in maximum adhesion of the laminated coating to the metal surface.

The texturing is defined as a set of technological operations resulting in surface structure changes and thus reaching the required surface profile parameters including roughness.

The following coating formation provides the material with predetermined metal – coating phase boundary texture. This approach results in maximum adhesion of the ceramic coating to the metal surface and durability of the coating in terms of exfoliation and destruction under different external impacts.

In this paper we offer texturing by microplasma deposition with the following etching of the coating as a new method to obtain a given metal surface texture. The method contains ceramic coating formation on processed valve metal by microplasma oxidation in solution. Texturing of the metal-ceramic coating phase boundary takes place due to multiple microplasma discharges and related processes during the coating formation.

It is evident that one can adjust the texture parameters by adjusting the parameters of electric impact during the coating formation. The following ceramic layer etching provides a metal surface with predetermined texture. The process can be carried out repeatedly achieving the required texture parameters. Thus, the number of formation – etching operations can also be an operation factor of the processed material texture.

Investigations of NIN coatings destruction process under mechanical loads provides conclusions about the nature of cracks formation and their growth on the coating surface and the way these processes are affected by phase boundary texturing.

2. Materials and methods

Flat square samples made of MA 2-1 magnesium alloy were investigated. Each side of the square was 3 cm.

2.1 Sample preparations

All samples were preliminarily subjected to polishing up to 10 roughness class in order to remove scratches and the oxide film. Then organic contaminants were removed from the surface. The cleaning operation consisted of rinsing the samples with 1 M Na_3PO_4 solution at a temperature not lower than 60 °C for 10 minutes. Then samples were rinsed with distilled water, alcohol and distilled water again. Samples were dried in desiccator for 30 minutes at 60 °C.

2.2 Microplasma texturing

Microplasma texturing of magnesium alloy was performed using the following fluoride phosphate solution: NaF – 5 g/l, $Na_2B_4O_7 \cdot 10H_2O$ – 7 g/l, HBO_3 – 10 g/l, Na_2HPO_4 – 18 g/l.

This electrolyte composition provided porous coatings forming on magnesium mainly consisting of magnesium oxide and phosphate:

$$Mg - 2e \rightarrow Mg^{2+},$$
$$Mg^{2+} + PO_4^{3-} \rightarrow Mg_3(PO_4)_2,$$
$$Mg^{2+} + O^{2-} \rightarrow MgO.$$

Such coatings could be easily removed afterwards by chemical methods, for example, by processing in phosphoric acid solution or concentrated alkali solution.

Samples were immersed in 1 M H_3PO_4 solution directly before the coating formation for 10-20 seconds in order to remove natural oxide layer from the surface.

The processed sample of magnesium alloy served as anode, a tub made of stainless steel with water cooling was cathode.

The temperature in the tub did not exceed 40 °C. Electroinfluence was implemented in the form of voltage impulses with 400 V amplitude and 200 microseconds duration. Entire duration of the process was 20 minutes. The obtained coatings had almost white uniform colour (Figure 1) and were quite dense and fitting well to the substrate. Thickness of the coatings determined by eddy currents method was (20±5) micrometers.

Figure 1. Coating formed on MA 2-1 magnesium alloy by microarc oxidation method.

The coating was etched by 1 M H_3PO_4 solution for 5 minutes. Reaction products were removed from the surface of the metal sample by momentary immersion in 30 % nitric acid solution after rinsing the samples with distilled water. The process was carried out repeatedly in order to determine how the formation – etching operations number influenced on metal texture. The M0 sample was etched without preliminary coating formation to reveal the separate effect of chemical etching on the surface texture. This approach was determined by high magnesium activity in chemical reactions with acids. Designations of samples and the number of cycles are shown in Table 1.

Table 1. Designations of samples and processing stages.

Sample	M0	M1	M2	M3	M4	M5
Number of coating formation operations	0	1	2	3	4	5
Number of coating etching operations	1	1	2	3	4	5

Roughness and profiles after texturing were studied by means of three-dimensional contactless profilometer (Micro Measure 3D Station) by STIL.

Results were processed by Mountains Map Universal software (Version 2.0.13). The software ensured formation of surface profile projections and determination of roughness parameters according to ISO 25178 and ISO 4287.

3. Investigations of samples under mechanical tensile loads

All samples were made of MA2-1 magnesium alloy in the form of flat trowels. Preparation of samples was carried out according to GOST 1497-84 (ISO 6892-84).

Metal with a special texture could be re-coated owing to the features of NIN coatings formation by microplasma deposition, and two-layered material with defined texture of the metal - ceramic coating phase boundary would be formed taking into account the additional stage of microplasma impact.

Treatment of the samples was carried out as described above.

Preparation of the samples and NIN coatings formation in microplasma mode with predetermined phase boundary texture also were carried out as described above.

Designations of samples and the number of deposition – etching stages during the formation of two-layered NIN coating on magnesium with predetermined phase boundary texture are shown in Table 2.

Table 2. Designations of samples and processing stages.

Sample	M1	M2	M3	M4	M5
Number of coating formation operations	1	2	3	4	5
Number of coating etching operations	0	1	2	3	4

3.1 Tensile tests

Tensile tests were carried out in accordance with ISO 21180-201 by means of INSTRON 5948 testing machine together with Bluehill software. The software and its application for metal testing provided simple user interface and full control over all testing parameters in accordance with requirements of ISO 6892-1.

Edges of the sample were fixed in clamps of INSTRON 5948 testing machine so that the sample was straight without application of force and was not slipping in clamps during the test.

Constant tensile load was applied to the sample at a rate of 5 mm/min.

Coating surface analysis was carried out after deformation by means of raster scanning electron microscope SEM 515 without preliminary preparation of the sample.

4. Results and discussion

Figures 2 - 7 show profilograms of magnesium samples after microplasma texturing.

Figure 2. Profilogram of M0 magnesium sample.

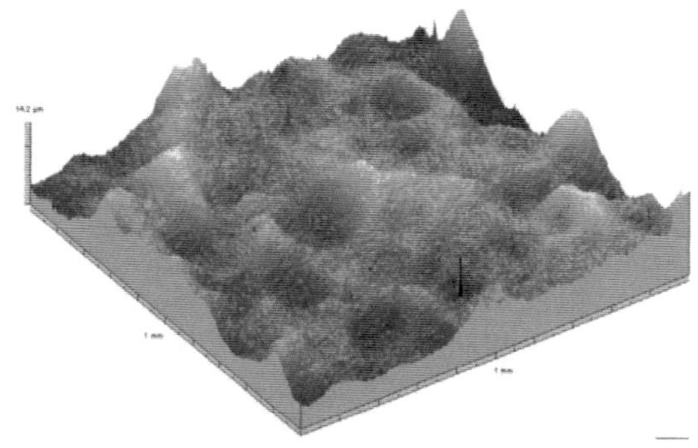

Figure 3. Profilogram of M1 magnesium sample.

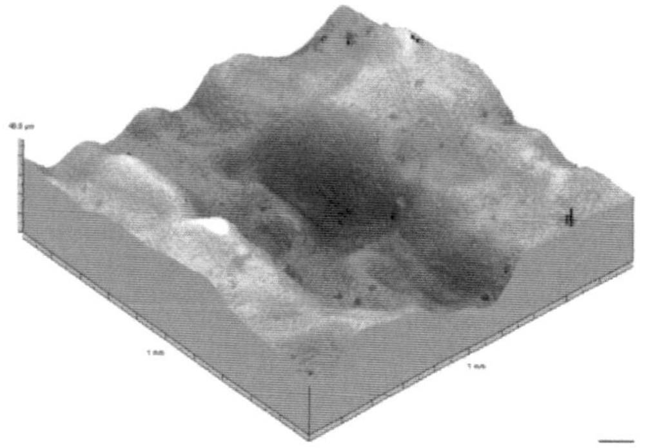

Figure 4. Profilogram of M2 magnesium sample.

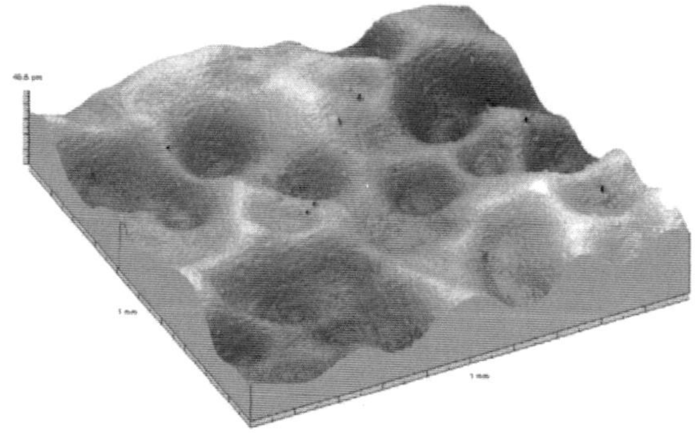

Figure 5. Profilogram of M3 magnesium sample.

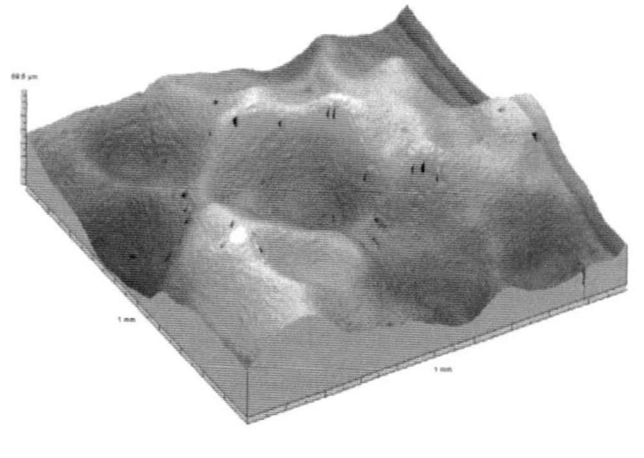

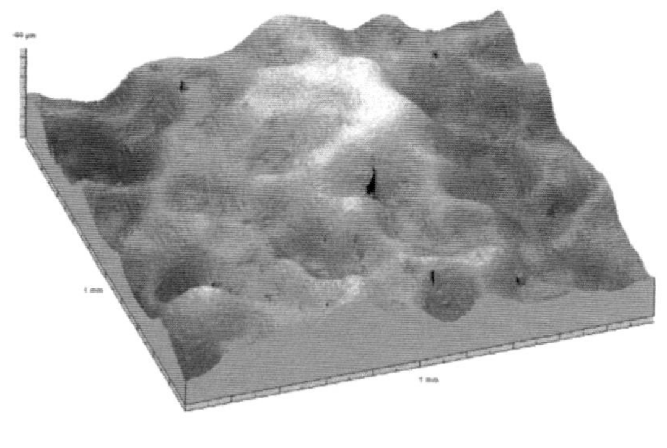

Figure 6. Profilogram of M4 magnesium sample. **Figure 7.** Profilogram of M5 magnesium sample.

Table 3 shows Ra and Rz averaged over 3 definitions.

Table 3. Roughness of magnesium samples textured by microplasma method.

Sample	M0	M1	M2	M3	M4	M5
Ra av., μm	0.818	1.123	4.180	5.523	3.977	2.883
Rz av., μm	3.75	5.88	19.83	20.53	23.57	13.50

The obtained data showed that the value of Ra roughness increased after every stage of microplasma texturing from stage 1 to stage 3. Then the surface smoothened, and the value of Ra roughness decreased. Thus, it was advisable to carry out three consecutive stages in the described mode of microplasma texturing that provided Ra values up to 5.5 μm.

The value of Rz roughness increased until the fourth stage of microplasma texturing and then decreased. Profilograms analysis showed that during the microplasma texturing of magnesium submicrotexturing took place aside from microtexturing. Submicrotexturing is rather difficult to quantify now but it will be useful in order to increase the adhesion with subsequent coating layers.

Mechanical characteristics obtained by tensile tests of the samples are presented in Table 4.

Table 4. Mechanical characteristics of coated MA2-1 magnesium alloy samples with different phase boundary texturing stages under mechanical loads.

Sample	M1	M2	M3	M4	M5
Maximum load, N	1230.00	900.00	900.00	700.00	490.00
Tension, MPa	224.59	190.57	165.01	130.91	104.79
Deformation (elongation), %	21.40	20.76	20.16	14.88452	10.91
Deformation (elongation) in total creep, %	7.13	1.87	5.77	0.83	0.69
Elongation, mm	3.27	1.62	2.47	0.69	0.37
Elongation in total creep, mm	1.14	0.29	0.92	0.13	0.11
Modulus of elasticity, MPa	2 612.87	4 061.91	3 976.41	5 195.19	4 526.07

The obtained data showed that maximum tensile load decreased, modulus of elasticity increased and the value of creep decreased with increasing of texturing stage number. This fact was connected to the thinning of the sample as a result of NIN coatings formation features in microplasma mode with the following etching. Plasticity reduction was proportional to the thickness reduction of the sample.

Thicknesses of samples are presented in Table 5.

Table 5. Thicknesses of samples.

Sample	M1	M2	M3	M4	M5
Thickness, mm	1.10	1.00	0.95	0.85	0.75

Figures 8 – 12 show micrographs (magnification 500 and 1000) of coatings before and after application of mechanical load. Micrographs under magnification 500 appeared to be more informative for most samples.

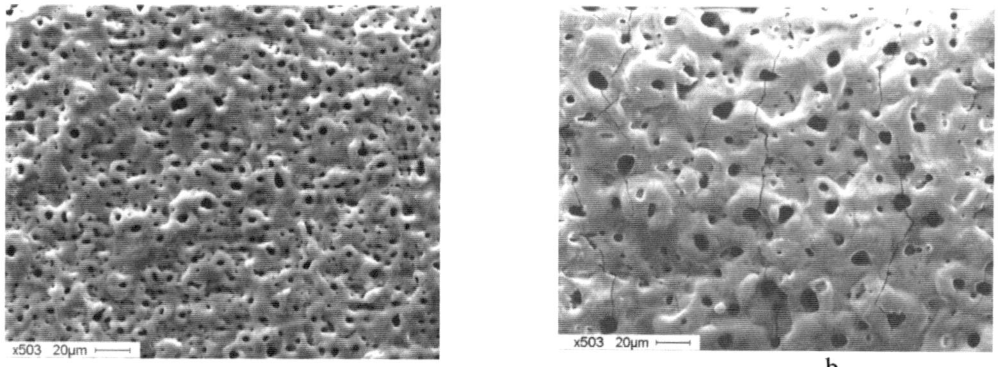

a b

Figure 8. Surface micrographs of coatings formed on magnesium alloy with predetermined phase boundary textures – one texturing stage, before (a) and after (b) application of mechanical load.

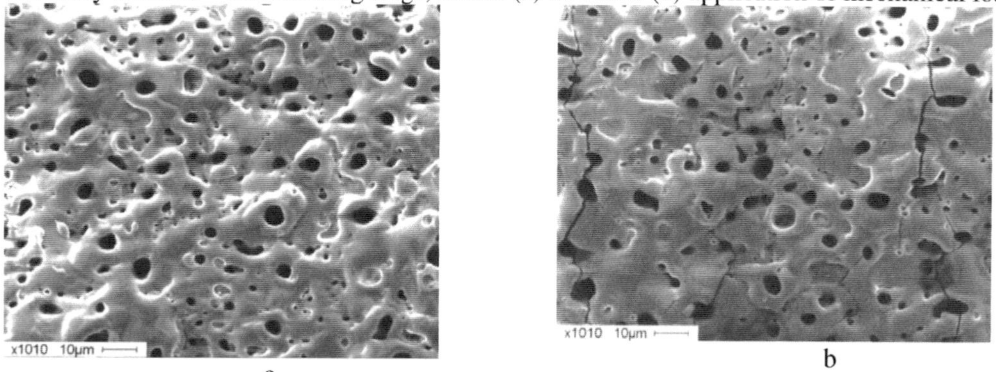

a b

Figure 9. Surface micrographs of coatings formed on magnesium alloy with predetermined phase boundary textures – two texturing stage, before (a) and after (b) application of mechanical load.

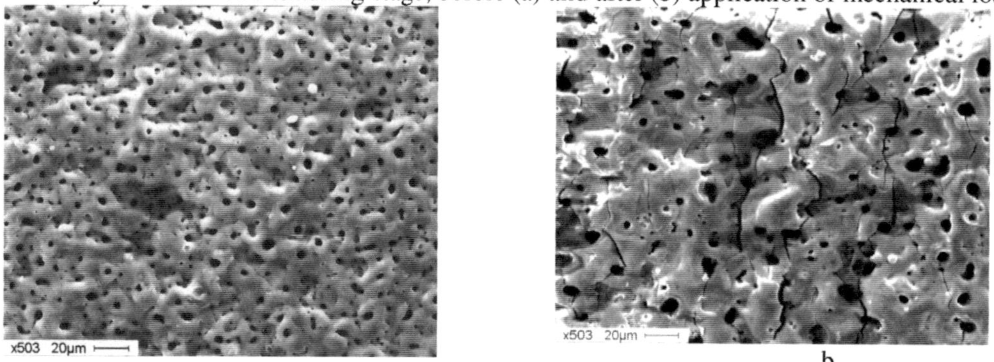

a b

Figure 10. Surface micrographs of coatings formed on magnesium alloy with predetermined phase boundary textures – three texturing stage, before (a) and after (b) application of mechanical load.

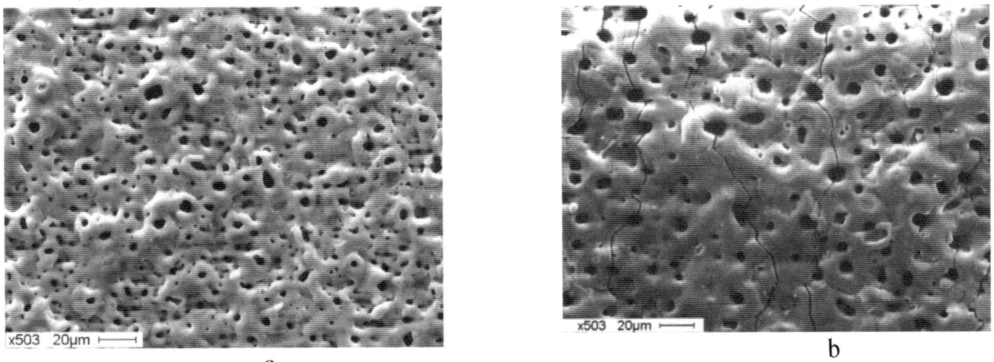

a b

Figure 11. Surface micrographs of coatings formed on magnesium alloy with predetermined phase boundary textures – four texturing stage, before (a) and after (b) application of mechanical load.

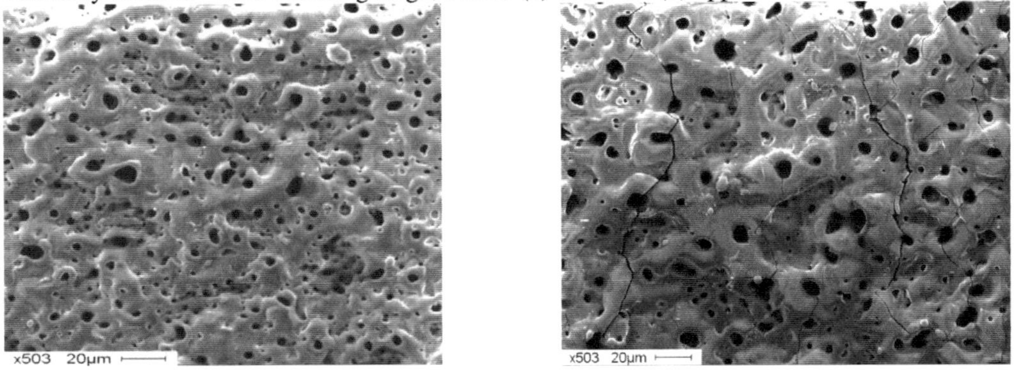

a b

Figure 12. Surface micrographs of coatings formed on magnesium alloy with predetermined phase boundary textures – one texturing stage, before (a) and after (b) application of mechanical load.

Micrographs showed that the coating was porous with pore size 1 - 5 μm, nanoscale pores were also observed. Visible cracks directed normal (parallel) to the loading line were apparent on the coating surface after mechanical load application. Cracks were wavy and characteristically curving around pores.

Cracks distribution character did not change for each cycle of texturing unlike the average distance between cracks. Thus, the average distance between cracks for samples after one texturing stage was 40 - 50 μm that corresponded to average texturing step at that stage (Figure 3), two texturing stages – 35 - 40 μm (Figure 4), three texturing stages – 25 - 35 μm (Figure 5), four texturing stages – 30 - 35 μm (Figure 6), five texturing stages – 35 - 45 μm (Figure 7).

Analysis of micrographs obtained after tensile tests for MA 2-1 magnesium alloy sample with NIN coating – one texturing stage under maximum load 1287 N (Figure 13) with deformation value 26.61 % – showed that there was a partial removal of upper coating layer. This fact indicated that the load was concentrated inside the coating. Since the coating was inelastic, the net of microcracks relaxing at the pores appeared.

Deformation value was 26.61 % that resulted in macrocracks formation. Exfoliation of the coating was not observed.

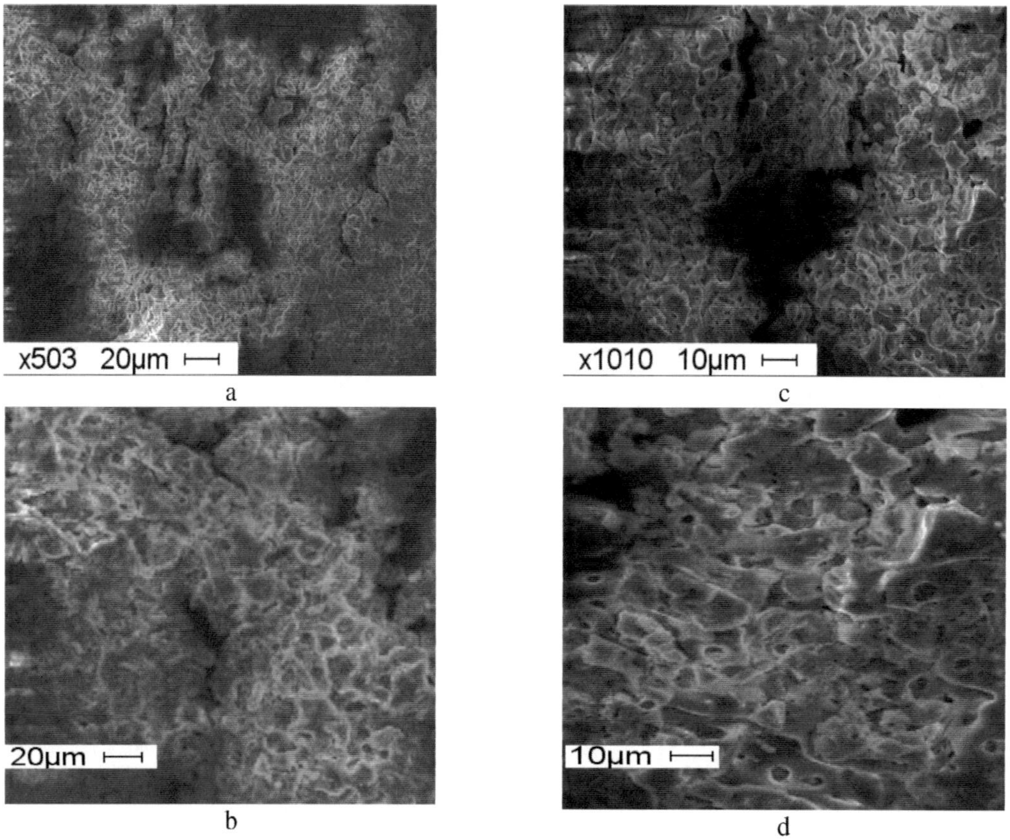

Figure 13. Surface micrographs of coatings formed on magnesium alloy after 26.61% deformation value under different magnifications: a – magnification 500, b – separate section on the photos a, c – magnification 1000, d – separate section on the photos c.

5. Investigations of NIN coatings destruction under critical tensile load

Figure 10 shows surface micrographs of NIN coating produced by microplasma oxidation after critical tensile load application.

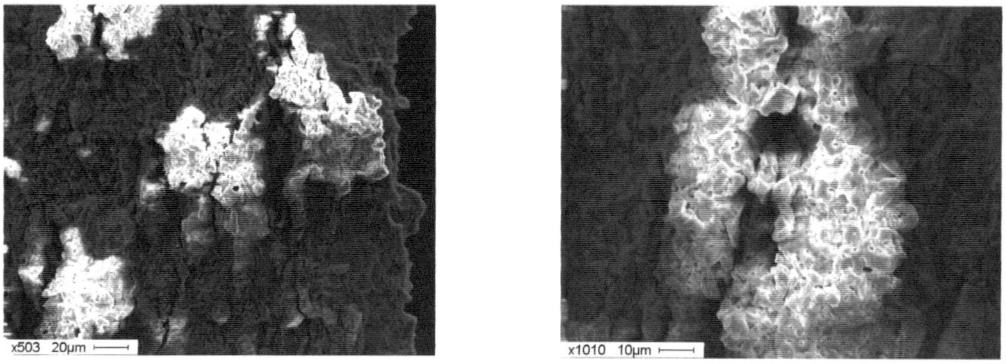

a b

Figure 14. Surface micrographs of the coating formed on magnesium alloy after critical tensile load application: a – magnification 500, b – magnification 1000.

Micrographs showed that circular microcracks were strongly marked around pores after exfoliation of upper coating layer, and circular cracks formation was also observed.

6. Conclusions

Step by step microplasma texturing method for metal surfaces was proposed and considered.

Special feature of magnesium alloys microplasma texturing was increase of Ra roughness value after every stage of microplasma texturing from stage 1 to stage 3 with the following surface smoothening characterized by decrease of Ra roughness value. Rz roughness value also increased until the fourth stage of microplasma texturing and then decreased.

Thus, the proposed method of step by step microplasma texturing can be considered as one of the best ways of valve metal surface texturing. Hence, microplasma texturing is an operation factor of phase boundary structure in laminated materials.

Destruction and deformation processes in two-layered NIN coatings with predetermined metal-coating phase boundary texture were investigated.

Metal thickness decreased during the coating formation with increase of texturing stage number, and the two-layered material became elastic and resilient wherein plasticity reduction was proportional to the thickness reduction of the sample.

Micrographs showed that the microcracks obtained during deformation processes were parallel to each other and normal to the loading line with wave nature of distribution.

Linear microcracks appeared at the thinnest parts of the coating, namely, on texture tops (Figure 8), and then relaxed at the pores during their extension. Since the load increased in that point, some new microcracks appeared starting from the previous relaxation point. New microcracks grew up to the next pores and so on.

With increase of texturing stage number, i.e. with increase of the surface roughness, the average distance between cracks decreased, in its turn the distance between cracks was inversely proportional to deformation values.

Exfoliation of upper coating layer and circular microcracks around pores were revealed after critical tensile load application, and circular cracks formation was also observed.

Samples with textured coatings were able to withstand substantial elongations up to 26 %.

Thus, investigations showed that insular circular microcracks were formed after moderate tensile loads. These microcracks did not result in coating exfoliation as the load was removed (Figures 8 - 10). Since the microcracks were circular, they relaxed on themselves during their extension. Circular cracks emergence removed the load within the coating and prevented the main crack leading to exfoliation of the coating. The obtained data corresponded to theoretical review from section 3.

Acknowledgement

This work was financially supported by the Ministry of Education and Science of Russian Federation, Assignment # 16.461.2014 / K for performing research work in the framework of project of the state assignment in the field of scientific activity.

References

[1] Mamaev A I, Mamaeva V A, Beletskaya E Yu, Chubenko A K and Konstantinova T A 2013 A teory of a collective microplasma process for formation of nanostructural inorganic nonmetallic coatings through localization of high-energy flows in the nanolayers of the metal-electrolyte interface. mathematical modeling. Part 1. *Russian physics journal* (Springer New York Consultant Bureau) **56** pp 959–969

[2] Dey A, Mukhopadhyay A K, Gangadharan S, Sinha M K and Basu D 2009 Development of hydroxyapatite coating by microplasma spraing. *Materials and manufacturing processes.* **24** pp 132 –1330

[3] Xin S, Jiang Z, Wang F, Wu X and Zhao L 2001 Effect of current density on Al alloy microplasma oxidation *J. Mater. Sci. Technol* **17** pp 657–660

IOP Conf. Series: Materials Science and Engineering **116** (2016) 012033 doi:10.1088/1757-899X/116/1/012033

Research of functional properties of nitride ion-plasma coatings

O Yu Vaulina, B B Ovechkin, A V Papchenko, E V Shvagrukova
National Research Tomsk Polytechnic University, Institute of High Technology Physics, 30, Lenin ave., Tomsk, 634050, Russia

E-mail: volgay@tpu.ru

Abstract. This paper considers the influence of ion-plasma coatings with the use of nitrogen (N), zirconium nitride (ZrN), titanium-aluminum nitride (Ti,Al)N and titanium nitride and zirconium nitride by-layer (TiN+ZrN – eight layers) on the properties of steel 65X13. The main functional properties of the coatings are determined: microhardness, nanohardness, Young's modulus and corrosion resistance. It is shown that all the types of coatings allow increasing the physical and mechanical characteristics of instrument steel 65X13. Hardness and wear-resistance, depending on the type of the deposited coating, increase from 1, 5 to 4 times, corrosion resistance increases by tens times.

1. Introduction

Development of mechanical engineering determines, mainly, the possibility to create and obtain instrument materials, available in sufficiency, which correspond to the range of requirements in the modern industry and new technologies. Creation of instrument materials, having the increased wear-resistance and high strength is a topical issue.

A traditional method of increasing the cutting tools properties by alloying is suppressed currently due to the deficit of some alloying elements. Therefore, an important applied task has arisen, which is to find new modes of additional tool processing. Ion-plasma coatings on the surface of cutting tools allow increasing their wear-resistance, when being in service, broadening the area of technological capabilities and decreasing the consumption of high-priced alloying materials as well as improving the quality of the machined surface of components. Wear-resistant coatings have become an integral part of the modern instrument materials, and the majority of modern plates are provided with one or another coating [1].

The aim of the work is to increase the microhardness and wear-resistance for steel 65X13, using the method of ion-plasma saturation of the surface N, ZrN, (Ti,Al)N and TiN+ZrN (8 layers).

2. Materials and methods

The samples of the investigation were the plates with the size 10×15×1 mm. The steel surface was prepared by abrasing machining and polishing. Then, in the conditions of the innovation industrial company LLC "Composite coatings" (Tomsk, Russia) the surface of the samples was saturated with nitrogen (N), zirconium nitride (ZrN), titanium-aluminum nitride (Ti,Al)N and titanium nitride and zirconium nitride by-layer (TiN+ZrN – eight layers).

Structural investigations were conducted, using a metallographic laboratory microscope LaboMet-I. Microhardness was measured with the help of a microhardness tester PTM-3M. Nanohardness was measured with a nanoindenter MTS G200. Tests for abrasion wear were conducted at a standard laboratory setup.

Nitrided materials have high surface hardness and strength, unchangeable at heating to 500…600°C. Nitrided steel is used for the production of cutting tools and the other components, working till wear as well as at high temperatures and corrosive environment.

The coating of zirconium nitride (ZrN) keeps a good combination of properties: high corrosion resistance and a low friction coefficient. It is used in various branches of industry.

The coating (Ti, Al)N increases wear-resistance, has good resistance to corrosion wear and high heat resistance. The instrument with the coating (Ti, Al)N is used in the situations, when heavy thermal loads are needed [1, 2]. During the work the deposition of the composition has been performed in a titanium-aluminum ratio 50/50.

It is known that the most effective method to prevent cracking in the surface layers of the instrument is to use multi-layer coatings, where "hard" layers interchange with the "soft" ones. For example, a central layer is iron (steel) and aluminum layers are disposed peripherically. At moving a crack will relax stresses in a soft aluminum layer, consequently, a life span of the given composition increases. Is it possible to make a similar kind of combinations for cutting tools coatings?

Evidently, a complete analog of the combination aluminum-iron-aluminum for cutting tools is impossible as far as the extreme layers of aluminum cannot withstand high cutting temperatures, which may reach 1200°C. But there are many coating materials that allow choosing the necessary interchangeability of the properties in a broad range [3].

In the frame of the multi-layer coating research the combination (TiN+ZrN – 8 layers) is offered, which, according to authors' opinion, allows increasing the wear-resistance and corrosion wear of the instrument.

3. Results and Discussion

The research was conducted, using instrument steel 65X13. The structure in the initial condition (condition as supplied) was a granular perlite with inclusions of carbides. Average microhardness was 220…240 N/mm^2.

The microhardness of the samples with deposited coatings was measured at the end surface from the surface to the centre of the sample, some tracks for one coating. The curves of microhardness dependencies according to the depth of the sample are presented in Figure 1.

For all the deposited compositions it can be seen that the saturated surface layer has a higher microhardness than the centre. The microhardness near the surface along the ends of the machined samples is 300…420 N/mm^2, and it decreases in the distance 400…600 μm to the centre to 200…220 N/mm^2. Thus, microhardness of all the samples with ion-plasma coatings has increased in 1.3…2 times in comparison with the microhardness of the initial steel. The character of a transient area has a smooth junction from the reinforced layer to the matrix metal. The thickness of the modified layer in all the cases is, approximately, equal to ~ 500 μm. The TiN+ZrN (8 layers) coating has the highest microhardness.

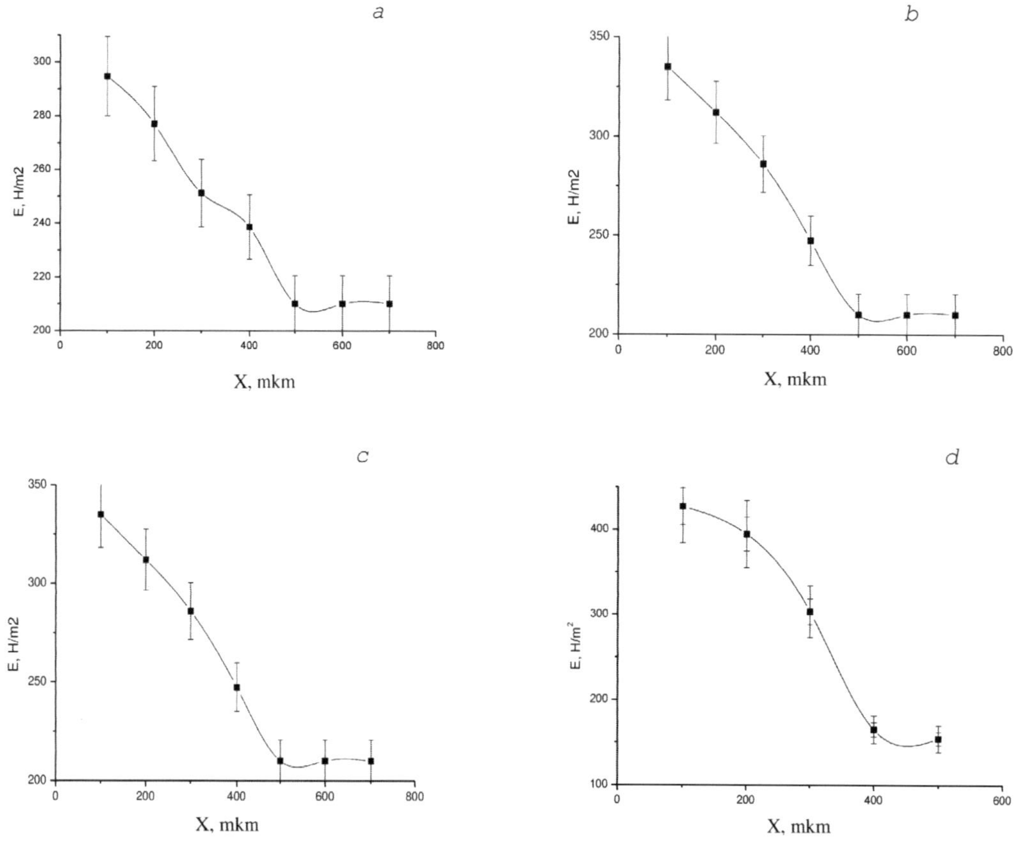

Figure 1. Microhardness dependence on the thickness of the samples with coatings: a – ZrN; b – N; c – (Ti, Al)N; d – TiN+ZrN

Hardness measuring of a thin surface layer at small loads was conducted with the help of the nanoindenter MTS G200. For each coating several measurements were made, and the curves of nanohardness dependence on the indenter penetration strength were plotted (Figure 2).
Also, the technology for nanohardness definition allows calculating Young's modulus automatically, which characterizes the material properties connected with the resistance to elastic yield. The obtained data of an average nanohardness and values for Young's modulus are presented in summary Table 1. It can be seen that the samples with the TiN+ZrN (8 layers) coating have the highest nanohardness, while the sample with the nitrided layer has the lowest one (after the initial steel).
Abrasion wear of the material in the work was defined at friction on the floated abrasive particles. Tests on wear-resistance were conducted, using the initial steel and steel 65X13 with all the types of investigated coatings. Wear tracks demonstrated the similar character for the initial steel with N-, ZrN-coatings. The surface of the tracks had a "streamy" structure; it had a metallic luster, what was typical for a brittle fracture. Wear tracks for (Ti, Al)N- and Tin+ZrN-coatings differed from the abovementioned tracks: a distinctive "cup-type" structure was revealed, the brushed character without any luster, which was usual for a ductile fracture. This fact was explained by the presence of lamellar constituents in these coatings.

IOP Conf. Series: Materials Science and Engineering **116** (2016) 012033 doi:10.1088/1757-899X/116/1/012033

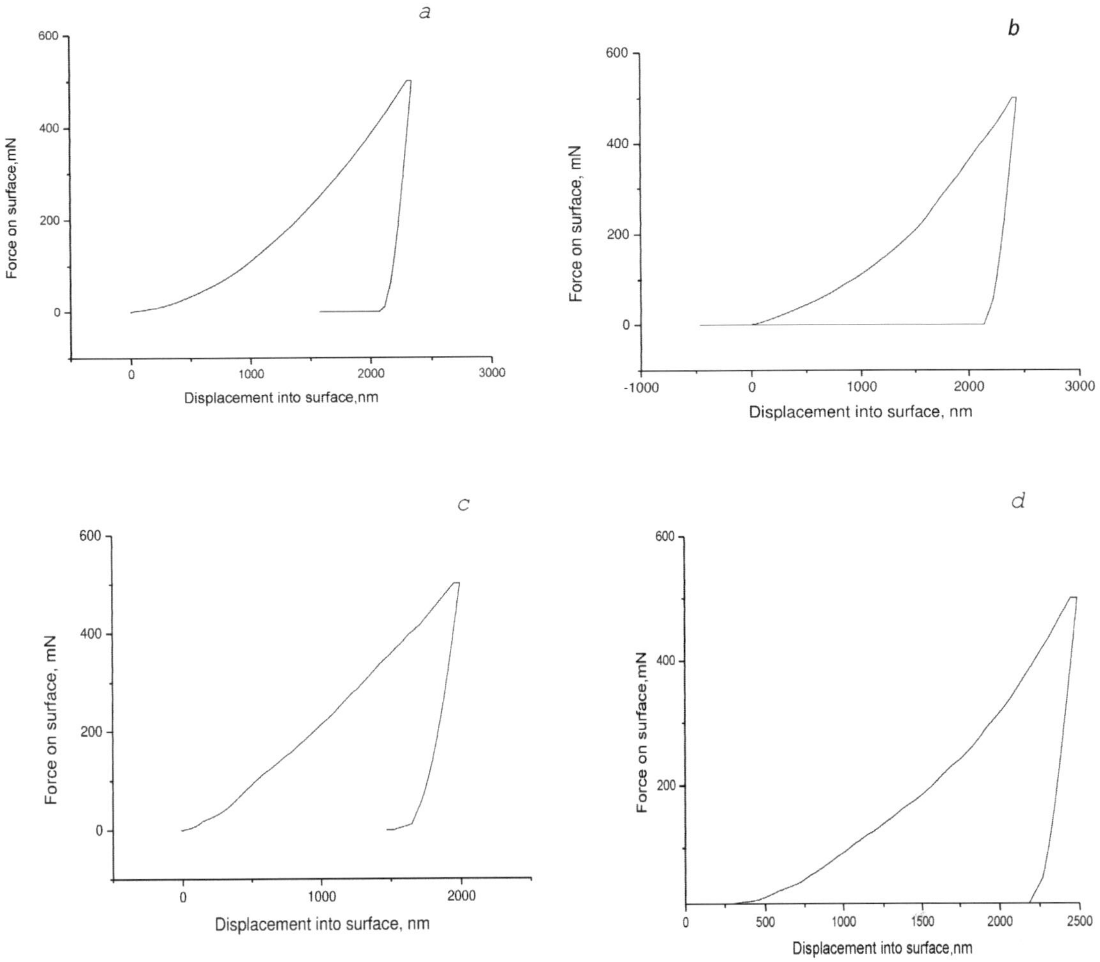

Figure 2. Nanohardness dependence on the indenter penetration strength into the sample: a – N; b – ZrN; c - (Ti, Al)N; d - TiN+ZrN

The obtained results of the tests for wear-resistance are presented as bar graphs in Fig.3 and in Table 1. It can be seen that the best results for wear-resistance a complex eight-layered TiN+ZrN coating has demonstrated.

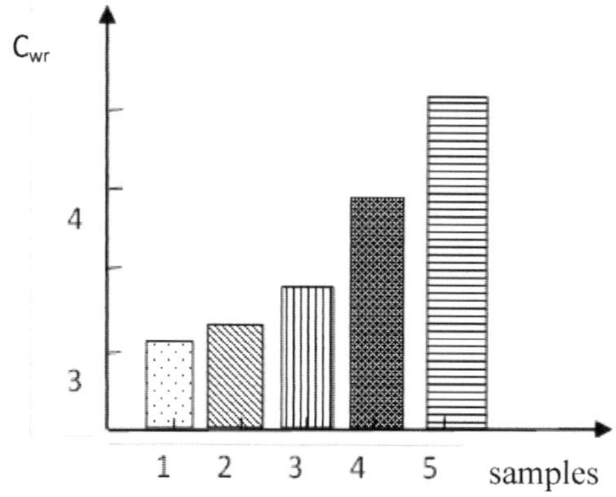

Figure 3. Wear-resistance coefficient (C_{wr}) of the investigated samples: 1 - 65X13, 2 – nitrided, 3 – ZrN, 4 - (Ti,Al)N; 5 - Tin+ZrN (8 layers)

Table 1. Summary data obtained as a result of research (E_{av} – Young's modulus; H_{av} – average nanohardness; R_w – relative wear; h_i – relative mass change as a result of corrosion in Δti=60min)

Material	E_{av}, H/mm^2	H_{av}, H/mm^2	C_{wr}	h_i
65X13	232024	4066	1.1	0.33
N	232111	4352	1.3	1.92
ZrN	233927	4361	1.9	0.36
(Ti,Al)N	235723	5833	2.9	0.87
TiN+ZrN	235983	5935	4.2	0.03

Corrosion tests were carried out in the laboratory conditions in a 10%-solution of a nitric acid in water. The nitric acid is a corrosive environment for the majority of metals. The scheme of testing may be different: dipping and storage in a basin with the high humidity or dipping and storage in the atmospheric humidity. The last is the most suitable for testing the instrument steels with coatings.
Nitride coatings testing for corrosion resistance to the static effect of liquid was carried out according to GOST 9.308-85. By this method the samples were deposited in the corrosive environment for 5 minutes of each hour with the following storage in the atmospheric humidity during 50 minutes. The tests were conducted at a room temperature. At every turn we conducted the estimation of a sample mass loss per the unit of area, the mass was determined by weighing. Moreover, the external examination of samples had been performed before the first attributes of corrosion appeared. The obtained results are demonstrated in summary Table 1.
The best result of corrosion testing was shown by the TiN+ZrN coating, which for the whole time of corrosion testing did not lose the mass. The next result was the stainless steel 65X13. This steel is known by its high ability to resist influencing of various corrosive media, which include the influence of nitric acid. The samples with the coating ZrN (Table 1) showed reasonably good results, but the coating itself had disappeared after 20 minutes, and the other time only steel worked. The nitrided samples demonstrated the worst corrosion resistance in a water solution of nitric acid.

4. Conclusion

The possible methods of increasing the hardness, wear-resistance and corrosion resistance of instrument materials by means of alloying have a significant disadvantage connected with the deficit and expensiveness of alloying elements, and they have already been worked out, to a great extent. The paper shows that the application of new progressive technologies of ion-plasma coatings deposition may increase the operational characteristics of instrumental materials and obtain a significant economic effect.

The coatings N, ZrN, (Ti, Al)N and a multi-layer TiN+ZrN (8 layers) deposited on steel 65X13 by the ion-plasma method are investigated in the paper. It is observed that all the types of coatings allow increasing the physical and mechanical characteristics of instrument steel 65X13. Hardness and wear-resistance, depending on the type of the deposited coating, increase from 1.5 to 4 times, corrosion resistance in the case of a multi-layer coating increases by tens times.

The complex TiN+ZrN (8 layers) coating has demonstrated the best values of working properties. However, the other investigated coatings have showed the results, which are much higher than the initial steel 65X13 has.

Therefore, it is shown that the use of modern technologies and materials for ion-plasma coatings deposition may significantly improve the physical and mechanical characteristics of instrument steels and provide the high technical and economic indicators of machine building.

References

[1] Tabakov V P and Chihranov A V 2010 Determination of mechanical characteristics of wear-resistant ion-plasma coatings based on titanium nitride *Proceedings of the Samara Scientific Centre of the Russian Academy of Sciences* **12** 4 292-97 (in Russian)

[2] Brozdnichenko A N, Gorchakov S A, Ryazancev S S, Senkevich S V, Tilte K L and Hinich I I. 2000 Influence of deposited TiN- and AlN-layers structure on the properties of a nanocomposite TiAlN coating. *Izvestia: Herzen University Journal of Humanities and Sciences* **15** 64-68 (in Russian)

[3] Cirkin A V 2003 Development of a multi-layer coating for end cutters *Reporter of Ulyanovsk State Technical University* **3-4** 33-35 (in Russian)

Deformation stages of technical aluminum at reverse

O Yu Vaulina, A N Durnovtseva, E V Shvagrukova
National Research Tomsk Polytechnic University, Institute of High Technology Physics, 30, Lenin ave., Tomsk, 634050, Russia

E-mail: volgay@tpu.ru

Abstract. Durability and reliability of machines and mechanisms are determined, mainly, by their fatigue resistance as far as, in the most cases, variable load impacts on machine components. Accordingly, the problem of fatigue failure is extremely topical, still. Its complexity is connected with a wide range of factors. First of all, at cyclic load the compatibility relations of a material surface layer, which is loaded over the yield point and the elastic-loaded substrate layer, play a very important role. This fact determines involvement into plastic flow and failure of all the scale hierarchy of deformation structural levels. Reverse loading under the condition of the elastic-loaded substrate layer causes strong localization of plastic deformation in the surface layers. In the deformation localization areas the material reaches its limit state, when fatigue cracks arise and expand. The paper presents the mechanisms of fatigue deformation for technical aluminum at various fatigue stages.

1. Introduction

Undiminishing interest towards the fatigue problem is determined, firstly, by the applied significance of the utmost importance for understanding the physics of this phenomenon as far as it determines the operational life for the majority of work-pieces and structures in machine building industry. Secondly, this is an extremely complicated phenomenon, thereby, the insufficiency of the knowledge about the nature of the fatigue process is kept, in spite of a great number of scientific studies in this direction. The most known investigations connected with the fatigue mechanism have been carried out at the microscale level, using the analysis of forming dislocation structures.

At the moment, the problems of fatigue are considered in a broad fashion. But, nevertheless, the mechanisms of surface deformation at a meso-level [1] stay underinvestigated. Therefore, this paper is devoted to the determination of fatigue processes stages for technical aluminum.

The aim of the present study was to investigate the behaviour of laminated Al alloy-Al-Mg alloy joints under compression and tensile loads.

2. Materials and methods

The technical aluminum of A7 grade was investigated in the course of work with the following composition: 0.25%Fe, 0.18%Si, 0.05%Zn, with 0.03% Mn and Mg, 0.02%Ti, 0.01%Cu, and other elements, the rest aluminum ~ 99.6%.

The aluminum surface was prepared, using elctropolishing. Load distribution – reverse bending, a frequency of 7 Hz, an amplitude ± 1.5 mm. Structural changes while loading were observed with the help of metallographic microscopes LaboMet-I and Axiovert-25CA with an add-on device DIC for getting contrast. Microhardness was carried on GOST 9450-76 with the use of a PMT-3 microhardness tester.

3. Results and Discussion

One of the most important tasks in the physics of strength and plasticity is to understand the physical nature of plastic yield and deformation. The mechanism of polycrystal plastic deformation is extremely complicated, and it depends strongly on the load distribution, conditions and the material itself [1, 2].

The method of microhardness testing, characterizing the resistance of material local volumes to plastic deformation, allows studying the changes of properties in the fatigue process. Therefore, it is widely used for fatigue processes investigations. For classification of structural changes at fatigue the dependence of microhardness on the number of loading cycles is chosen (Figure 1), which shows how the microhardness of a thin surface layer is changing with the increase of loading cycles.

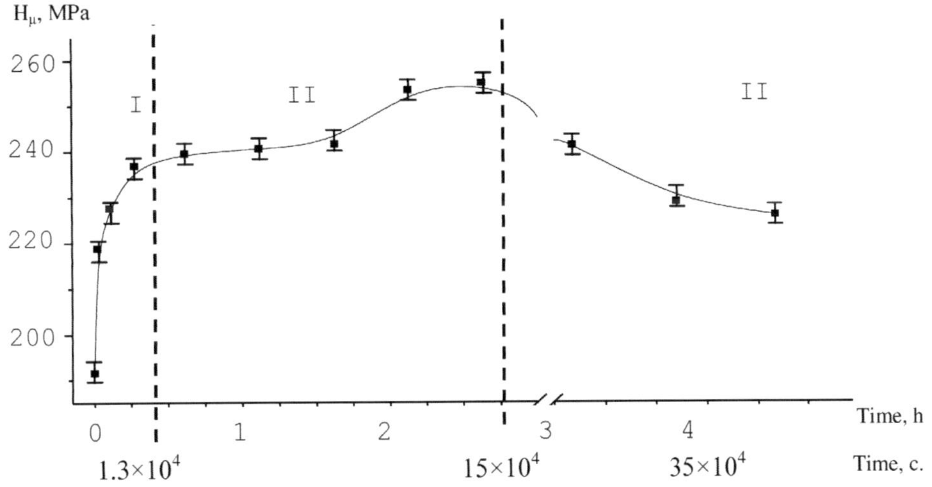

Figure 1. Dependence of microhardness (Hμ) on the time of technical aluminum loading

3.1. Fatigue Stage I

At Fatigue Stage I for aluminum A7 the microhardness increases with the increase of loading cycles during a rather small number of cycles $N=0\div1.3\cdot10^4$. Only in separate grains of technical aluminum near the fixed grip the individual, rather thin, sliding lines in one system are formed, which cause stress fields on the boundaries, influencing on the given element from the area of the surrounding grains. They appear, first of all, in the grains, which are the most favorably pointed towards the applied stress. It is known that at cyclic load sliding occurs along the same atomic planes and in the same directions as under the influence of static load. However, at cyclic load the number of sliding systems, operating in the grain, is limited.

In polycrystal grains maximum two sliding systems operate. As a rule, the primary sliding in one grain is matched with the secondary one in the adjacent grain. Deformation in the reverse direction, initiating the matched system in the adjacent grain, leads to relaxation on the stress grain boundaries, which prevent a shear in the straight direction. As a result, when the sign changes, sliding may develop in the primary system again.

3.2. Fatigue Stage II

Fatigue Stage II corresponds to the number of loading cycles $N=1.3\div15\cdot10^4$. It is characterized by the abrupt decrease of microhardness growth rate as a result of cyclic load. First of all, distinguishable expanding of sliding traces is observed (Figure 2a).

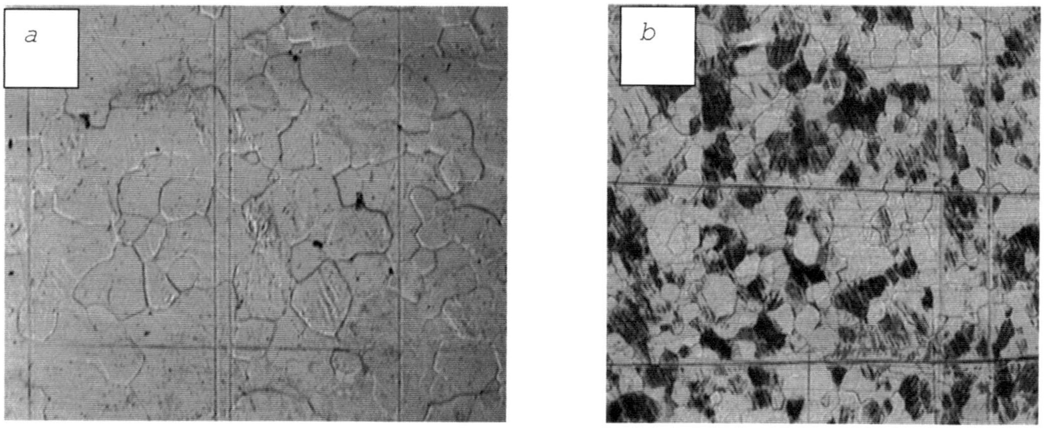

Figure 2. Structural changes in the technical aluminum A7: a – Stage II: $N=11\cdot10^4$ cycles, ×200; b – Stage III: $N=26.7\cdot10^4$ cycles, ×80

For reversal load it is typical to have heterogeneous deformation distribution in the sample, on the whole, as well as in the grains, particularly [1-4]. In aluminum A7 many grains of the surface layer stay non-deformable. A new deformation quality at Fatigue Stage II is an experimental fact, confirming that these grains are distributed along the polycrystal not randomly, but they are grouped in the couples, triples, etc. of the grains with self-consistent sliding. The origin for self-consistent shears in the adjacent grains of a polycrystal is as follows.

Individual sliding bands, leading to the shear of crystallographic planes, are accompanied by grain material rotation in the active grains. Due to the reaction of the adjacent grains the stress raisers appear on the boundaries. The interaction of short-range raisers leads to fine adjustment of active stressed adjacent grains in self-consistent deformable conglomerates.

3.3. Fatigue Stage III

Fatigue Stage III is the main part of the process according to its duration. A new cyclic deformation quality in aluminum A7 is the beginning of the locking process for separate groups of self-consistent deformable grains into loop deformation conglomerates (Figure 2b), i.e. formation of new structural deformation elements on a much greater scale in comparison with the original grains. Rotation accommodation of these large structural deformation elements is carried out, initially, as fragmentation of the grains, composing a conglomerate (Figure 3a). Fragmentation is one of the most efficient mechanisms for stress meso-raisers relaxation. However, in this case, in the grains with coarse shear bands of one system only two fragments are formed, as a rule. Further sample loading is accompanied by the increase of a fragment turn. Consequently, this accommodation along with stress meso-raisers relaxation caused by the rotation of grain conglomerates forms the secondary stress meso-raisers connected with a strong turn of large-scale fragments.

When the opportunity to have fragmentation as an accommodative process of a rotation type is exhausted, a relaxation mechanism of cracking comes into action.

Consequently, while in the surface layer the self-consistent adjustment of plastically deformable grains is taking place, the compensation of material rotations of plastic shears occurs in each of these grains with keeping the continuity of the material.

Cracks arise in the area of the maximum bending, as a rule, on the boundaries of self-consistently deformable grains conglomerates. This makes perfect sense as far as at the rotation of such a large-scale structural element as a grain conglometrate, the most powerful stress raisers are formed on its boundary. It could be observed in Figure 3b. This process is completed with a sample destruction.

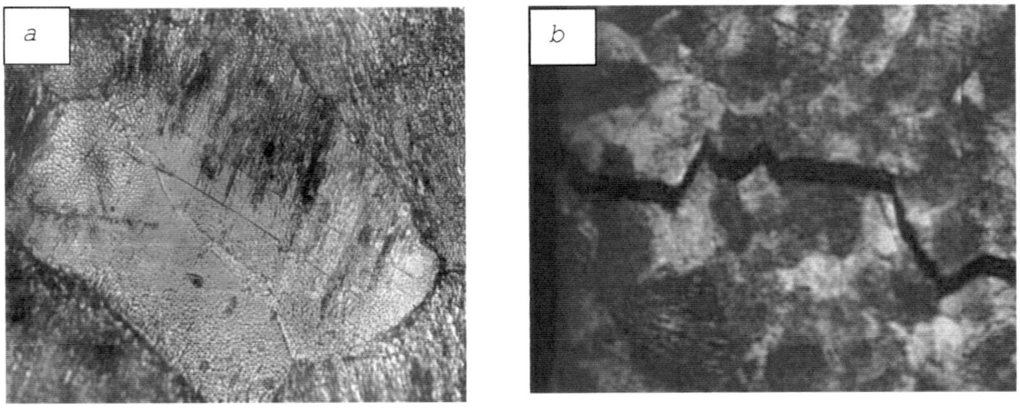

Figure 3. Structural changes: A7: a - N=$50 \cdot 10^4$ c., ×2000; b - N = $56 \cdot 10^4$ c., ×200

4. Conclusion

Fatigue process development stages are determined, using the method of microhardness testing, as well as investigations connected with the character of mesosubstructure formation at various fatigue stages of technical aluminum. It is shown that microhardness dependence on the number of loading cycles consists of three (four) stages, which are distinct from each other according to the rate and the sign of microhardness value changes.

At Fatigue Stage I for technical aluminum the microhardness value increases abruptly while loading, the picture of deformation is presented by separate thin sliding lines in a small amount of grains.

Fatigue Stage II is characterized by the abrupt decrease of microhardness growth rate. In aluminum A7 in the favorably pointed grains the sliding bands are accumulated, which are grouped in the grain couples, triples with self-consistent deformation according to the increase of cycles' number.

At Fatigue Stage III, which is the main part of the process, the decrease of microhardness occurs. A new quality in the meso-structure nature at this stage is the formation of self-consistently deformable grains conglomerates, simulating the loops with the centers of non-deformable grains in technical aluminum.

Fatigue failure of technical aluminum at reverse bending has a mixed character: in some areas of its path a crack is propagating only along the grain or conglomerate boundary, in the other areas it propagates by the grain body along the direction of single sliding. This experimental fact proves that in the given conditions of loading in aluminum A7 the quantities of volume hardness and grain boundaries are commensurable.

References

[1] Panin V E, Elsukova T F, Egorushkin V E, Vaulin O Yu , Pochivalov Yu I 2008 Nonlinear wave effects of curvature solitons in surface layers of high-purity aluminum polycrystals under severe plastic deformation. I. Experiment. *Physical Mesomechanics* **11** 1-2 63-72

[2] Kuzina O Yu, Elsukova T F, Panin V E 2005 Mesoscopic structural levels of deformation in surface layers and the character of polycrystal fatigue failure at reverse bending. Part II. Multilevel approach *Physical Mesomechanics* **8** 4 13-26

[3] Vaulina O Yu, Sergeeva E A 2013 Influence of the surface layer condition on the mechanisms of cyclic deformation and titanium polycrystal fatigue failure *Russian Journal of Physics* **56** 7/2 195-199

[4] Vaulina O Yu, Sergeeva E A 2014 Investigations of commercial titanium polycrystal deformation and failure under the condition of variable loading. *Modern materials science: materials and technologies of new generations: proceedings of Russian national seminar with international participation, Tomsk, June, 9-11, 2014, National Research Tomsk Polytechnic University (TPU)* 49-52

IOP Conf. Series: Materials Science and Engineering **116** (2016) 012035 doi:10.1088/1757-899X/116/1/012035

Theoretical analysis of features of dipole and quadrupole configurations of partial disclinations in nanocrystals of metals

I I Suhanov[1,4], I A Ditenberg[1,2,3], A N Tyumentsev[1,2,3]

[1] National Research Tomsk State University, Tomsk, 634050, Russia
[2] Institute of Strength Physics and Materials Science SB RAS, Tomsk, 634055, Russia
[3] Siberian Physical-Technical Institute, Tomsk, 634050, Russia

E-mail: suhanii@mail.ru, ditenberg_i@mail.ru, tyuments@phys.tsu.ru

Abstract. The results of theoretical analysis of fields of local internal stresses and their gradients of nonequilibrium substructures in metal materials during the formation of nanostructural states in conditions of intensive deformation are presented. Possibility of use of disclinational approach within the continual theory of defects for description of the above mentioned states at nanoscale structural level is shown. Dipole and quadrupole configurations of wedge partial disclinations are considered. It's shown that the size of local internal stresses and their gradients can reach $P \approx E/50$ and $\partial P/\partial x \approx 0.02\ nm^{-1}$ respectively. The calculating experiment demonstrates that high values of local internal stresses and their gradients in the field of elastic distortions can be described by multipole configurations of partial disclinations like dipole/quadrupole. The problem of the nanodipole configuration of partial disclinations and its disintegration on quadrupoles system is considered.

1. Introduction

One of the intensively developing directions of physics of strength and plasticity is creation and study of nanostructural metal materials. By now the experimental results show that the main factor in charge of the formation of unique physical and mechanical properties in such materials is features of their nonequilibrium structure [1 - 4]. Such structure is characterized by existence of the high-defect states differing in elastic fields with the considerable values of local internal stresses [5, 6], and in some cases essential increase of the specific atomic volume [7].

Unfortunately, practically there are no works on the theoretical description and modeling of such non-equilibrium states, which have a direct influence on the evolution of the microstructure, as well as the level of physical and mechanical properties of nanostructured metal materials under different conditions of external impact. The above mentioned problem, in our opinion, is the result of insufficient study of representations of the relationship between the nanograin's sizes and the processes of evolution of defective substructures. Moreover, the development of the most complete model of evolution in metal crystal grains within multilevel approach assumes consideration of

[4] To whom any correspondence should be addressed.

mechanisms of deformation at the different scale levels [8]. It's important that in contrast to the macro-, meso- and submicrocrystalline scale levels to describe the processes of the transformation of the microstructure on the nanoscale level classical (dislocation) approaches often can't be applied. So in the papers [9 - 10] it's shown that in some materials grains practically don't contain a dislocations after decrease of their sizes to several tens nanometers. At the same time it's established that in conditions of intensive plastic deformation the rotational mode appears in the metal nanocrystals. Such mode leads to the formation of a substructure with high curvature of a crystal lattice, zones of the braked shifts and turns, nanostrips of the reorientation [5, 6]. The existence of this rotational mode indicates the possibility of using disclination models and representations which were applied only to the description of structural conditions of metal materials on meso- and submicrocrystalline scale levels.

In this paper the simulation of the dipole and quadrupole disclination configurations in the field of elastic distortions and the subsequent theoretical analysis of fields of local internal stresses and energies of these configurations are conducted.

2. Experimental materials and procedures

Computing experiment was conducted in computer algebra system Maple version 15. Simulation of tension fields is carried out with using of the explicit form of components of the stress tensor [12, 13]. For the convenience of calculation were considered only wedge components of disclinations. Visualization of stress fields and the spatial distribution of elastic energy is produced using graphical tools of Maple 15. Analysis of the dependence of the energy of disclination configurations (dipole / quadrupole) is performed using numerical integration by Gauss method in the adaptive 30-point variation, at estimates of values of an integration the Kronrod's rule on 61 points was applied.

In the simulation of the crystal representations of the homogeneous isotropic continuum in which the Hooke's law is carried out are used. Independent elastic constants were the shear modulus (G) and Poisson's ratio (ν). The value of the Frank's vector (power of disclinations) ω during computing experiments didn't change and amounted to $\omega \approx 1°$. The last condition is necessary for a geometrical linearization of displacement from disclinations, i.e. in the formulas for the elastic fields are only saved by line ω members. It allows to receive closed analytical relations for elastic fields and energies of rectilinear disclinations [13].

3. Results and discussion

As a result of the conducted electron-microscopic research in works [5, 6] existence of nanostrips of reorientation tens long and few nanometers wide, characterized by high (hundreds of degrees / micrometer) values of elastic curvature of the crystal lattice and high ($\partial P / \partial x \approx 0{,}02$ nm^{-1}) local stress gradients is revealed. On the basis of the analysis results is obtained misorientation scheme (figure 1 c) [6]. The important feature of the found nanostrips is that they break off in a crystal (figure 1 c) and thus they are becoming the zones of the braked shifts and turns. The typical misorientation angles (φ) have values $\approx$ 0.5° - 2° (figure 1 a, b).

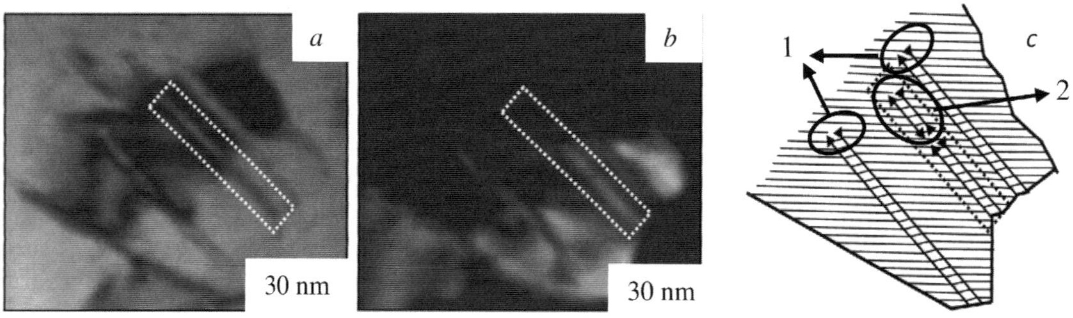

Figure 1. Nanostrips of reorientation in submicrocrystalline nickel after intensive plastic deformation a - bright-field electron - microscope image; b - dark-field image; c - misorientation scheme[5, 6].

On the above mentioned scheme (figure 1 c) the configuration of identified partial disclinations are shown. In particular, dipole (the areas 1 in figure 1 c) and quadrupole (the area 2 in figure 1 c) configurations are presented. From the experiment were found that the characteristic parameters of disclination configurations are the size l = 3 – 5 nm and the value of the Frank's vector defined angles of misorientation, $\omega\approx$ $\omega\approx$ 0.5 ° - 2 °.

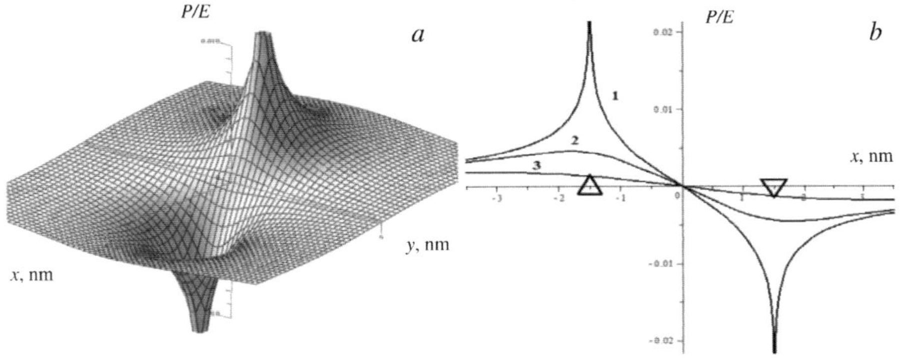

Figure 2. The character of change of pressure $P = (\sigma_{11} + \sigma_{22} + \sigma_{33})/3$ for the nanodipole of partial wedge disclinations with the size l = 3 nm and the value of the Frank's vector $\omega\approx$ 1 °. a - spatial distribution; b - at different distances from the bedding plane of the dipole y = 0 nm (curve 1): at the distance y = 1 nm (curve 2); at the distance y = 3 nm (curve 3).

Within continual approach using described earlier formalism the analysis of character and magnitude of the stress field of nanodipole with parameters l = 3 nm and$\omega\approx$ 1 ° in the field of elastic distortions is produced. In figure 2 the spatial distribution of local internal stresses is presented. As can be seen, the stress field is characterized by a complex form with localization in the region of the partial disclinations. It's shown that the maximum stresses and pressure gradients are observed in the bedding plane (y = 1 nm) of dipole (figure 2 b). And their values reach the order of $P\approx$ E / 50 and $\partial P / \partial x \approx$ 0,02 nm^{-1}. The presented curves in figure 2 b curves indicate rapid decrease of the interaction forces with increasing distance from the bedding plane, due to the effect of shielding of the individual disclinations fields. Figure 2 b shows that, in contrast to the stress field of individual disclination at a distance of x = 6 - 9 nm (2 l - 3 l) almost complete relaxation of the elastic fields of the nanodipole configuration is observed. Thus, the presented nanodipole of partial disclinations really is the concentrator of high tensions in the localized area of the elastic continuum.

In the case of the quadrupole configuration modeling of stress fields was conducted on the equilibrium quadrupole with the size $l = m$ = 3 nm, and the value of the Frank's vector$\omega\approx$ 1 °. Figure 3 a shows that the field of local internal stresses is symmetric and has four zones of increased stress concentrations corresponding to the locations of partial disclinations. Similarly dipole configurations in the quadrupole are also achieved high (up to $P = E$ / 50) values of internal stress fields. In addition, the distinctive feature of the quadrupole configuration is the increase of the values of stress gradients in the bedding plane of dipoles in contrast to the above presented lonely dipole (figure 3 b). It's found that in the quadrupole bounding planes gradients and internal stresses are maximum and reach values $\partial P / \partial x$ = 0.03 nm^{-1}. This fact testifies to the presence of an additional screening of interaction between the individual disclination dipole.

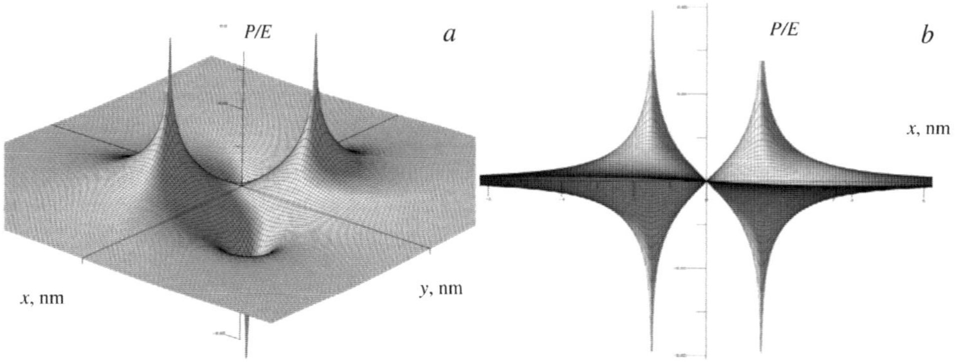

Figure 3. The character of change of pressure $P = (\sigma_{11} + \sigma_{22} + \sigma_{33})/3$ for the quadrupole of partial wedge disclinations with the size l = 3 nm and the value of the Frank's vector $\omega \approx$ 1.a - spatial distribution; b - projection on the plane $x0P$ °

Thus, the above presented analysis of the stress distribution demonstrates the possibility of existence of the considered configurations partial disclinations as concentrators of high tensions in the localized area of the elastic continuum. However, to assess probability of the existence of certain disclination configuration requires knowledge of its energy. Due to above told, the analysis of distribution of specific resilient energy in a continuum of dipole and quadrupole configurations (using the formulas [13]) is produced. The results of this analysis are presented in figure 4. It should be noted that the elimination of a logarithmic divergence of energy at large distances is the result of the screening and, thus, energy of dipole and quadrupole systems become independent of the sizes of grain and are defined only by their size. However, the maximal error of calculation of elastic energy distribution is localized in the region of the disclinations, that is seen by the presence of singularities along the line defect (figure 4). A distinctive feature of the quadrupole is that a significant part of the elastic energy is concentrated in its physical volume (figure 4 b), while in the case of the dipole configuration, the area of distribution of elastic energy exceeds its physical size.

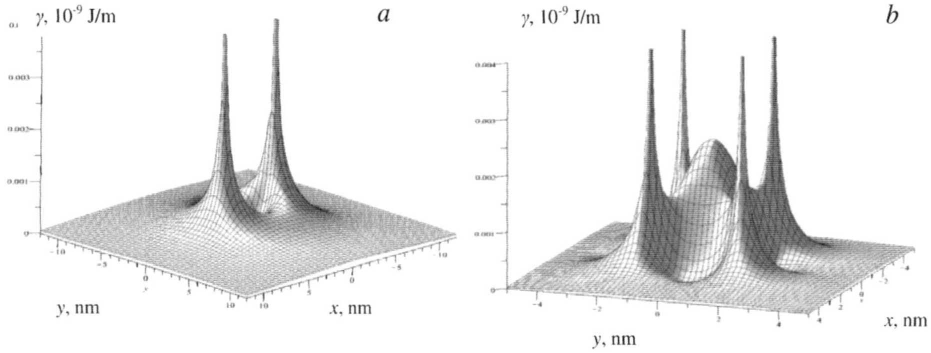

Figure 4. The distribution of the specific elastic energy a - the nanodipole of partial wedge disclinations with the size l = 3 nm and the value of the Frank's vector $\omega \approx$ 1; b - quadrupole partial wedge disclinations with parameters $l = m$ = 3 nm, $\omega \approx$ 1 °.

The analysis of stability considered disclination configurations performed on the basis of estimates of the elastic energy depending on the its size (l). It's shown that after increasing l the proportion of the exchange interaction is reduced that leads to increase in fields of tension and energy (figure 5). As can be seen, for l > 20 cm (figure 5 curve 1) the value of energy for nanodipole system becomes comparable with energy of the separate disclination that testifies to disintegration of system. At the same time, for small values of l < 6 nm in the continuum approach the dependence of nanodipole energy has square character and agrees well with analytical expression for the linear energy of the edge dislocation with Burgers vector $\boldsymbol{b} \approx \boldsymbol{l}\,\boldsymbol{w}$ (figure 5 curve 1).

In comparison with the quadrupole configuration (figure 5 curve 2) the value of the dipole energy is order of magnitude larger and more dependent on the its size(figure 5). Thus, in a free elastic continuum (without the presence of internal and external sources of tension), the quadrupole configuration is represented energetically more preferable. This fact indicates the possibility of transformation (disintegration) of the dipole configuration of partial disclinations in the quadrupole.

It's important to note that in the absence of screening by other disclinations systems dipole configuration may be more stable. One of these variants is presented in figure 1. While the central nanodipole is transformed to a quadrupole other dipole configurations near border remain steady.

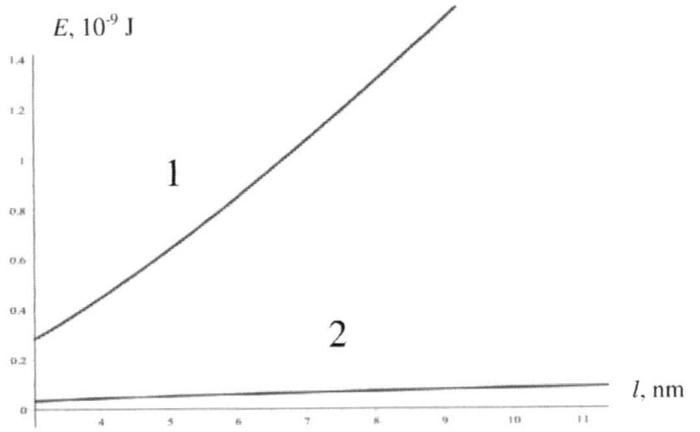

Figure 5. The dependence of the energy for the disclination configuration from the size: 1 - the nanodipole of partial wedge disclinations with the value of the Frank's vector $\omega \approx$ 1; 2 - quadrupole partial wedge disclinations with parameters m = const = 3 nm, $\omega \approx$ 1 °.

4. Summary

The possibility of using disclination approach within the continuum theory of defects for describing the above presented states on the nanoscale structural level is shown. Based on the study of the force characteristics of dipole and quadrupole configurations of wedge partial disclinations it's established that these configurations characterized by high gradients ($\partial P / \partial x \approx$ 0.02 nm^{-1}) of the local stresses in the spatial scales l < 20. It's shown that in the energy relation the existence of a quadrupole configuration is more favorable.

References

[1] Valiev R Z and Alexandrov I V 2007 *The Bulk Nanostructural Metal Materials* (Moscow: Academkniga) [in Russian]
[2] Valiev R Z, Mulyukov R R, Ovchinnikov S V 1990 *Metalophisika* **12(5)** 124 [in Russian]
[3] Gleiter H 1995 *NanoStruct. Mat.* **6** 3
[4] Horita Z, Furukawa M, Nemoto M, Valiev R.Z. and Langdon T G 2000 *NATO Sci. Series* **3(80)** 155
[5] Tyumentsev A N and Ditenberg I A 2011 *Phys. Mesomech.* **14(3)** 55 [in Russian]
[6] Tyumentsev A N and Ditenberg I A 2011 *Rus. Phy. Jour.* **54(9)** 26
[7] Zolnikov K P, Psakhie S G and Panin V E 1986 *J. Phys.* **16(8)** 1145
[8] Panin V E and Egorushkin V E 2011 *Phys. Mesomech.* **14(5-6)** 207 [in Russian]
[9] Kozlov E V, Popova N A and Koneva N A 2010 *Izvestiya RAN* **74(5)** 630 [in Russian]
[10] Romanov A E 1995 *Nanostr. Mat.* **6** 125
[11] Koneva N A , Kozlov E V, Popova N A 2008 *et al Mat. Sci. Forum* **584-586** 269
[12] Lihachev V A 1975 *The Introduction to the Theory of Disclinations* (Leningrad University Pub lishing) [in Russian]
[13] Vladimirov V I and Romanov A E 1986 *The Disclinations in Crystals* (Leningrad: Nauka) [in Russian]

Improvement of the heat resistance of carbon steels by thermocycling thermochemical treatment with self-protective pastes based on boron carbide and aluminum

I Polyansky, I Sizov, U Mishigdorzhiyn, V Butukhanov
East Siberian State University of Technology and Management, Ulan-Ude, Russian Federation

E-mail: i.polyansky@mail.ru

Abstract. The positive effect after thermocycling thermochemical treatment on the structure and properties of formation of boroaluminized layer on steel 20 was established. The research results of metallographic analysis, x-ray spectrometry microanalysis and microhardness were given. The modes of thermocycling treatment (temperature, speed of heating and cooling) were carried out.

1. Introduction

One of the effective methods of metals and alloys surface hardening is thermochemical treatment. Thermochemical treatment is widely used in mechanical engineering to get surface protective coatings with high service properties. Protective coatings received by the thermochemical treatment on steel machine parts and tools (which work under high temperatures, impact loads, wear by friction, corrosion) can increase the service life up to 10-30 times.

There are many methods of thermochemical treatment [1]: saturation in powders, in liquid, with the use of gaseous phases and in plasma. The process of diffusion saturation in pastes is technologically and economically advantageous and is widely used.

Depending on the saturation element the processes of thermochemical treatment are recognized: single-component and multi-component saturation by metal or non-metal layers with the wide range of properties can be found.

One of the effective methods of increasing heat resistance and durability is boroaluminizing. Diffusion saturation by boron and aluminum can be carried simultaneously or one after the other.

The diffusion saturation is carried at high temperature 900-1000 °C and long duration of the process. This causes the growth of the structure, hogging and other defects. Such disadvantages can be eliminated with the use of thermocycling treatment [2]. Thermocycling treatment has positive effect on metal structural changes.

2. Materials and methods

We chose steel 20 (1020 AISI steel analogue) as object of our study. Paste was applied on the samples which were then dried at 200 °C. Paste consisted of (in weight %): boron carbide 78, aluminum 18, sodium fluoride 4.

Diffusion saturation was conducted at isothermal mode and thermocycle mode (4, 8 and 16 cycles).

Isothermal mode was conducted at 950 °C. Thermocycle mode was performed in the range of temperatures, which coincident with the temperature of phase transformations (650-950 °C). Rate of heating and cooling between cycles was controlled with thermocouple. Thermocouple was fixed on a sample. The time for cooling and heating was 3 and 12 min., speed of heating – 0.4 °C/sec., speed of cooling – 1.6 °C/sec.

Metallographic analysis was performed with the use of equipment in Institute of Materials Technical University of Dresden. Microhardness was determined using "PMT-3M" microhardness tester. X-ray spectrometry microanalysis system in Centre for collective use "Progress" East Siberian State University of Technologies and Management. The heat resistance was determined after holding the specimens in the furnace for the specified time (50 hours) at a constant temperature (1000 °C) using the method of weighing for determining the growth in the mass of the specimens (GOST 6130-71). The change in the mass was detected in every 5 hours using an analytical balance.

3. Results and discussion

The influence of the paste composition on structure and properties of boroaluminized layer is shown in work [3]. It is known [4] that phases with high content of aluminum (Fe_2Al_5 and $FeAl_3$) cause an increase in heat resistance. Simultaneous saturation with boron and aluminum happens while treatment with the following composition (in weight %): 78 B_4C; 18 Al; 4 NaF (figure 1). Boride crystals formed in the shape of separate needles in aluminized part of the layer. There is no grain growth of the base metal during the thermocycling treatment.

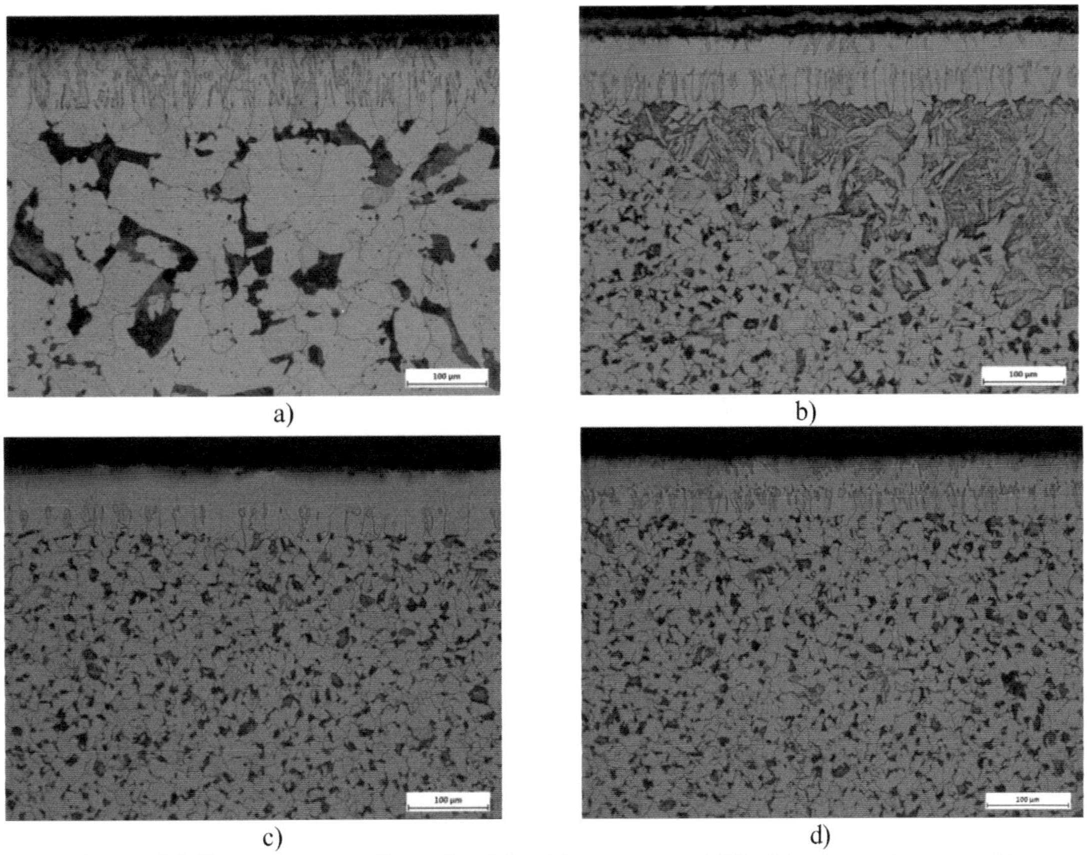

Figure 1. Microstructures of boroaluminized layers on steel 20: a) isothermal process; b) thermocycling process after 8 cycles; c) thermocycling process after 16 cycles

X-ray spectrometry microanalysis has shown (figure 2) that there is the maximum content of boron (17 %) after isothermal treatment. There is minimal content of boron after the process with 8 cycles. After process with 16 cycles the content of boron is 9.5 %.

The aluminum content gradually decreases from the surface towards the base metal after all modes. Medium content of aluminum after the isothermal process is 5.6 %, after the process with 4 cycles is 3.1 %, after the process with 8 and 16 cycles increases from 6.5 to 6.8 %.

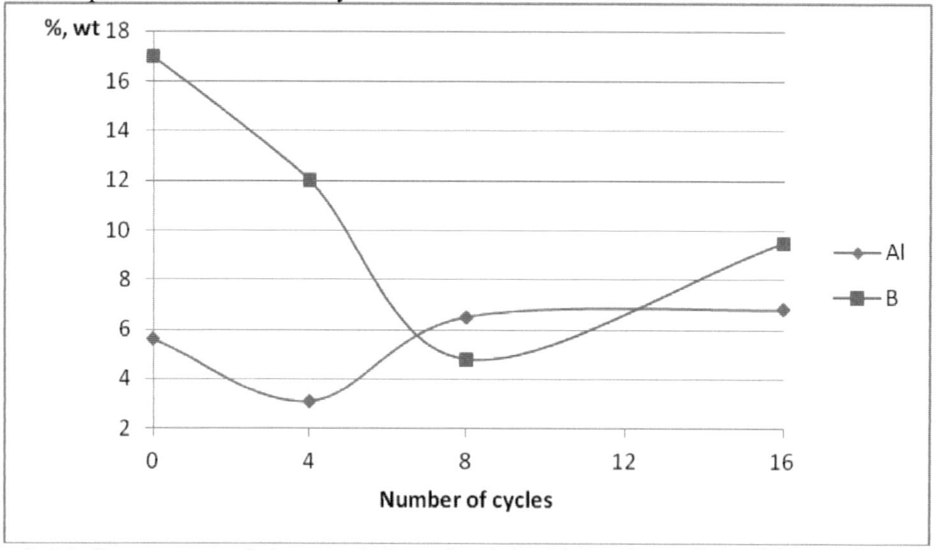

Figure 2. Medium content of elements in boroaluminized layer depending of the number of cycles

Microhardness after isothermal and thermocycling treatment with 4 cycles is 500-550 HV. It is known that $FeAl_3$ and Fe_2Al_5 have such microhardness [4]. Microhardness of needles under base layer is 2500-2700 HV. Such microhardness relates to iron boride FeB [5].

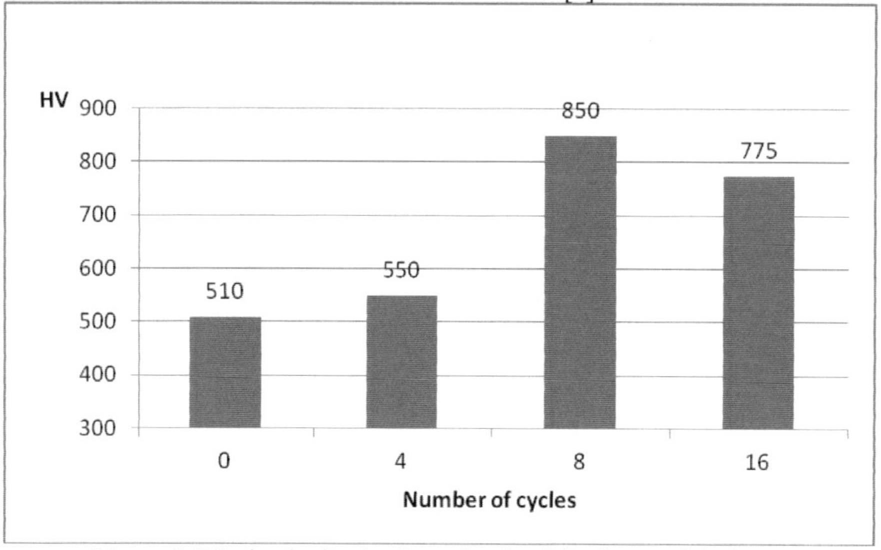

Figure 3. Maximal microhardness in aluminized part of the layer

The thickness of diffusion layers depends on process duration. Increasing the number of cycles leads to a decrease of layer thickness. This is because an increasing of cycle numbers causes the growth of total time for cooling and heating. The time (when samples are at high temperatures) decreases.

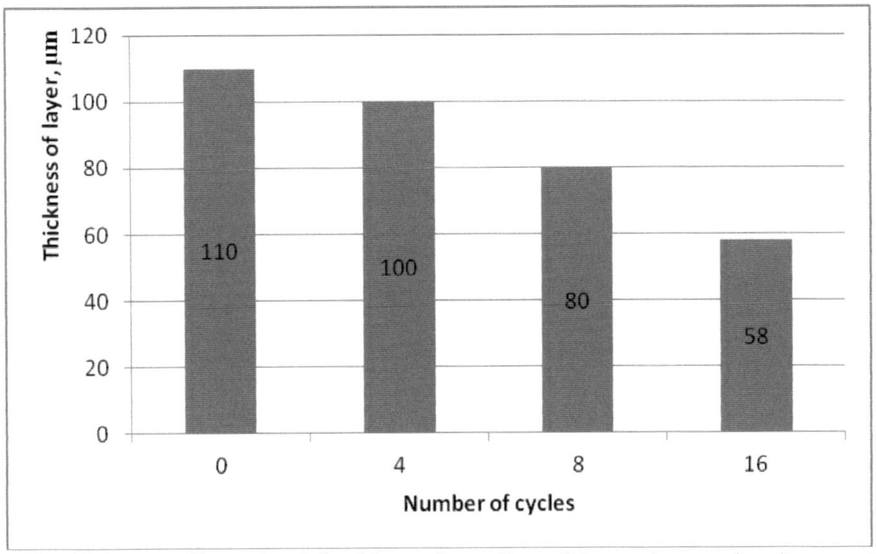

Figure 4. The thickness of boroaluminized layer depending of the cycles number (temperature 1000 °C, 50 hours)

The tests for high temperature strength have shown considerable growth in the resistance of samples after treatment with 8 and 16 cycles (figure 5).

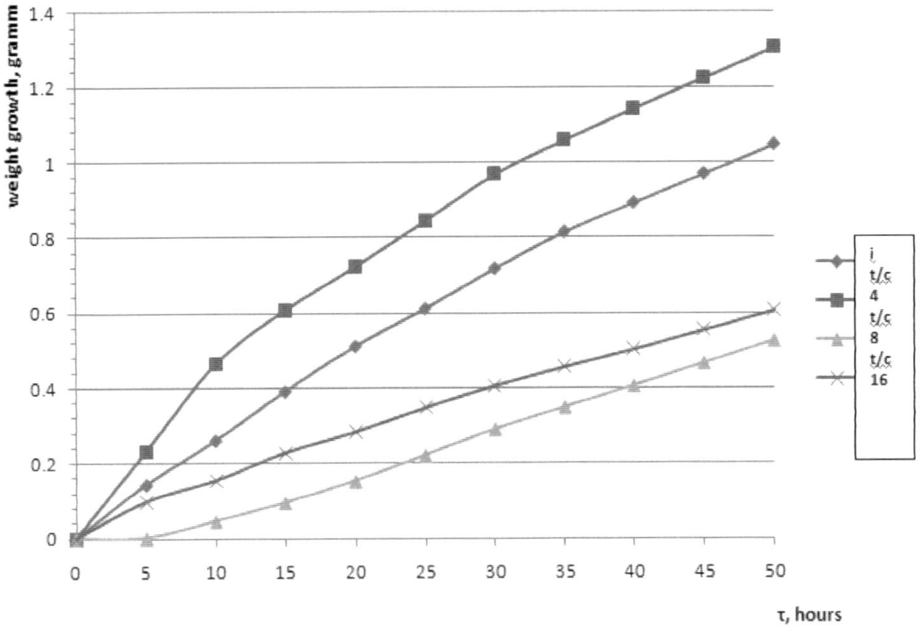

Figure 5. The results of the tests for high temperature strength, increase in weight

3. Conclusion

Thermochemical treatment with thermocycling has a positive effect on a base metal structure. Heat resistance and microhardness are increased by appearance of high aluminum content phases in diffusion layer. Structure and properties of boroaluminized layer depends on the number of cycles.

Small number of cycles does not lead to the formation of boroaluminized layer with necessary properties.

Acknowledgement
This work was supported by Grant of Russian Federation President # 14.Z56.15.3112-MK, 2015.

References

[1] Voroshnin L, Mendeleeva O, Smetkin V 2010 *Theory and technology of thermochemical treatment* (Minsk: Novoe znanie) p 304 (in Russian)
[2] Fedyukin V, Smagorinskiy M 1989 *Thermocycling treatment of metals and machine parts* (Leningrad: Mashinostroenie) p 255 (in Russian)
[3] Sizov I, Polyansky I, Mishigdorzhiyn U, Makharov D 2014 *Paste composition effect on structure and properties of boroaluminized layer* (Metal processing: technology, equipment, tools vol. 1(58) (Novosibirsk: NSTU) pp 22-26 (in Russian)
[4] Ryabov V 1973 *Aluminizing of steel* (Moscow: Metallurgiya) p 240 (in Russian)
[5] Krukovich M, Prusakov B, Sizov I 2010 *Plasticity of boronized layers* (Moscow: FizMatLit) p 384 (in Russian)

IOP Conf. Series: Materials Science and Engineering **116** (2016) 012037 doi:10.1088/1757-899X/116/1/012037

Features of formation of nanocrystalline state in internal-oxidized V-Cr-Zr-W and V-Mo-Zr system alloys during deformation by torsion under pressure

I V Smirnov[1,2,4], I A Ditenberg[1,2,3], K V Grinayev[1,2,3] and V L Radishevsky[1]

[1] National Research Tomsk State University, Tomsk, 634050, Russia
[2] Siberian Physical-Technical Institute, Tomsk, 634050, Russia
[3] Institute of Strength Physics and Materials Science SB RAS, Tomsk, 634055, Russia

E-mail: smirnov_iv@bk.ru, ditenberg_i@mail.ru, kvgrinyaev@inbox.ru, vladrad95@mail.ru

Abstract. The results of investigation of features of nanostructural state formed during deformation by torsion under pressure in high-strength vanadium V-Cr-Zr-W and V-Mo-Zr systems alloys are presented. It was found that after deformation at number of revolutions N = 1, samples are characterized by high anisotropy of defect and grain structure. Inside grains, limited by high-angle boundaries, the formation of two-level structure states was revealed: fragmentation of the above grains on nanofragments from 5 to 20 nm in size with a dipole nature of low-angle misorientations and high (hundreds of degrees per micron) elastic curvature of crystal lattice. Formation of the above structural states leads to a 3-fold increase in microhardness values. Further increase in deformation degree leads to fracture of samples of vanadium alloy V-Mo-Zr with a high volumetric content of fine-disperse oxide phase. At the same time V-Cr-Zr-W-system alloy with a lower concentration of Zr and, as a result, a lower volume fraction of fine particles remains ductile.

1. Introduction

Investigation of the specificity of nanostructural states formation in metallic materials of different classes under severe plastic deformation today is still one of the rapidly developing areas of modern physical materials science. According to modern concepts, the behavior of nanostructural metallic materials in the fields of external influences and their physical and mechanical properties are largely determined by features of nonequilibrium structural states formed inside and regularities of their evolution in the process of generation and propagation of defects at different scale levels [1 – 5].

Analysis of published data shows that the study of such issues is advantageously carried out on pure FCC metals and alloys. Much smaller number of experimental works is devoted to studying of features of formation of nanostructural states in metals and alloys with BCC lattice during large plastic deformations. Meanwhile BCC metals are the basis for creation of a broad class of structural materials for use in extreme conditions (high temperatures and pressures, aggressive media, etc.). Due to high chemical activity of these materials relative to interstitial impurities, their alloys often become

[4] To whom any correspondence should be addressed.

heterophase [6] with opportunities for implementation of solid-solution, dispersion, precipitation, etc. types of strengthening. So, one of the actual trends is the investigation of features of formation of nanostructural states in metallic materials with BCC lattice with different strengthening types.

In current work the investigation of features of nanostructural states and parameters of microhardness of high-strength internal-oxidized vanadium V-Cr-Zr-W and V-Mo-Zr systems alloys with different volume fraction of fine-disperse second phase particles after severe plastic deformation by torsion on Bridgman anvils was carried out.

2. Experimental materials and procedures

In this work vanadium alloys of V-Cr-Zr-W (V - 1.17 Zr - 8.75 Cr - 0.14 W - 0.01 C - 0.02 O - 0.01 N) and V-Mo-Zr (V - 9 Mo - 1.5 Zr - 0.07 C) (wt. %) were used. Before plastic deformation in Bridgman anvils V-Cr-Zr-W alloy was processed by thermo-mechanical treatment (TMT) by scheme [7] which provides the dispersion of initial coarse phase. Further, oxygen doping during chemical-heat treatment (CHT) [8, 9] was conducted on alloys V-Cr-Zr-W and V-Zr-Mo to achieve oxygen concentration of 1.34 and 2.7 (at. %) respectively. Then one-hour annealing was conducted at 1400 °C. Deformation of disc samples with thickness $h = 0.2$ mm and diameter of 8 mm was produced by torsion under high pressure (7 GPa) at room temperature and the number of revolutions of the anvil $N = 1$. Thickness of specimens after deformation was 0.15 mm. Values of shear ($\gamma \approx 2\pi NR/h$) and true logarithmic ($e \approx \ln\gamma$) deformation depending on the distance from the center of the deformable disk (R) changes from $\gamma \approx 16$, $e \approx 2.8$ at $R = 0.5$ mm to $\gamma \approx 110$, $e \approx 4.7$ at $R = 3.5$ mm.

Structural studies were conducted by transmission electron microscope Philips CM-30 STEM-TWIN (300 kV). Thin foils were prepared in sections normal to anvils plane. For their preparation, copper layer was electrolytically precipitated on samples after deformation. Flat samples in the above sections were cut with electrospark machine and mechanically thinned to a thickness of 100 microns. Further thinning was carried out by bilateral sputtering by argon ions at accelerating voltage of 5 kV.

Methods of analysis of high continuous and discrete misorientations [10, 11] were used to study the characteristics of defect structure of volume and boundaries of grains. Microhardness (H_μ) was measured by Vickers method on «Neophot 21», with a load of 0.5 N and 15 seconds of exposure in section perpendicular to anvils plane (AP) at different distances from torsion axis (TA).

3. Results and discussion

Figure 1 shows the bright-field (figure 1 a) and dark-field (figure 1 b) electron microscopic images of V-Cr-Zr-W alloy microstructure after CHT and subsequent deformation on Bridgman anvils at N = 1.

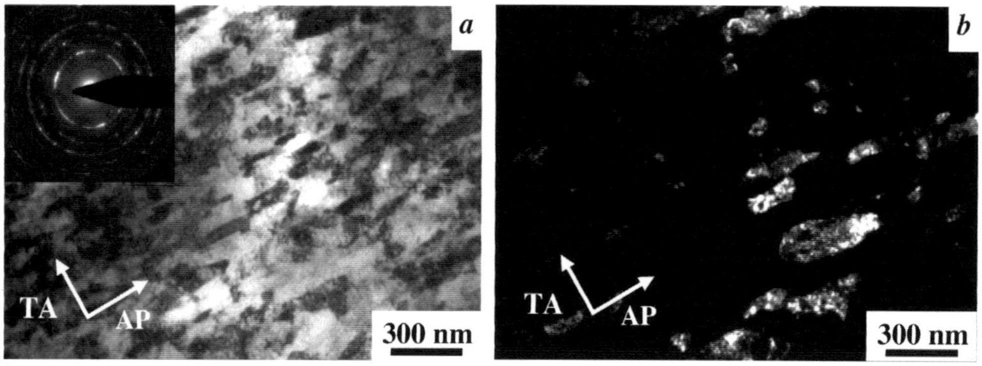

Figure 1. V-Cr-Zr-W alloy microstructure ($N = 1$, $e \approx 4.7$). Bright-field (a) and dark-field (b) images taken in cross section perpendicular to anvils plane.

Formation of similar structural state was also found in V-Mo-Zr-system alloy after such treatment (figure 2 a). The analysis revealed that characteristic dimensions of crystallites of both alloys in directions parallel to the anvils plane are in the range from 150 to 600 nm, in rare cases reaching 1 micron or more in the alloy V-Mo-Zr, whereas in perpendicular to them (parallel to the torsion axis) crystallites sizes does not exceed 100 nm (figure 1 and figure 2 a). Thus, in the studied alloys the formation of structural state with pronounced grain structure anisotropy is observed.

In addition to anisotropy of grain sizes, high anisotropy of discrete low-angle and high-angle misorientations was revealed that determine [10, 11] characteristics of grain and subgrain structure, as well as continuous misorientations that characterize parameters of dislocation-disclination substructure of volume and boundaries of crystallites. A schematic representation of anisotropic structural state is presented on figure 2 b. It was found that values of crystal lattice curvature tensor components of χ_{31} and χ_{21} type reach 20 – 30 degrees/μm (with maximum up to 40 degrees/μm) which two or three times exceed values of χ_{13} and χ_{23} type components (5 – 15 degrees/μm). Thus, the intensity of all these misorientation types in the directions lying in anvils plane greatly exceeds crystal lattice misorientations in the direction normal to this plane, parallel to the torsion axis (figure 2 b).

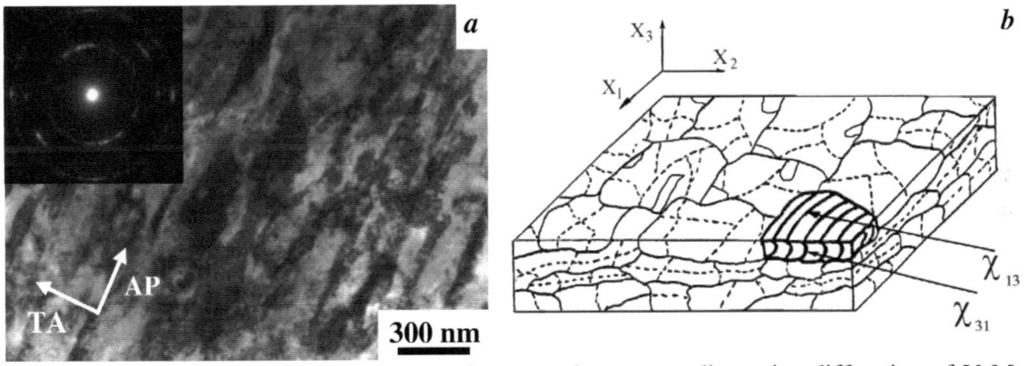

Figure 2. a – Bright-field microstructure image and corresponding microdiffraction of V-Mo-Zr- system alloy (N = 1, $e \approx 3$). Section perpendicular to anvils plane. b – Scheme of anisotropic structural state of material after deformation by torsion under pressure [11].

Previous, similar structural features were detected in V-Ti-Cr-system alloy after deformation by torsion under pressure [11], and it was shown that observed microstructure anisotropy is a consequence of high anisotropy of fields of displacements and rotations due to scheme of the method.

An important feature specific to studied in this paper vanadium alloys of V-Zr-Cr-W and V-Mo-Zr systems is the fact that after deformation by torsion at N = 1 in both central and peripheral parts of specimens the formation of two-level structural states is observed. Figure 3 is an example of electron microscopic dark-field images obtained at different goniometer tilt angles.

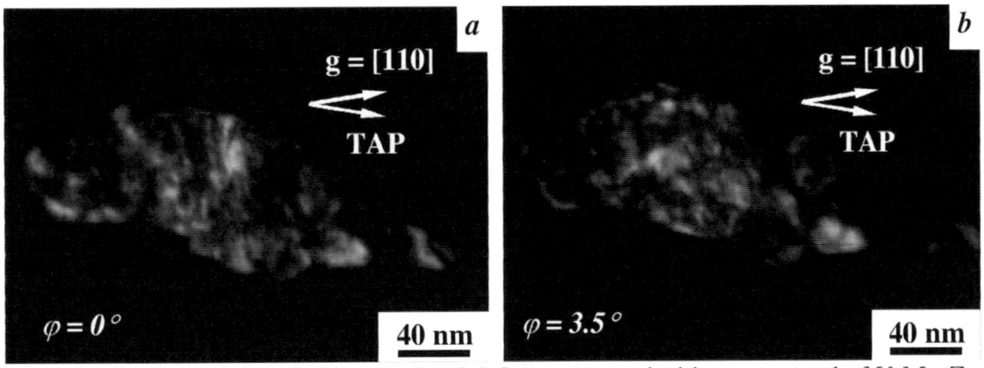

Figure 3. Example of dark-field analysis of defect structure inside nanocrystal of V-Mo-Zr-system alloy (N = 1, $e \approx 3$). Section perpendicular to anvils planes. φ - specimen tilt angle in goniometer. TAP – goniometer tilt axis projection.

This analysis revealed that characteristic dimensions of nanofragments are 5 – 20 nm and values of crystal lattice curvature reaches up to several tens and sometimes even hundreds of degrees/μm.

Formation of a similar two-level structural state in V-4Ti-4Cr-system alloy after deformation by torsion at N = 1 was detected only at peripheral area, in which $e \approx 4.7$, and filling of the entire volume of disk sample by structural state of this type was observed only when N = 5 [12]. Figure 4 a is a diagram of defect substructure with a dipole character of misorientations, reflecting the main features of the two-level structural state.

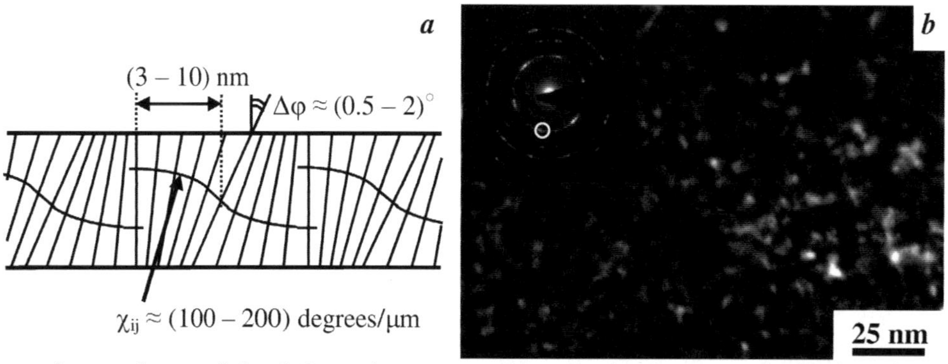

Figure 4. a – scheme of the defect substructure with a dipole character of misorientations [12]. b – dark-field image of the fine-disperse phase of Zr-(O-N-C)-based complex oxycarbonitride in V-Cr-Zr-W alloy (N = 3, $e \approx 4.7$)

In addition to differences in values of plastic deformation at which begins the formation of two-level structures, vanadium alloys with internal oxidation are-characterized by more intense strengthening, which manifests in microhardness measurements (table 1), before the deformation values were 1.5 GPa for all these alloys.

Table 1. Values of vanadium alloys microhardness at N = 1 subject to distance from center of disks (R), H_μ, GPa

R, mm	V-Cr-Zr-W	V-Mo-Zr	V-Ti-Cr [13]
0.5	3.17	3.63	3.21
3.5	4.01	4.17	3.44

Figure 4 b shows an example of dark field electron microscopic image of nanoscale (from 3 to 10 nm) FCC particles in V-Cr-Zr-W alloy. It was found that these particles are Zr-(O-N-C)-based complex oxycarbonitride with a lattice parameter of ≈ 4.28 Å. An important feature of identification of these particles in BCC alloys is the realization of Bain orientation relationship [14], according to which the microdiffraction pictures reflexes of [200]-type of FCC particles and [110]-type of the BCC matrix coincide or are close enough.

Vanadium alloys used in current work are characterized by different content of the volume fraction (f) of second phase disperse particles. The concentration of Zr in V-Mo-Zr-system alloy allows oxidation with the achievement of a significantly greater f. The results of evaluations conducted on the basis of data of the elemental composition shown that as a result of chemical-heat treatment the maximum values of f, mainly represented by fine particles of ZrO and/or ZrO_2 (figure 4 b), reaches about 0.027 and 0.054 respectively in alloys of V-Cr-Zr-W and V-Mo-Zr systems.

In our opinion, the high density of distributed over the volume Zr-based fine particles, and possibly, a high concentration of oxygen in solid solution, are the main reasons of so intense evolution of microstructure and a sharp changes in H_μ. These factors combine to provide a high localized stress concentration at the nanoscale level while reducing the relaxation ability of the material.

It should be noted that a further increase in deformation degree on Bridgman anvils leads to intense embrittlement and destruction of samples of internal-oxidized V-Mo-Zr-system alloy: material crumbled into fine powder. At the same time V-Cr-Zr-W-system alloy, with a smaller volume fraction of fine particles remains plastic and deforms to 5 and more revolutions by torsion under pressure.

Unfortunately, in the framework of this article even rough estimates of contribution of any particular type of strengthening, just as was done for fine-crystalline V-4Ti-4Cr alloy [15], are not possible to be made. The reason is the necessity to integrate a large number of interrelated factors (grain and subgrain size, particle size and density, dislocation density, etc.), which, to date, have not been fully studied in conceptions of strengthening of nanostructural states of heterophase systems.

4. Summary

Comparison of presented data with the results of [13] shows that the volume fraction of fine-disperse second phase particles, apparently influences the intensity of two-level-type structural states formation. The evolution of these states actually determines microhardness of analyzed vanadium alloys, including V-4Ti-4Cr. Furthermore, the maximum values of microhardness of these alloys are achieved when the whole bulk of material is presented by two-level-type state at deformation degree typical for each alloy.

Acknowledgements

Authors express gratitude to professors A.N. Tyumentsev, V.M. Chernov and A.V. Korznikov.

The work was conducted within the framework of the Program for Fundamental Scientific Research of State Academies of Sciences for 2013 – 2020 (deformation treatments) at partial financial support of Tomsk State University Competitiveness Improvement Program (transmission electron microscopy). Investigation was carried out using the equipment of the Tomsk State University.

References

[1] Panin V E and Egorushkin V E 2011 *Phys. Mesomech.* **14(5-6)** 207
[2] Korotaev A D, Tyumentsev A N and Pinzhin Yu P 2001 *Theor. and Appl. Fract. Mech.* **35** 163
[3] Antolovich S D and Armstrong R W 2014 *Prog. Mater Sci.* **59** 1
[4] Valiev R Z 2013 *Nature Mater.* **12(4)** 289
[5] Andrievski R A and Glezer A M 2009 *Physics-Uspekhi* **52(4)** 315
[6] Fromm E and Gebhardt E 1976 *Gases and Carbon in Metals* (Berlin: Springer-Verlag)
[7] Tyumentsev A N, Korotaev A D, Pinzhin Yu P, Ditenberg I A, Litovchenko S V, Shuba Ya V, Shevchenko N V, Drobishev V A, Potapenko M M and Chernov V M 2004 *JNM* **329-333** 429
[8] Tyumentsev A N, Korotaev A D, Pinzhin Yu P, Ovchinnikov S V, Ditenberg I A, Shikov A K, Potapenko M M and Chernov V M 2007 *JNM* **367-370** 853
[9] Chernov V M, Potapenko M M, Drobishev V A, Kravtsova M V, Tyumentsev A N, Ovchinnikov S V, Ditenberg I A, Pinzhin Yu P, Korotaev A D, Smirnov I V, Grinyaev K V and Sukhanov I I 2015 *Nucl. Mater. Energ.* **3-4** 17
[10] Tyumentsev A N and Ditenberg I A 2012 *RUPJ* **54(9)** 977
[11] Ditenberg I A, Tyumentsev A N, Korznikov A V, Chernov V M and Potapenko M M 2012 *Phys. Met. Metallogr.* **113(2)** 160
[12] Ditenberg I A, Tyumentsev A N, Grinyaev K V, Chernov V M, Potapenko M M and Korznikov A V 2011 *Tech. Phys. The Russ. Jour. of Appl. Phys.* **56(6)** 815
[13] Ditenberg I A, Tyumentsev A N, Korznikov A V and Vins S A 2009 *Jour. Perspektivnye materialy* **7** 103 [in Russian]
[14] Korotaev A D, Tyumentsev A N and Sukhovarov V F 1989 *Dispersion Hardening of Refractory Metals* (Novosibirsk: Nauka) [in Russian]
[15] Ditenberg I A, Tyumentsev A N, Chernov V M and Potapenko M M 2004 *VANT ser. Nuclear Fusion* **2** 28 [in Russian]

Electromagnetic properties of texture composite materials based on hexagonal ferrites/multiwalled carbon nanotubes

O A Dotsenko[1], K O Frolov[1] and D V Wagner[1]

[1] Department of Radio Electronics Tomsk State University, 36, Lenina Ave., Tomsk, 634050, Russian Federation

E-mail: FrolovKirill.O@yandex.ru

Abstract. In this article, the frequency dependence of the absorption coefficient and electromagnetic losses of the composite based on ferrite powder and / or multi-walled carbon nanotubes are presented. The dielectric and magnetic losses in the composite were measured in the range of 0.01 – 20 GHz. It has been found to increase the absorption of electromagnetic radiation and increased losses in the samples containing multi-walled carbon nanotubes.

1. Introduction

Nowadays microwave electronics devices are used in many areas of industry, health, education. It is a reason to a problem of electromagnetic compatibility of radio electronic devices. Interference leads to a deterioration of the nominal characteristic and the stability of the devices. The high power of the microwave radiation is dangerous to biological objects. [1, 2]

Composite materials with unique physical properties: a low weight, strength, plasticity, resistance to mechanical loads and high radar-absorbing properties are developed for the decision of this problem. They can be used as radioabsorbing materials and radioabsorbing coatings. Ferrites and multiwall carbon nanotubes (MWCNT) as materials with controlled electromagnetic characteristics were investigated in the articles [3-7]. It was shown that the change of concentration of the active fraction leads to a change of electromagnetic properties of the experimental samples. Moreover, this change also depends on the type of extender. The present article is devoted the possibility of creating of multi-functional composite materials with controlled electromagnetic characteristics in during production. The coaxial method was used to investigate electromagnetic characteristics. The study was conducted in the frequency range from 0.01 GHz to 20 GHz. Micran P2M-18 (Micran, Research & Production Company, Tomsk, Russia) was used to measure the modulus of the reflection coefficient and the transmission coefficient.

2. Results and Discussions

Standard ceramic techniques were used to the fillers creation. The initial materials for the synthesis were barium oxide BaO powders, cobalt (II) oxide CoO, titan oxide TiO_2 and iron (III) oxide Fe_2O_3. Prior to synthesis, the oxides powders were dried for 3 h at temperature of 200 °C. After that powders were weighed according to a ratio

$$3BaO + 2{,}4CoO + 0{,}4\,TiO_2 + 11{,}6\,Fe_2O_3 = Ba_3Co_{2,4}Ti_{0,4}Fe_{23,2}O_{41}.$$

Powders used for sintering experiments were ground for 4 h in a vibration ball mill. The ball-to-powder mass ratio was 5:1. Powders were die-pressed at about 1000 atm. Solid samples were heated up in 10–15 min to the sintering temperature (1150 °C), fired for 4 h and were cooled to the room temperature. After that samples were crushed. Powders were ground for 35 min in a vibration mill. Powders were die-pressed repeatedly at about 1000 atm. Solid samples fired in oxygen for 6 h 1200 °C finally. After that samples were ground once again. The powders were sifted by separator. The sizes of powders don't exceed 80 microns.

Multiwalled carbon nanotubes with a diameter of 9.4 nm were produced by the Institute of Catalysis SB RAS [4]. Epoxy resin EDP 20 was prodused by Gmbh «ChK FEM». Fillers and epoxy resin were weighed in the required proportions of mass on the scales Shimadzu AUX-320 (error ±0.5 mg) were placed into the vessel and thoroughly mixed for 15 minutes until homogeneous. The compositions of samples are presented in Table 1. The resulting mixture was poured into three identical shapes. The magnetic field intensity was 1 kOe. Duration of stay in the magnetic field of the samples was 5 hours at room temperature.

Table 1 – Descriptions of test samples

Number	Concentration (%)			Magnetic field (kOe)
	Ferrite	MWCNT	Epoxy	
1	65	1	34	1
2	66	-	34	1
3	66	-	34	0

The prepared samples were representing washers with a thickness of 1.35 mm, an outer diameter of 7 mm and inner diameter of 3.04 mm. The results of experimental measurements of the electromagnetic characteristics are shown in Figure 1 and Figure 2.

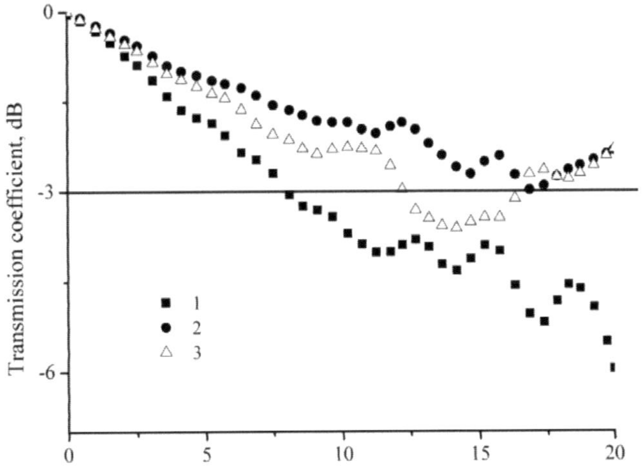

Figure 1. Coefficient of transmission experimental samples

The comparison of the samples line was held at the level 3 dB in Figure 1. This line corresponds to the absorption capacity of 50% of the incident wave. Figure 1 shows that in the frequency range from 8 to 20 GHz 1 wt.% of nanotubes sample absorbs EMR in 2-2.5 times more compared to the same sample without nanotubes. For a sample that is polymerized without external magnetic field at all similar parameters absorption power is worse in 1.2-1.8 times. The sample number 1 has a several distinct areas of the dispersion. They are from 12 to 15 GHz with a peak of absorption minus 4.3 dB, from 16 to 18 GHz with a peak of absorption equal minus 5.2 dB and also from 19 to 20 GHz, but the peak of absorption is not observed in this range. The maximum value of the absorption power is minus 6 dB in the range from 19 to 20 GHz. The sample number 3 has a clearly expressed dispersion region. It is in the range from 12 to 16 GHz. The sample number 2 has nothing special in this frequency range.

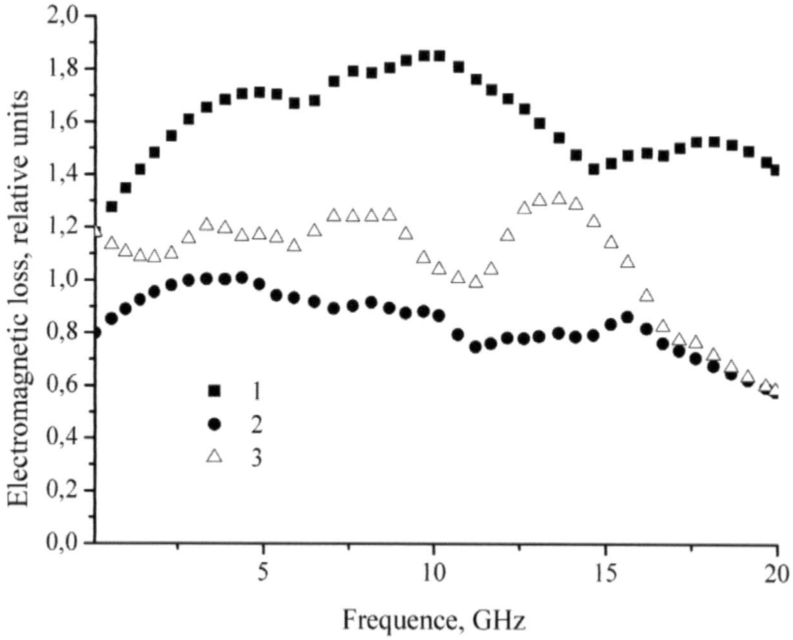

Figure 2. Coefficient of electromagnetic loss

The results of calculation of cumulative electromagnetic losses for thin dielectric layer are shown in Figure 2. For sample number 1 in the range from 0.01 to 14 GHz dispersion electromagnetic losses is observed with a maximum value of 1.85 rel. units at a frequency of 10 GHz and a small peak of absorption between 15 and 20 GHz at 1.5 rel. units. Comparison with samples of composite materials, fillers which are the MWCNTs [4, 7], and carbonyl iron [8] shows that result obtained can be explained by large dielectric losses in the MWCNT. Comparison samples number 2 and number 3 showed that the effect of an external magnetic field to the sample during the polymerization resulted in a reduction of losses in the material, which worsened its absorbent properties.

3. Conclusion

The study showed that the addition of MWNTs into the mixture until fabrication of textured materials increases the absorbing properties of the composite material. The effect of external magnetic field during the polymerization of samples and the addition of MWNT leads to a broadening of the absorption spectrum.

Acknowledgements

Authors are thankful to PhD V.A. Zhuravlev of National Research Tomsk State University PhD V.I. Suslyaev of National Research Tomsk State University for help and support in our research.

This Research is supported by Tomsk State University Competitiveness Improvement Program.

References

[1] Bespalova E E, Belyaev A A and Shirokov V V. Radio-absorbing materials for high power microwave radiation, Available at: http://viam-works.ru/ru/articles?art _id= 788 (accessed 10 April 2015)
[2] Bespalova E E, Belyaev A A, Romanov A M and Shirokov V V. Study of the dielectric layers of the multilayer radar absorbing material for anechoic chambers on the basis of foam asbestos filled with carbonized fiber, Available at: http://viam-works.ru/ru/articles?art_id=703 (accessed 10 April 2015)
[3] Wagner D V and Dotsenko O A 2014 Electromagnetic properties of Z-hexaferrities composites with magnetic texture, *EDM.* 132
[4] Mazov I N, Kuznetsov V L, Moseenkov S I, Ishchenko A V, Romanenko A I, Anikeeva O B, Buryakov T I, Korovin E Yu, Zhuravlev V A and Suslyaev V I. 2010 Electrophysical and Electromagnetic Properties of Pure MWNTs and MWNT./PMMA Composite Materials Depending on Their Structure. *Fullerenes, Nanotubes and Carbon Nanostructures* **18** **4** 505
[5] Dotsenko O A and Ulyanova O A 2014 24nd Int. Magnetic insulator with texture: electromagnetic properties. *Crimean Conf. "Microwave & Telecommunication Technology"(CriMiCo'2014)* Sevastopol. 696.
[6] Zhuravlev V A, Suslyaev V I, Korovin E Yu and Dorozhkin K V 2014 Electromagnetic waves absorbing characteristics of composite material containing carbonyl iron particles. *MSA.* **11** 803
[7] Kachusova A O 2015 The study of the spectra of the dielectric constant of composite materials based on carbon nanotubes *Eurasian Scientific Association* **2** 22

Energy Approach-Based Simulation of Structural Materials High-Cycle Fatigue

A F Balayev[1], A V Korolev[1], A V Kochetkov[2], A I Sklyarova[1], O V Zakharov[1]
[1] Yuri Gagarin State Technical University, Saratov, Russia,
[2] Perm National Research Polytechnic University, Perm, Russia

E-mail: zov17@mail.ru

Abstract. The paper describes the mechanism of micro-cracks development in solid structural materials based on the theory of brittle fracture. A probability function of material cracks energy distribution is obtained using a probabilistic approach. The paper states energy conditions for cracks growth at material high-cycle loading. A formula allowing to calculate the amount of energy absorbed during the cracks growth is given. The paper proposes a high-cycle fatigue evaluation criterion allowing to determine the maximum permissible number of solid body loading cycles, at which micro-cracks start growing rapidly up to destruction.

1. Introduction

Durability is one of the most important product performance criterion. That is why the crucial task is to evaluate durability under high-cycle loading subject to structural materials high-cycle fatigue. High-cycle fatigue was studied in many papers [1-8] which deal with destruction of solid bodies due to development of micro-cracks. These studies proposed durability evaluation criteria based on deterministic approach to description of destruction process considering random distribution of micro-cracks. In fact, not only distribution of micro-cracks in materials but the process of their growth and distribution of internal energy is random as well.

In our opinion, in order to determine the conditions of micro-cracks occurrence and development and to choose a criterion of evaluation of residual life of a real solid body in high-cyclic loading it is necessary to use a probabilistic approach to describe energy process of cracks development.

2. Problem Statement

Let's consider a plate of volume V stretched out and contracted alternately under a load P (Figure 1). The plate includes multiple cracks distributed at random. When approximating the crack for the first time, one can take it as flat, elliptical.

Let's take the theory of A. Griffiths as a basis for energy criterion of brittle body fracture. According to the paper [1] the energy of a crack can be determined using the following formula:

$$w_l = 2 \cdot \gamma_s \cdot s = 2 \cdot \pi \cdot \gamma_s \cdot l^2 , \tag{1}$$

where γ_s is the crack material specific surface energy; s is the crack surface area; l is the crack equivalent length equal to the average D.

The material includes a large number of random micro-cracks of different size. Therefore, the Rayleigh law can describe the distribution of micro-crack length, and the exponential law can describe the density of the cracks energy distribution:

$$f_l(w_l) = \lambda \cdot e^{-\lambda \cdot w_l} = \frac{e^{-w_l/(8\cdot\gamma_s\cdot l_o^2)}}{8\cdot\gamma_s\cdot l_o^2}, \tag{2}$$

where l_o is the expected value l.

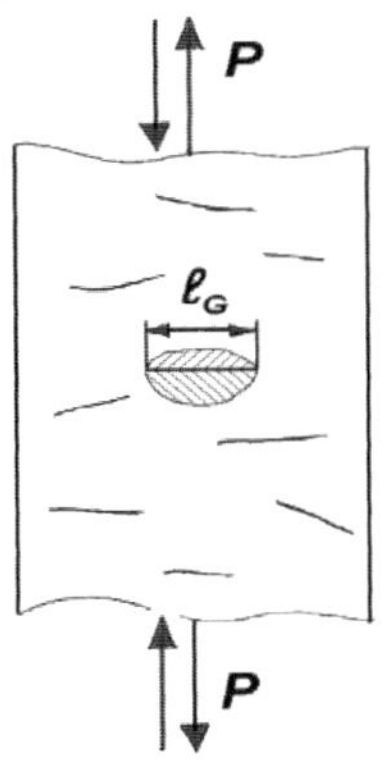

Figure 1. *P* plate cyclic loading diagram.

The considered volume V of the material contains a random integral nonnegative n number of cracks, which distribution is described by the Poisson law. Then, a probable occurrence of micro-cracks with total energy W_n in the volume V of the material is described with the probability function:

$$f_V(W) = \sum_{n=1}^{\infty} f_W(W_n)\cdot H_n = \lambda\cdot e^{-\lambda\cdot\left(\sum_{k=1}^{n} w_{lk}\right)\cdot V\cdot\rho_o} \sum_{n=0}^{\infty} \frac{\left(\lambda\cdot\left(\sum_{k=1}^{n} w_{lk}\right)\cdot V\cdot\rho_o\right)^n}{(n!)^2}, \tag{3}$$

where H_n is the possible presence of n number of micro-cracks (n = 1, 2, 3, ...) with the total energy W_n in the volume V of the material; ρ_o is the average density of the cracks distribution in the volume V of the material; w_{lk} is the energy of the k-th crack.

Let's consider the plate stretched out and contracted alternately under the symmetrical cyclic load P (Figure 1). The plate stretching-out requires the amount of work determined with the following formula:

$$U = \frac{\sigma^2}{2\cdot E}\cdot V, \tag{4}$$

where σ is the stretching-out stress occurring in the plate under the P force.

The energy gained by the plate is concentrated in the material defects and is distributed among them. If the energy gain portion is proportional to the area of the crack, then the energy gain of the crack of l size is equal to:

$$u_l = \varphi_s\cdot s = \pi\cdot\varphi_s\cdot l^2, \tag{5}$$

where φ_s is the specific portion of the external energy U falling to a unit of free micro-cracks surface area.

Similarly with the equations (1) and (2) for the crack energy, distribution of the crack energy gain is expressed with the following formula:

$$f_l(u_l) = \nu\cdot e^{-\nu\cdot u_l} = \frac{e^{-u_l/(8\cdot\varphi_s\cdot l_o^2)}}{8\cdot\varphi_s\cdot l_o^2}. \tag{6}$$

Assuming that the energy required for stretching-out (4) will be absorbed by the micro-cracks, the expected value of the probability function of the total energy gain is as follows:

$$\varphi_s = \frac{\sigma^2}{16 \cdot E \cdot \rho_o \cdot l_o^2}. \tag{7}$$

A crack can grow under the impact of the energy u_l gain if energy is sufficient for this. The crack may remain unchanged if the energy gain is insufficient to form a new free surface. Hence, if, as a result of external impact on the material, the energy u_l gained by the crack is less than the standardized value $u_g (u_l < u_g)$, the crack does not grow. If the energy gained by the crack is larger than this value ($u_l \geq u_g$), the crack starts growing. The value of this energy u_g does not depend on the size of the crack.

In this case, external forces impact on the material is as follows. Only a portion of all the micro-cracks being under the external load in the volume V can grow. To ensure the growth this portion of the cracks will absorb the additional energy u_l attributable to them. The remaining part of the cracks will not grow, and, consequently, after unloading and elastic recovery of the material the part of the additional energy attributable to them dissipates into the surrounding space.

3. Simulation results

Based on the equation (6) the part of the external energy absorbed by the micro-cracks can be described as follows:

$$P_g = \int_{u_g}^{\infty} f_l(u_l)du = e^{-v \cdot u_g}. \tag{8}$$

According to the formula (1) upon receiving additional energy $u_l \geq u_g$ (equation (5)) the micro-cracks gain value with initial energy w_l, will be equal to:

$$\Delta l = l_1 - l = \sqrt{\frac{w_l}{2 \cdot \pi \cdot \gamma_s}} \cdot \left(\sqrt{1 + \frac{u_l}{w_l}} - 1 \right) = l \cdot \left(\sqrt{1 + \frac{\varphi_s}{\gamma_s}} - 1 \right). \tag{9}$$

During cyclic deformation a stretching-out half-cycle is followed by a contraction half-cycle, which energy is calculated in the same way.

Occurrence and development of the cracks in structural materials are accompanied by cracks setting under the loads opposite to those favoring cracks growth. This issue was studied in detail in the following papers [2, 3]. According to these papers, the crack setting requires much more energy than its development, therefore cracks development process will outpace their setting. Theoretical and experimental studies [4, 5] have shown that high hydrostatic pressure or substantial biaxial contraction in the elliptic pore causes reduction of a minor axis, while the major axis stays unchanged in terms of dimensions.

Thus, the micro-cracks setting requires certain conditions, which are difficult to ensure under a simple cyclic load. Therefore, we can assume that the micro-cracks can grow in the cyclic alternating contraction and stretching-out, if the energy concentrated in them exceeds a certain threshold, while in the micro-cracks contraction half-cycle nothing happens.

If in one loading cycle the micro-crack increased in size by a value determined with the equation (9), then in the i-th loading cycle the micro-crack length gain will be:

If microcrack at one loading cycle has increased in size by an amount determined by the equation (9), then the i-th cycle of loading length of microcracks increment will be:

$$\Delta l_i = l \cdot \delta^i - l = l \cdot \left(1 + \frac{\varphi_s}{\gamma_s} \right)^{\frac{i}{2}} - l \text{ при } u_l \geq u_g, \tag{10}$$

while the micro-crack length will increase δ-fold.

Since the external deformation energy is absorbed only by the cracks, which energy exceeds the u_g, then the cracks, which did not start growing in the first loading cycles, would not grow in the subsequent cycles too. Therefore, the value of the external energy absorbed by the micro-cracks is the sum of the energy gains of all the micro-cracks in the i-th cycle which original energy exceeded the u_g:

$$\sum u_{li} = \pi \cdot \varphi_s \cdot \sum l_i^2 = \pi \cdot \varphi_s \cdot \delta^{2i} \cdot \sum l^2 . \quad (11)$$

It appears from the equation (11) that the gain of the energy absorbed with the micro-cracks is proportional to the square in parentheses. Initial energy consumption to deform the stretching-out forms the U value from the formula (4). Then the total energy loss to deform the micro-cracks in the k cycles is defined with the following formula:

$$\sum \Delta U_i = U \cdot P_g \cdot \sum_{i=1}^{k} \delta^{2i} . \quad (12)$$

According to the formula (12) the relative energy loss in the i-th cycle will be equal to:

$$\Delta_i = \delta^{2i} . \quad (13)$$

If we know the permissible length of the micro-crack l_G, which dramatically increases probable destruction of an item, then using the formulas (10) and (13) we can determine the number of maximum permissible cycles of elastic loading:

$$k = \frac{\ln l_G - \ln l}{\ln \delta} = \frac{2i \cdot (\ln l_G - \ln l)}{\ln \Delta_i} . \quad (14)$$

The resulting formula (14) can be used as a criterion for the residual life of the item: k-i.

4. Experimental studies

We carried out experimental studies of vibromechanic relaxation of residual stresses of the item shaped like a rod with dia. = 4 *mm* and length of L = 120 mm made of structural steel 70 GOST 14959-79.

The rod was bent under the concentrated load, which resulted in generation of internal stresses added to the initial residual stresses. The cyclic changes of the internal stresses were obtained by uniform rotation of the rod, which resulted in harmonic oscillations of the rod bent axis relative to its rotation axis. Rotation speed, treatment time and bending moment determining the internal stresses were taken as variable parameters.

During the experiment, it was found that if the number of loading cycles is less than the value determined with the formula (14), the stress is relieved from the item only partially. The number of the loading cycles was defined by the duration of treatment and rotation speed. If the vibromechanic treatment time exceeds the specified value, new stresses can appear in the item and material cyclic strength decreases. The results obtained are shown in Figure 2. The figure shows dependence of the residual stresses in the item $\sigma_\tau(k_\sigma, t)$ at the item rotation speed of t =160 min^{-1} on the treatment time t for three values of relative stress $k_\sigma = \sigma_t/\sigma_m$ (0.6; 0.8; 0.9), where σ_m is the maximum stress of the item elastic deformation, σ_t is the yield strength of the item material.

The Figure 2 shows that during the vibromechanic treatment the residual stresses first decrease down to zero, and then increase. The decrease of the residual stresses is due to the fact that during the vibromechanic stabilization the energy generated in the treatment process gradually absorbs the potential energy generating residual stresses in the item. As soon as the potential energy of the item material becomes equal to zero, further vibromechanic treatment causes increase in the potential energy of the item material and generates internal stresses.

5. Findings

The mathematical simulation of cracks occurrence mechanism with the use of probabilistic and energy approaches resulted in the following findings:

- brittle fracture energy is exponentially dependent on a solid body destroyed volume;
- increase in the volume of a body increases the probability of its destruction;
- crack length gain per one loading cycle is characterized by power-law dependence on the value of the energy gained during such cycle;

- increase in the energy absorbed by the micro-cracks is proportional to the square of the micro-cracks size gain.

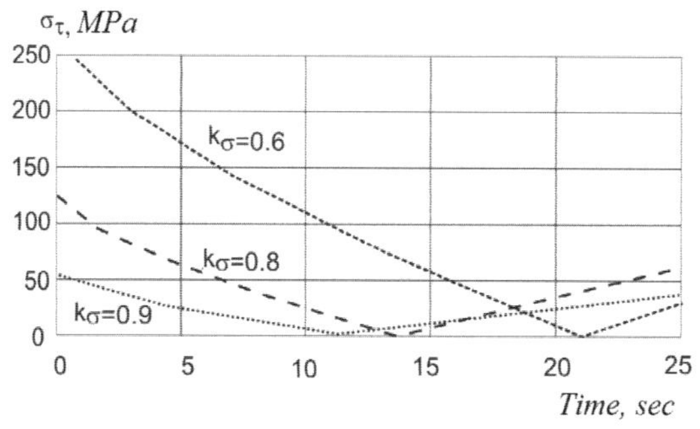

Figure 2. Residual stresses dependence on vibromechanic treatment time.

According to the results of the experimental studies, it was found that increase in the number of loading cycles causes relaxation of internal stresses that characterizes accumulation of internal energy in the cracks. After complete relief of the residual stresses, while continuing the cyclic loading process, internal stresses start to increase. It characterizes crossing of critical energy threshold and subsequent rapid growth of the micro-cracks followed by the destruction of the material.

The experimental results confirm the adequacy of the mathematical models obtained and criteria of evaluation of the material residual life.

6. Conclusion

The above method allows to evaluate high-cycle fatigue and residual life of structural materials based on the energy approach to description of the cracks occurrence mechanism. The results can be used in designing of steel structures and in development of new technological methods of residual stresses relaxation.

Acknowledgments

This paper has been developed under the state order of the Ministry of Education and Science of the Russian Federation (No. 9.896.2014/K).

References

[1] Gumbel E I 1958 Statisticks of extremes (New Jork: Columbia University Press) p 375
[2] Bezobrazov Yu A, Kolbasnikov N G and Naumov A A 2014 Tension-compression method in the simulation of multistage plastic deformation *Steel in Translation* **44** pp 71-79
[3] Betekhtin V I, Veselkov S Yu, Dal Yu M, Kadomtsev A G and Amosova O V 2003 Theoretical and Experimental Investigation of the Effect of an Applied Load on Pores in Solids *Physics of the Solid State* **45** **4** 618-624
[4] Betekhtin V I and Kadomtsev A G 2005 Evolution of microscopic cracks and pores in solids under loading *Physics of the Solid Stat* **47** 825-831
[5] Petukhov A N 2005 High-Cycle Fatigue of Materials and Parts of Gas-Turbine Engines *Strength of Materials* **37** 231-242
[6] Petukhov A N 2010 Problems of the high-cycle fatigue of the materials intended for the parts of modern gas-turbine engines and power plants *Russian Metallurgy* **10** 973-978
[7] Xiaojing Cai and Jinquan Xu 2015 Effective Fatigue Stress and Criterion for High-Cycle Multi-axial Fatigue *Journal of Materials Engineering and Performance* **24** 158-166

[8] McDiarmid D L 1991 A General Criterion for High Cycle Multiaxial Fatigue Failure *Fatigue Fract. Eng. Mater. Struct.* **14** 429-453